ECOLOGY, BIODIVERSITY AND CONSERVATION

The world's biological diversity faces unprecedented threats. The urgent challenge facing the concerned biologist is to understand ecological processes well enough to maintain their functioning in the face of the pressures resulting from human population growth. Those concerned with the conservation of biodiversity and with restoration also need to be acquainted with the political, social, historical, economic and legal frameworks within which ecological and conservation practice must be developed. The new Ecology, Biodiversity, and Conservation series will present balanced, comprehensive, up-to-date, and critical reviews of selected topics within the sciences of ecology and conservation biology, both botanical and zoological, and both 'pure' and 'applied'. It is aimed at advanced final-year undergraduates, graduate students, researchers, and university teachers, as well as ecologists and conservationists in industry, government and the voluntary sectors. The series encompasses a wide range of approaches and scales (spatial, temporal, and taxonomic), including quantitative, theoretical, population, community, ecosystem, landscape, historical, experimental, behavioural and evolutionary studies. The emphasis is on science related to the real world of plants and animals rather than on purely theoretical abstractions and mathematical models. Books in this series will, wherever possible, consider issues from a broad perspective. Some books will challenge existing paradigms and present new ecological concepts, empirical or theoretical models, and testable hypotheses. Other books will explore new approaches and present syntheses on topics of ecological importance.

Ecology and Control of Introduced Plants
Judith H. Myers and Dawn Bazely

Invertebrate Conservation and Agricultural Ecosystems
T. R. New

Joint Species Distribution Modelling

Joint Species Distribution Modelling (JSDM) is a fast-developing field and promises to revolutionise how data on ecological communities are analysed and interpreted. Written for both readers with a limited statistical background and those with statistical expertise, this book provides a comprehensive account of JSDM. It enables readers to integrate data on species abundances, environmental covariates, species traits, phylogenetic relationships and the spatio-temporal context in which the data have been acquired. Step-by-step coverage of the full technical detail of statistical methods is provided, as well as advice on interpreting results of statistical analyses in the broader context of modern community ecology theory. With the advantage of numerous example R-scripts, this is an ideal guide to help graduate students and researchers learn how to conduct and interpret statistical analyses in practice with the R-package Hmsc, providing a fast-starting point for applying JSDM to their own data.

OTSO OVASKAINEN is Professor of Mathematical Ecology at the University of Helsinki, Finland. He has conducted research in mathematical, statistical and empirical ecology, with a particular focus on metapopulation ecology, movement ecology, population genetics, molecular species identification and community ecology.

NEREA ABREGO is a postdoctoral researcher at the University of Helsinki, Finland. After obtaining her PhD in fungal ecology, she expanded her research to general community ecology. She has conducted research in empirical, theoretical and statistical ecology, including recent developments in JSDM.

Risks and Decisions for Conservation and Environmental Management
Mark Burgman

Ecology of Populations
Esa Ranta, Per Lundberg and Veijo Kaitala

Nonequilibrium Ecology
Klaus Rohde

The Ecology of Phytoplankton
C. S. Reynolds

Systematic Conservation Planning
Chris Margules and Sahotra Sarkar

Large-Scale Landscape Experiments: Lessons from Tumut
David B. Lindenmayer

Assessing the Conservation Value of Freshwaters: An international perspective
Philip J. Boon and Catherine M. Pringle

Insect Species Conservation
T. R. New

Bird Conservation and Agriculture
Jeremy D. Wilson, Andrew D. Evans and Philip V. Grice

Cave Biology: Life in darkness
Aldemaro Romero

Biodiversity in Environmental Assessment: Enhancing ecosystem services for human well-being
Roel Slootweg, Asha Rajvanshi, Vinod B. Mathur and Arend Kolhoff

Mapping Species Distributions: Spatial inference and prediction
Janet Franklin

Decline and Recovery of the Island Fox: A case study for population recovery
Timothy J. Coonan, Catherin A. Schwemm and David K. Garcelon

Ecosystem Functioning
Kurt Jax

Spatio-Temporal Heterogeneity: Concepts and analyses
Pierre R. L. Dutilleul

Parasites in Ecological Communities: From interactions to ecosystems
Melanie J. Hatcher and Alison M. Dunn

Zoo Conservation Biology
John E. Fa, Stephan M. Funk and Donnamarie O'Connell

Marine Protected Areas: A multidisciplinary approach
Joachim Claudet

Biodiversity in Dead Wood
Jogeir N. Stokland, Juha Siitonen and Bengt Gunnar Jonsson

Landslide Ecology
Lawrence R. Walker and Aaron B. Shiels

Nature's Wealth: The economics of ecosystem services and poverty
Pieter J.H. van Beukering, Elissaios Papyrakis, Jetske Bouma and Roy Brouwer

Birds and Climate Change: Impacts and conservation responses
James W. Pearce-Higgins and Rhys E. Green

Marine Ecosystems: Human impacts on biodiversity, functioning and services
Tasman P. Crowe and Christopher L. J. Frid

Wood Ant Ecology and Conservation
Jenni A. Stockan and Elva J. H. Robinson

Detecting and Responding to Alien Plant Incursions
John R. Wilson, F. Dane Panetta and Cory Lindgren

Conserving Africa's Mega-Diversity in the Anthropocene: The Hluhluwe-iMfolozi Park story
Joris P. G. M. Cromsigt, Sally Archibald and Norman Owen-Smith

National Park Science: A century of research in South Africa
Jane Carruthers

Plant Conservation Science and Practice: The role of botanic gardens
Stephen Blackmore and Sara Oldfield

Habitat Suitability and Distribution Models: With applications in R
Antoine Guisan, Wilfried Thuiller and Niklaus E. Zimmermann

Ecology and Conservation of Forest Birds
Grzegorz Mikusiński, Jean-Michel Roberge and Robert J. Fuller

Species Conservation: Lessons from islands
Jamieson A. Copsey, Simon A. Black, Jim J. Groombridge and Carl G. Jones

Soil Fauna Assemblages: Global to local scales
Uffe N. Nielsen

Curious About Nature
Tim Burt and Des Thompson

Comparative Plant Succession among Terrestrial Biomes of the World
Karel Prach and Lawrence R. Walker

Ecological-Economic Modelling for Biodiversity Conservation
Martin Drechsler

Joint Species Distribution Modelling

With Applications in R

OTSO OVASKAINEN
University of Helsinki

NEREA ABREGO
University of Helsinki

CAMBRIDGE
UNIVERSITY PRESS

CAMBRIDGE
UNIVERSITY PRESS

University Printing House, Cambridge CB2 8BS, United Kingdom

One Liberty Plaza, 20th Floor, New York, NY 10006, USA

477 Williamstown Road, Port Melbourne, VIC 3207, Australia

314-321, 3rd Floor, Plot 3, Splendor Forum, Jasola District Centre, New Delhi - 110025, India

103 Penang Road, #05-06/07, Visioncrest Commercial, Singapore 238467

Cambridge University Press is part of the University of Cambridge.

It furthers the University's mission by disseminating knowledge in the pursuit of education, learning and research at the highest international levels of excellence.

www.cambridge.org
Information on this title: www.cambridge.org/9781108492461
DOI: 10.1017/9781108591720

First published 2020

A catalogue record for this publication is available from the British Library

Library of Congress Cataloging in Publication data
Names: Ovaskainen, Otso, author.
Title: Joint species distribution modelling : with applications in R / Otso Ovaskainen, University of Helsinki, Nerea Abrego, University of Helsinki.
Description: Cambridge, United Kingdom ; New York, NY : Cambridge University Press, 2020. | Series: Ecology, biodiversity and conservation | Includes bibliographical references and index.
Identifiers: LCCN 2019042580 (print) | LCCN 2019042581 (ebook) | ISBN 9781108492461 (hardback) | ISBN 9781108716789 (paperback) | ISBN 9781108591720 (epub)
Subjects: LCSH: Biogeography–Statistical methods. | Ecology–Statistical methods. | R (Computer program language)
Classification: LCC QH84 .O93 2020 (print) | LCC QH84 (ebook) | DDC 577.2/2–dc23
LC record available at https://lccn.loc.gov/2019042580
LC ebook record available at https://lccn.loc.gov/2019042581

ISBN 978-1-108-49246-1 Hardback
ISBN 978-1-108-71678-9 Paperback

Contents

Preface

Species distribution modelling has become one of the most widely used tools in ecology, conservation biology and wildlife management. While methods for species distribution modelling are continually being developed, it is fair to say that the field itself is well established. Thousands of research papers have developed and applied statistical methods to map how the occurrence or abundance of species depends on environmental and spatial predictors. These methods have been summarised in several influential reviews and books, some of which are part of the Ecology, Biodiversity and Conservation series of Cambridge University Press (Franklin 2009; Guisan et al. 2017). However, the largest body of species distribution modelling literature concerns single-species models in which the response variable is the occurrence or abundance of a focal species. Compared to single-species distribution modelling, the methodological advances in multiple-species distribution modelling have lagged behind. When applying single-species models to data on multiple species, a separate model needs to be developed and validated for each species, making it challenging to model real communities consisting of many species. This is particularly difficult with regard to rare species, which are inherently common in most ecological communities. Furthermore, species do not live in isolation from each other, and thus viewing a community as a set of species that respond individualistically to environmental variation represents a major simplification. From the perspective of assembly theory in community ecology, biotic filtering is ignored when treating each species independently.

While species distribution modelling is routinely applied in single-species studies, the reasons outlined above make it less ideal for modelling species-rich ecological communities. Instead, the most widely applied methods in community ecology are ordination-based methods. Ordination methods were developed to enable the patterns in community composition to be summarised along spatial and environmental gradients. This is done by simplifying the high-dimensional structure of community data into few axes that explain the dominating part of the variation. While ordinations are very powerful for summarising the patterns in complex community data, they have limitations as well. Most importantly, ordination methods have been criticised for being of descriptive rather than predictive nature.

Both species distribution models and ordination methods are used to achieve the same general aim, namely to better understand the drivers controlling biodiversity across environmental gradients, space and time. Consequently, there is no reason why these two methodological fields should continue to develop independently of each other; rather, they could each learn from each other and eventually merge to combine their strengths. In recent years, statistical ecologists have taken the first steps in this direction, by developing the so-called joint species distribution modelling (JSDM) approaches. JSDMs build more heavily on single-species distribution models than on ordinations, as they involve a single-species distribution model for each of the species comprising the community. However, they are not a mere collection of single-species models: the species are not modelled independently of each other, but jointly, as with ordination methods. The 'joint' aspect of JSDM relates to both environmental filtering and biotic filtering. The responses of the species to environmental variation (i.e. environmental filtering) are assumed to have a joint structure that can depend on e.g. species' traits or phylogenetic relationships. This is achieved by a hierarchical model structure that involves both community-level and species-level parameters. The species' responses to each other (i.e. biotic filtering) are modelled through residual association matrices that describe the co-occurrence or co-abundance patterns that are not explained by environmental filtering.

The first JSDM approaches modelled species associations separately for each pair of species, and were thus feasible only for communities with few species (the number of pairs of species – and hence model parameters – becomes otherwise too large to be estimated). To overcome this limitation, the next generation of JSDMs applied latent variable approaches, making it possible to estimate association matrices also for communities with many species. This is where joint species distribution models have approached ordination methods. Namely, the latent variable approach is used to reduce the high dimensionality of community data. In fact, it can actually be viewed as model-based ordination. Therefore, JSDMs involve both species-specific distribution models and ordinations in their machinery, bringing these two fields closer to each other.

JSDM is currently one of the fastest developing fields in statistical ecology. While several kinds of JSDMs have already been implemented and successfully applied, the field is still in its infancy, especially compared to single-species modelling and ordination techniques. Consequently, the field of JSDM is currently experiencing much turbulence, with new approaches emerging at a fast rate and parallel developments of related approaches being simultaneously undertaken by different research groups. Some of these approaches may prove to be viable in the long run, while others may become superseded by improved approaches. While the ongoing rapid turnover of JSDM provides exciting possibilities, it also makes it difficult for their users to keep track of the pros and cons of the different approaches, and to gain an adequate understanding of their underlying assumptions and limitations. For these reasons, we considered it timely to devote an entire book to joint species distribution modelling, as this provided the possibility to present the conceptual, statistical and implementation aspects of JSDM in a much more profound and collective way than would be possible in focal research papers or software tutorials.

While several JSDM approaches and software implementations have been developed over the past decade, this book develops the argument of joint species distribution modelling from the point of view of one specific framework, namely Hierarchical Modelling of Species Communities (HMSC). However, as many of the existing JSDM approaches are closely related and can be considered as different branches of the same tree, we hope that this book will help deepen the readers understanding of the fundamentals of JSDM in general. In addition to presenting the conceptual, theoretical and statistical foundations of JSDM, this book also provides 'hands on' examples of how JSDM can be applied in practice. To this end, we build heavily on the R-package Hmsc; its use is demonstrated through R-scripts, and it has also been used to generate the majority of the figures/illustrations. Furthermore, we note that writing this book motivated us to implement some new features and extensions of HMSC, so some of the material here has not been published yet in research papers. We hope that the many R-scripts presented in this book (and the related online resources at www.helsinki.fi/en/researchgroups/statistical-ecology/hmsc) will provide a convenient starting point for a reader who wishes to apply JSDM for his or her own purposes.

Acknowledgements

This book builds on the development of Hierarchical Modelling of Species Communities (HMSC) that has continued over the past 10 years. Thus, we are thankful for the many researchers who have contributed to the work. One influential event that took place in the early phase of HMSC development was a research seminar in Helsinki in 2008, in which OO presented an approach that he had developed for species co-occurrence analyses in the context of fungal interactions. In this seminar, Janne Soininen asked whether the method could also be applied if the community matrix was transposed, to model joint responses of the species to environmental covariates instead of the responses of the species to each other. This resulted in Ovaskainen and Soininen paper entitled 'Making more out of sparse data: Hierarchical modelling of species communities', from which the HMSC approach derives its name. In this early phase of HMSC development, Guillaume Blanchet joined OO's group as a postdoctoral researcher, making many valuable conceptual and technical contributions.

Another influential event was a research visit to Duke University by OO and Guillaume Blanchet in 2013, hosted by Alan Gelfand. After a seminar given by OO, David Dunson pointed out some developments in modern Bayesian statistics that could be utilised to improve the applicability and computational efficiency of HMSC. This started a critically important and still continuing collaboration, which has resulted in the implementation of latent variable approaches to HMSC, as well as many other aspects that have made HMSC applicable to much bigger data than was possible before.

In 2013, Gleb Tikhonov started as a PhD student in OO's group. Gleb quickly became a key developer of HMSC, and defended his PhD thesis on this topic in 2018, with Alan Gelfand as the opponent. In addition to his numerous conceptual and statistical developments, Gleb made the very important contribution of leading the implementation of the R-package Hmsc (Tikhonov et al. 2020b). While the early versions of

HMSC were implemented first with Mathematica and then with Matlab, it became increasingly clear that an R-implementation would be needed for most ecologists to apply the method. The existence of the R-package is mainly thanks to the major efforts by Gleb. Another PhD student who made many contributions to HMSC was Anna Norberg, who also defended her thesis in 2018. While Gleb focused on developing the statistical approaches, Anna's main focus was on applying HMSC. This greatly aided the development and software implementations. In particular, Anna made the heroic effort of comparing the predictive performances of 33 single-species and joint species distribution models (Norberg et al. 2019), thus helping researchers assess the strengths and weaknesses of the many available approaches. Another key person who contributed to both the conceptual and implementation aspects of HMSC is Øystein Opedal, who joined the development team as a postdoctoral researcher with OO. More recently, Melinda de Jonge and Jari Oksanen also took part in the development of the R-package Hmsc, with major contributions in making the software more user-friendly and improving documentation. We also thank the many participants of the HMSC courses (organised in the context of the International Biometric Society meeting in Hobart in 2015, the European Congress of Conservation Biology in 2018 in Jyväskylä, the International Statistical Ecology Conference in 2018 in St Andrews and the Kaamos Symposium in Oulu in 2019) for their feedback, which has greatly contributed to the development of the approach itself, as well as the material presented in this book.

The participation of NA in the development of HMSC would have not been possible without the encouragement and support of her advisors. Back in 2013, Panu Halme promoted the collaboration, which resulted in some of the key papers in the development of HMSC (Abrego et al. 2017a; Ovaskainen et al. 2016a). Bernt-Erik Sæther gave valuable support while NA worked on the development of HMSC during her first postdoc, especially in the context of developing a time-series version of HMSC in collaboration with Steinar Engen and Vidar Grøtan (Ovaskainen et al. 2017a). Since 2017, NA has got the unconditional support of Tomas Roslin to continue collaborating on HMSC as her 'side project'; Tomas has also contributed to the development of HMSC himself (Ovaskainen et al. 2017b).

Writing this book was greatly facilitated by the support that we received from the publisher. Michael Usher, the series editor to Ecology, Biodiversity and Conservation, originally approached us about our interest in writing this book and encouraged us to do so. The senior commissioning editor Dominic Lewis helped us to develop the more detailed plan for the book. We are especially grateful to the editorial assistant Aleksandra Serocka, who made the process of writing the book as painless as possible by providing clear instructions and being very pragmatic and supportive during the entire process.

During the writing process, we received excellent feedback from many of our colleagues. In particular, we would like to thank Laura Antão, Melinda de Jonge, Janet Franklin, Elina Kaarlejärvi, Tuomas Kankaanpää, Janne Koskinen, Øystein Opedal, Iñaki Odriozola, Isabella Palorinne, Federico Riva, Bernt-Erik Sæther, Panu Somervuo, Tomas Roslin, Marjaana Toivonen, Jarno Vanhatalo and Helena Wirta for their invaluable comments on earlier versions of the book. Furthermore, we thank Aleksi Lehikoinen for providing the bird community data used as a case study in Chapter 11, and Atte Moilanen for performing the Zonation analyses presented in Section 11.3. We are especially grateful to Bess Hardwick for her major assistance in formatting the figures of this book and to Jacquelin De Faveri for her excellent work in revising the English language.

The writing of this book coincided with an especially busy period of our lives. To this end, we are most thankful to Argia Abrego, without whom this book would not have happened. Argia allowed us to focus on the work by having a great time with our baby for those three critically important weeks during which we wrote the first draft.

Part I

Introduction to Community Ecology

Theory and Methods

1 · *Historical Development of Community Ecology*

In this first chapter we give a brief overview of the history of community ecology, starting from the early twentieth-century debates on how communities should be defined, and continuing until the modern conceptual frameworks. The aim is not to review every single theory, model or framework that has been developed in community ecology – that would call for an entire book! Instead, we give an overview of how this field has developed through history. Most importantly, this chapter is needed to introduce the concepts and ideas that underline the ecological assumptions behind species distribution models (SDMs) in general, and Hierarchical Modelling of Species Communities (HMSC) in particular. Here we will briefly mention how some of the theoretical concepts relate to HMSC, but more thorough discussions on how HMSC ties to ecological theory will be given later in the book, under each of the relevant chapters where the different components of HMSC are introduced.

The reader may wonder why a statistically orientated book starts with a historical tour of the development of community ecology. Many readers interested in figuring out how to fit a joint species distribution model (JSDM) in R might be tempted to completely skip this chapter and jump straight to where the equations and scripts start. While this is understandable, we strongly recommend that you keep reading. In our view, ecologists should think about the theoretical context in which their study questions are framed, before starting to fit any model. We start by recalling what community ecology is about (Section 1.1) and how an ecological community may be defined (Section 1.2). We then briefly review the developments in community ecology from the foundational ideas during the twentieth century up to the current frameworks (Sections 1.3–1.5).

1.1 What Is Community Ecology?

Community ecology is a cross-disciplinary field that aims to describe and understand the spatio-temporal structure and dynamics of ecological communities. Although nowadays community ecology is well rooted within the broader scope of ecology, this has only recently become the case.

One of the most influential papers in community ecology is Lawton (1999), which critically questions the entity of community ecology as a field. In his own words, 'community ecology is a mess with so much contingency that useful generalizations are hard to find'. What Lawton found problematic was that conclusions from studies in this field were mostly case-specific and lacked general or unifying conceptual frameworks. This was indeed the case, as the conceptual and theoretical developments in community ecology have lagged behind other fields, such as population ecology and population genetics. Since the influential 'community ecology is a mess' statement, the past two decades have experienced a proliferation of unifying theory and general conceptual frameworks for community ecology (for books on community ecology theory see Leibold & Chase 2018; Morin 2011; Vellend 2016).

In the next sections we will review the most important early debates that formed the basis for the current conceptual and theoretical frameworks in community ecology.

1.2 What Is an Ecological Community?

Nowadays, the term 'ecological community' is generally understood as the assemblage of at least two potentially interacting species at a given time and location. However, throughout history this term has acquired disparate meanings for different scientists (Fauth et al. 1996; Stroud et al. 2015). For some early ecologists, the basic feature of a community was that species must interact. Whittaker (1975) defined an ecological community as 'an assemblage of populations of plants, animals, bacteria and fungi that live in an environment and interact with one another, forming together a distinctive living system with its own composition, structure, environmental relations, development, and function'. Others did not put such emphasis on interactions, but rather on the spatial co-occurrence among species. Along these lines, for Krebs (1972) a community is 'an assemblage of populations of living organisms in a prescribed area or habitat', and for Ricklefs (1990) a community reflects 'associations of plants and animals that are spatially delimited and that are dominated by one or more prominent species or by a physical characteristic'.

Because of the tradition of studying different taxa separately, community ecologists often work with communities of species that are phylogenetically related (e.g. insects, birds, fungi, plants, etc.). Although we normally use the term 'community' to refer to these (e.g. insect community, bird community, fungal community, plant community), the technical word for referring to communities of taxonomically similar species is 'taxocene'. Other terms that are often used in place of 'ecological community' are 'guild' and 'assemblage'. The term 'guild' is used when the ecological community is formed by species that use resources in similar ways (Root 1967). For instance, all grazers (either mammals or insects) or saprotrophs (either fungi or bacteria) form their own guilds. The term 'assemblage' refers to species that exist in a given area, but do not necessarily interact. In the ecological literature, 'assemblage' usually refers to the species pool present in a large spatial area, and when the interspecific relationships among species are not so clear (Stroud et al. 2015). As an example, atlas data on species' distributions are considered 'assemblage' data rather than community data: information about a species' occurrence has often been recorded at different time points, and the size of the spatial unit at which the data are recorded (i.e. grid size) is not necessarily related to the spatial scale of the ecological processes, and is usually quite large (e.g. tens of km).

For the purpose of analysing data with HMSC, it does not matter whether the data are community data or assemblage data. In both cases, the input data matrices will have the same structure, and the results will look the same, in the sense that the output from the model will be in the same format. Yet, for the ecological interpretation, the distinction between these two can be critical. For example, empirical community ecologists are often interested in studying how species interact with each other, which can be described as interaction networks or food webs. Interaction networks are essentially communities in which all interactive relationships among the species are depicted, whereas food webs focus on the feeding relationships (i.e. food chains) among species (Elton 1927). In the case of assemblage data, the species-to-species association matrices (on which we focus in Chapter 7) may have nothing to do with species interaction networks, while for community data they might.

As seen from those pioneering definitions of ecological communities, some of the early scientists emphasised the taxonomical identity of the species as a characteristic to form an ecological community. Most early community ecologists worked on terrestrial plant and animal communities, as these contain the most conspicuous study organisms. Consequently, pioneering conceptual frameworks in community ecology were developed using terrestrial plant and animal communities as model systems. Many modern community ecologists consider it equally important to work with any taxonomical group from any environment, including for example microorganisms from the air (Barberán et al. 2015) or from the digestive tract (Burns et al. 2015). Molecular species identification methods now allow us to study many more kinds of communities than before. This is greatly facilitating the detection and identification of less conspicuous and highly diverse organisms.

In the context of HMSC, what we call an 'ecological community' follows the definition by Fauth et al. (1996): a collection of species occurring in the same place and at the same time, the species not being necessarily restricted by phylogeny or resource use, and allowing the spatial boundaries to be either natural (e.g. islands) or arbitrary (e.g. study plots).

1.3 Early Community Ecology: A Descriptive Science

In the beginning community ecology was a merely descriptive scientific field. After Linnaeus' work, naturalists began building species inventories, i.e. identifying and listing species from given localities. They soon started to realise that there are predictable differences in the numbers and abundances of species among localities that differ in their environmental conditions. This inspired scientists to classify communities according to the species composition patterns and environmental variation (Köppen 1884; Wallace 1876; Whittaker 1962).

Some of the community classifications developed in the 1960s and 1970s are still currently used. Perhaps the most remarkable example is Whittaker's (1975) classification of terrestrial communities according to the dominant plant species and environmental conditions. Whittaker borrowed from previous biome classifications (e.g. Clements 1916) and assigned them to annual precipitation and average temperature conditions. Although this classification has undergone several modifications since its original publication, it still represents a basic system for understanding biodiversity organisation globally.

Furthermore, Whittaker provided the first definition of one of the most popular concepts for assessing between-site variation in species composition: beta diversity (Whittaker 1972). Whittaker defined beta diversity as an index to measure the 'extent of differentiation of communities along habitat gradients'. Currently known as Whittaker's multiplicative law, he postulated that the total gamma diversity (total number of species) of a geographic area is a product of the alpha diversity (average number of species in a single locality) and the beta diversity (variation in species composition between localities). Since Whittaker's seminal work on beta diversity, this concept has been redefined in a number of ways, and a multitude of indices and methods for measuring beta diversity have been developed (Anderson et al. 2011; Tuomisto 2010).

In spite of the modernisation of the concept of beta diversity since its origin, community ecologists assessing beta diversity essentially aim to do what Whittaker did, i.e. to assess the variation in species composition among sites. Indeed, classifying communities according to the species composition patterns and environmental variation is still of central interest in community ecology. Novel sampling methods and species identification techniques are revolutionising the amount and accessibility of information about biodiversity, yet there is still a large gap in our knowledge about how communities are distributed on Earth. Describing the community composition patterns along environmental, spatial and temporal gradients is an indispensable step towards understanding the structure of species communities.

As mentioned above, the justification of community ecology as a proper discipline was highly debated in the end of the twentieth century. In the early twentieth century, the debate was centred on whether ecological communities are self-organised and delineable systems, or collections of populations with unclear boundaries. These contrasting views are known as the *organismic concept of communities* and the *individualistic continuum concept*, and were advocated by botanists Clements (1916) and Gleason (1926), respectively. Under the organismic view, Clements believed that ecological communities form static and definable units that can be classified, similar to the Linnaean taxonomical system for species. On the contrary, according to Gleason, a community can be seen as an assemblage of populations of different species whose traits allow them to persist in a given area. Therefore, opposed to Clements' view, Gleason thought that communities result from species-specific responses to the environment, rather than from the associations among species. Under Gleason's view, the spatial boundaries of ecological communities are not so sharp, and the composition of communities may change over time and space.

These two disparate views mainstreamed the avenue of plant community ecology. Clements' organismic view of communities represents the foundation of phytosociology, i.e. the science that aims at classifying plant communities into fixed units. Following Clements' idea, plant communities reach a steady state after the process of ecological succession occurs (Clements 1936). Phytosociology culminated in the beginning of the twentieth century, when botanists around the globe developed their own classification systems and most plant communities were assigned to vegetation types. The current view on how communities are structured is more dynamic, and therefore closer to Gleason's view. The current emphasis is not on classifying species assemblages into a discrete set of archetypal communities, but rather on understanding the mechanisms allowing species coexistence within communities (Götzenberger et al. 2012).

Another debate began in the twentieth century, about the spatial scale at which ecological communities should be described. Partially reflecting the Clementsonian vs. Gleasonian view of communities, the debate was focussed on the extent to which communities are spatially bound. The definitions of ecological community always implied a spatial aspect: an assemblage of populations of living organisms *in a prescribed area* (Krebs 1972); an assemblage of populations of plants, animals, bacteria and fungi that live *in an environment* and interact with one another, forming together a distinctive living system (Whittaker 1975); associations of plants and animals that are *spatially delimited* and that are dominated by one or more prominent species or by a physical characteristic (Ricklefs 1990). As an implicit consensus, communities were conceptually delimited at the spatial scale that interspecific interactions physically take place. But often the spatial scale at which an observational study is conducted is decided quite arbitrarily, partially because the true spatial scale at which species interactions operate (or even the interactions themselves) are usually unknown beforehand. As such, the uncertainty about the spatial scale at which communities should be defined continued gaining much attention, especially after Ricklefs' influential work on the importance of spatial scale on the processes structuring communities (Ricklefs 1987, 2008).

Another early line of research in community ecology focused on patterns of accumulation of species and individuals across space and time, such as the species–area relationship, species–time relationships and species abundance distribution (Arrhenius 1921; Fisher et al. 1943; Preston 1948, 1960). Since the first descriptions of these relationships, community ecologists and macroecologists have been fascinated by the high consistency of their shapes across ecosystems and taxonomical groups. For example, the species–area curve is often found to follow a power-law (Arrhenius 1921; Dengler 2009), whereas the species abundance distribution tends to show great variation among species, in particular a long tail of many rare species (Fisher et al. 1943; McGill et al.

2007; Preston 1960). The question of what mechanisms underpin these patterns has been a major inspiration for the development of theories about the drivers of community assembly (McGill et al. 2007).

1.4 Emergence of the First Theories

By the end of the twentieth century, two controversial theories about community assembly were formalised, namely the *Niche Theory* (Hutchinson 1959; MacArthur & Levins 1967) and the *Neutral Theory* (Hubbell 2001). The early ideas of the concept of ecological niche had already emerged in the beginning of the twentieth century, when an ecological niche was considered the place that a species occupies in an ecological community. During these early years, the concepts of Grinnellian and Eltonian niches originated, which were later formalised as the concepts of fundamental and realised niches. For Grinnell (1917), the ecological niche was 'the sum of habitat requirements and behaviours that allow a species to persist and produce offspring'. Elton (1966) defined the ecological niche as 'the place of an animal in the abiotic environment, its relations to food and enemies'.

These ideas persisted for decades, but it was not until the end of the twentieth century that the concept of ecological niche was formalised. Hutchinson (1959) developed a formal notion of the ecological niche as a n-dimensional hypervolume, and this concept has remained to the present day (Blonder 2018). The n dimensions of the hypervolume are the environmental and resource characteristics that the species requires to persist. Hutchinson also formally introduced the ideas of fundamental and realised niches. Specifically, the fundamental niche of a species is represented by the hypervolume defined by the environmental and resource characteristics that the species require to persist, whereas the realised niche is what remains from the hypervolume after interactions with other species are taken into account. Another important contribution to the Niche Theory was provided by MacArthur and Levins (1967), who implemented Hutchinson's niche concept into a mathematical model. The consumer-resource model of MacArthur and Levins (1967) illustrates the overlap in resource use among species.

The niche concept has been surrounded by confusion since its foundation, and the controversy about how exactly to define it still continues (Pocheville 2015). This problem arises from the fact that different ecologists have meant slightly different things when referring to a niche (Leibold 1995). Additionally, it is difficult to distinguish between the effect of the environment on a species and the effect of the species on the environment (Chase & Leibold 2003). In spite of this, the Niche Theory remains a central principle in ecology, and one of the fundamental theoretical pillars in species distribution modelling (Peterson et al. 2011). As in most SDMs, the species niche in HMSC is the relationship between species occurrence or abundance and the environmental conditions, and thus refers more to realised rather than fundamental niche. The niche of a particular species is thus measured by regression parameters that describe how the occurrence or abundance of that species depends on the environmental conditions that are included in the analyses (Chapter 5). The distribution of species-specific niches describes how the entire community responds to environmental variation (Chapter 6).

A milestone for the development of predictive community ecology research was the *Equilibrium Theory of Island Biogeography* by MacArthur and Wilson (1967). This theory was originally developed for explaining the species richness patterns in oceanic islands, and was later empirically validated by Simberloff and Wilson (1969). This theory predicts that on an island the number of species is determined by a balance between immigration and extinction. The 'equilibrium' part of the theory comes from the assumption that immigration rate decreases and extinction rate increases with an increasing number of species that already occupy the island. Thus, the number of species that can persist will converge to an equilibrium. The number of species on islands that are large or near the mainland is predicted to be larger than the number of species on distant small islands, because there is higher immigration to large islands near the mainland, and higher extinction on small islands. Many kinds of suitable habitats within a matrix of less suitable habitats can be conceptually viewed as an island. As such, this theory represents a baseline for understanding species diversity far beyond true island systems (Hanski 2016). For example, for forest-dwelling organisms, forest fragments embedded within an agricultural matrix would be analogous to islands distributed within the ocean. Similarly, for aquatic organisms, lakes embedded within terrestrial habitats could be considered as islands. For species with low tolerance to anthropogenic disturbance, the islands could be protected natural areas embedded within the matrix of human-modified areas.

By the end of the twentieth century, a new ground-breaking theory on how communities are assembled emerged: The *Unified Neutral Theory of Biodiversity and Biogeography* (Hubbell 2001). Inspired by the incredibly high plant diversity in tropical environments – which is very difficult to relate to variation in environmental conditions – Hubbell proposed that biodiversity arises and is organised at random. From the Neutral Theory perspective, all individuals are ecologically identical and niche differences are not needed to explain biodiversity patterns. Highly diverse communities of equivalent species (i.e. species with identical niches) arise solely because of random events (i.e. chance extinctions balanced by chance speciations). More specifically, stochastic random processes that include birth, death and immigration of individuals, as well as speciation, can lead to species-rich communities. Because of the extreme point of view that biodiversity originates solely from random processes, the Neutral Theory of biodiversity provoked a wave of criticism. This resulted in the development of tests in which the predictions of niche-based and the neutral theories were compared against empirical data, for example in terms of species abundance distributions (e.g. McGill 2003; McGill et al. 2006a; McGill et al. 2007; Wootton 2005). While these empirical tests failed to find general support for Hubbell's Neutral Theory, they did establish its position as a highly useful null model for evaluating the roles of non-neutral processes such as adaptation and natural selection in shaping ecological communities.

The proliferation of mathematical models in population ecology during the 1960s and 1970s (see Kingsland 1986) greatly influenced the field of community ecology. Single-species population models started incorporating the influences of other competing species, and linking the patterns of resource use to competitive abilities. Extensions of the original Lotka-Volterra two-species competition model and consumer-resource models allowed modelling networks of interacting species (e.g. Levine 1976; MacArthur 1972; Tilman 1994). The development of multi-species models of interacting species also raised one of the central issues in community ecology today: the relationship between network stability and complexity (May 1971). Furthermore, the emergence of Neutral Theory motivated the development of more complex and realistic niche-based modelling frameworks (e.g. Chave et al. 2002). One important conclusion from such modelling studies was that many contrasting types of community assembly processes can result in surprisingly similar patterns, for example in terms of species abundance relationships (Chave et al. 2002).

1.5 Current Community Ecology: Search for the Unifying Theory

The turn of the 21st century saw a change in community ecology research, where the interest switched from describing community patterns to understanding the processes underlying the patterns. In other words, community ecology shifted from a descriptive to a more predictive science. One key question in modern-day community ecology – especially relevant in the context of ongoing environmental change – is the following: given past and current environmental conditions, and the composition and characteristics of past and current species communities, can we predict future community compositions?

For researchers aiming to develop predictive community ecology, a major concern has been the lack of a general unifying conceptual framework. This gap is now rapidly filling, with the development of conceptual and theoretical frameworks remaining a major focus of current community ecology (McGill 2010). The three most renowned frameworks for understanding community assembly are the *Metacommunity Theory,* the *Assembly Rules Framework* and Vellend's *Theory of Ecological Communities.* All of these build on the ideas developed in the previous century, as well as from theoretical frameworks in other fields such as metapopulation ecology and population genetics.

As mentioned earlier, one of the problems that early community ecologists faced was how to decide on the spatial scale at which communities should be delineated. Since the most natural delineation was case-specific, the mechanisms that governed the communities were also considered case-specific. Consequently, it remained difficult to synthesise, from among studies, which processes and mechanisms drive community dynamics (Lawton 1999; Ricklefs 2008). For instance, a microcosm experiment may show that interspecific interactions are the main force driving community structure, whereas in a continental-scale observational study interaction effects might not be observed. It is nowadays well-recognised that the processes and mechanisms driving community dynamics are scale dependent (e.g. Chase et al. 2019; Jarzyna & Jetz 2018). Similarly, there was a hot debate in community ecology about whether neutral or niche models were more valid for explaining community structure and dynamics (McGill et al. 2006a). Merging the contrasting viewpoints called for frameworks that could integrate multiple processes, such as neutral and niche-based, operating at multiple spatial and temporal scales. In particular, these considerations led to the emergence of the metacommunity framework (Holyoak et al. 2005; Leibold et al. 2004) and the assembly rules framework (Keddy 1992).

1.5.1 The Metacommunity Framework

A metacommunity is defined as 'a set of local communities that are linked by dispersal of multiple potentially interacting species' (Holyoak et al. 2005). Metacommunity Theory explains how networks of local communities result from the interplay of stochastic and deterministic processes at both local and regional scales (Holyoak et al. 2005; Leibold et al. 2004). For this, Metacommunity Theory synthesises four perspectives, each arising from different – but not mutually exclusive – conceptual perspectives: neutral, patch dynamics, species sorting and mass effects perspectives.

With its roots in Hubbell's Neutral Theory (Hubbell 2001), the *neutral perspective* posits that individuals are considered to be equal in competitive capabilities and niche preferences, irrespective of which species they belong to. Thus, in the neutral perspective, any variation in species composition emerges solely from stochastic ecological drift.

The *patch dynamics perspective* assumes that species track ephemeral habitat patches through colonisation–extinction dynamics. One classical result arising from implementing this perspective into a mathematical model is that species coexistence can be facilitated by colonisation–competition trade-off (Tilman 1994). This happens because the species with higher colonisation ability are faster at colonising newly emerged habitats, but are later outcompeted by the species with lower ability, as the latter are assumed to be competitively superior ability.

Related to the Niche Theory, the *species sorting perspective* focuses on the differences in the species niche preferences, which lead different species to inhabit different parts of environmental gradients (Chase & Leibold 2003), thus reflecting Gleason's view on how communities are organised.

The *mass effects perspective* differs from the species sorting perspective by assuming a much greater dispersal rate between the local communities within the metacommunity. The high dispersal rate influences variation in community composition. For example, in source-sink dynamics, the "sink species" can be found outside their fundamental niche due to high immigration rate (Amarasekare & Nisbet 2001).

A core assumption of the Metacommunity Theory is that the four different perspectives discussed above are not mutually exclusive – some or all may be simultaneously relevant for a given metacommunity. Their relative roles depend on the degree of environmental heterogeneity and degree of dispersal in the metacommunity, as well as the scale at which it is observed. Therefore, this framework allows the coexistence of previously competing theories.

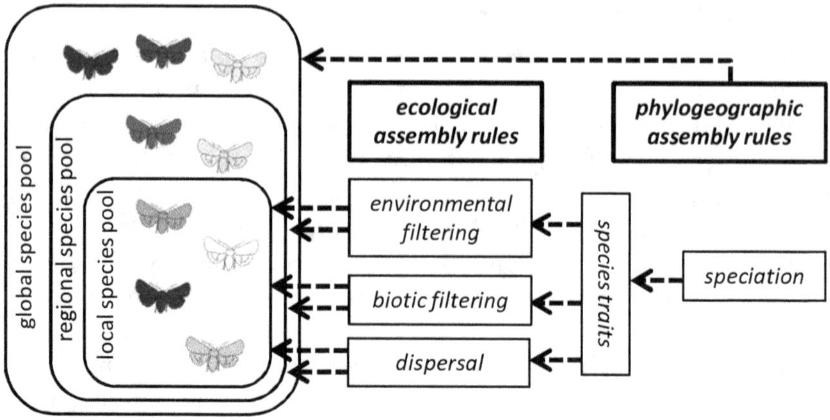

Figure 1.1 A conceptual diagram of the assembly rules framework, illustrating how assembly processes influence ecological communities at different spatio-temporal scales. The composition and dynamics of local, regional and global communities are influenced by the combined effects of phylogeographic and ecological assembly rules. The responses of the species to the biotic and abiotic environments depend on their traits, which are ultimately shaped by speciation and adaptation

1.5.2 The Assembly Rules Framework

Already in the 1970s, Diamond (1975) introduced the term 'assembly rule' to refer to the restricted species combinations to which competitive interactions can lead. But soon after Diamond's work, the meaning of assembly rule was broadened, from competition to any ecological process favouring or disfavouring the occurrence of a species (Keddy 1992). Assembly rules, which are more naturally called assembly *processes* to emphasise their dynamic and stochastic nature, can be conceptually viewed as 'filters' that act on scales ranging from the regional species pool to increasingly finer scales until the local community composition is determined (Zobel 1997). What is regional and local does not have a precise meaning beyond the fact that the latter is nested within the former. Thus a 'regional pool' can be simply defined as the set of all species that would in theory be able to colonise a given area, and the 'local species pool' defined as the set of species that are actually found from that area (Figure 1.1). As discussed further throughout the book, HMSC is conceptually linked to the assembly rules framework (Ovaskainen et al. 2017b).

The assembly rules framework distinguishes phylogeographic from ecological assembly processes (Figure 1.1). Phylogeographic assembly processes refer to the restrictions in species composition that result from historical patterns of speciation and large-scale migration, whereas ecological assembly processes refer to the restrictions due to smaller-scale dispersal (dispersal assembly rules), abiotic environment (environmental filtering), and biotic interactions (biotic filtering). While the original definition of biotic assembly rules by Diamond (1975) included only negative or competitive interactions, current community ecology includes all interactions, for example facilitative ones (Bruno et al. 2003). Of course, the influences of different assembly processes are not necessarily additive. Instead, they may be interactive, as for instance environmental variation may modify biotic interactions. Therefore, the assembly rules framework emphasises that different processes can act simultaneously at multiple spatial and temporal scales, as is also the case with the metacommunity framework.

Trait-based research has gained much popularity within the assembly rules framework (Cadotte et al. 2015; McGill et al. 2006b). Rather than focusing on species *per se*, current community ecologists acknowledge that a more profound understanding of assembly processes can be obtained by identifying the traits that influence the responses of species to the environment, and by linking assembly processes to speciation and adaptation by considering how these traits evolved. The traits that influence the responses of species to changes in environmental conditions are called response traits (Lavorel & Garnier 2002). For example, traits related to dispersal and competitive capabilities may determine which species reach and colonise a given area, and which species succeed in securing adequate resources. Yet, traits may appear linked to occurrences not only because of their adaptive significance *per se*, but also because phylogenetically related species can be expected to be similar in terms of both traits and occurrence patterns (Harvey & Pagel 1991). This is the result of a phenomenon known as phylogenetic niche conservatism, which refers to the fact that species tend to retain their ancestral traits, and consequently traits of related species tend to be similar. Even if only some traits causally influence species' occurrences, other traits will also appear to be associated with occurrence variation. Alternatively, some response traits may not be known, in which case their influences are seen in phylogenetic relationships in species niches. For these reasons, HMSC models species niches both as a function of their response traits as well as phylogenetic relationships (Chapter 6).

One of the long-standing principles in ecology has been that the coexistence of two species competing for a single resource type is not possible (Gause 1934), known as the 'competitive exclusion principle'. Since the resource use of a species is the trait that describes its fundamental niche, this poses an important question in functional community ecology: to what extent can similar species be found together? (Wiens et al. 2010). While niche conservatism and environmental filtering would suggest that species with similar traits are likely to be found together, competitive exclusion and other processes related to niche partitioning would suggest that only dissimilar species can be found together. A central evolutionary concept related to niche partitioning is that of adaptive radiation, where an ancestral species rapidly diversifies into a variety of species. As a classical example, Darwin's finches diversified their beak shapes to partition the niche space consisting of different food resources.

1.5.3 Vellend's Theory of Ecological Communities

Motivated by the proliferation of disparate conceptual frameworks in community ecology, and inspired by the theory in population genetics, Vellend proposed a unifying theory that he called the *Theory of Ecological Communities* (Vellend 2010; Vellend 2016). Vellend brought a more synthetic perspective to community ecology theory by integrating all processes of community dynamics into four fundamental or 'high-level' processes: selection, ecological drift, dispersal and speciation. Selection results from deterministic fitness differences between individuals of different species, and is expected to change community composition to the extent that species vary in their average relative fitness. Ecological drift refers to the random component that drives community dynamics when demographic events occur randomly with respect to species' identities. Dispersal refers to the movement of species. As movement brings individuals to locations where their respective species might not be able to persist, it is expected to increase both species richness and similarity in species composition across space. Speciation is the process that creates variation in community composition at larger scales by the emergence of new species, which obviously increases species richness.

All previously mentioned assembly processes can be grouped into Vellend's four high-level processes. Biotic and environmental filters both deterministically select against or in favour of species, depending on the traits that determine their fitness. Dispersal includes all events related to the movement or arrival of new species, including historical migrations. All stochastic events that create unpredictable pathways in community composition can be grouped in drift. Finally, speciation accounts for

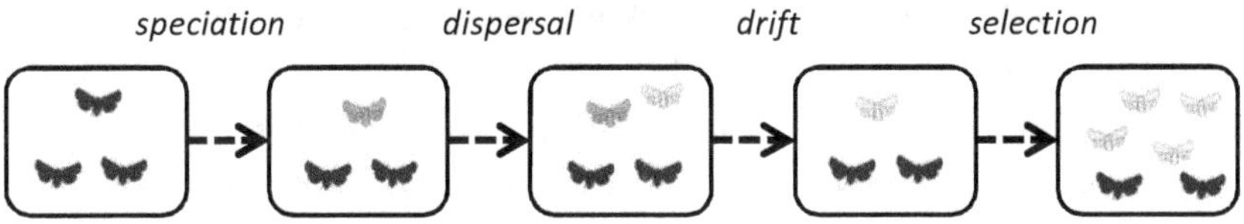

Figure 1.2 Conceptual diagram depicting the four high-level processes considered in Vellend's *Theory of Ecological Communities* (adapted from Vellend, 2016). After speciation, a subpopulation of the black species diverges into a grey species, increasing the number of species. By dispersal, an individual of a white species arrives from elsewhere, adding another species. Drift reduces the number of species, because stochastic events lead the grey species (that was at low population size) to become extinct. A selection process occurs because the white species is better adapted to the environment than the black species, and thus increases in abundance.

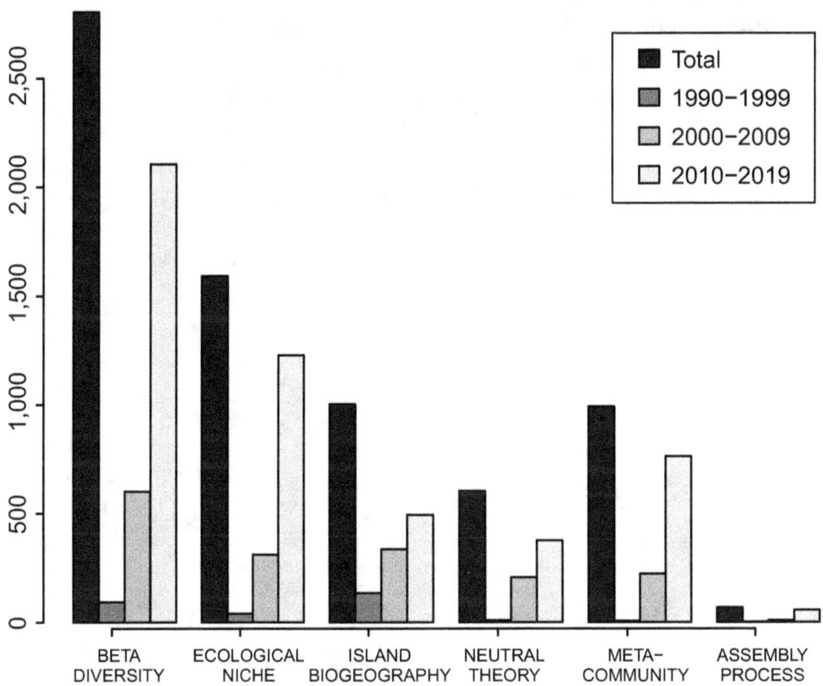

Figure 1.3 Number of scientific studies that have addressed central concepts in community ecology (Web of Science, 19.03.2019)

evolutionary processes. These four high-level processes can be further classified as those that add species to a community (dispersal and speciation) and those that decrease the number of species (selection and drift) (Figure 1.2).

1.5.4 Which Ecological Theories Are Prevailing in the Current Community Ecology Literature?

Figure 1.3 shows the results of a Web of Science search that counts the number of studies in ecological research that have addressed some of the central concepts of community ecology. The number of studies is continually rising. Even those concepts that originated in the 1960s and 1970s (e.g. beta diversity, ecological niche, island biogeography) are still hot topics in current ecological research. The total number of studies considering more modern frameworks (e.g. metacommunity, assembly process) is expectedly lower, but currently experiencing a drastic increase.

As in any other scientific field, community ecologists should be aware of the history of the field as well as its current state in order to make further progress. In particular, neglecting the theoretical foundations of community ecology when analysing data can lead to misinterpretation of the statistical results. In contrast, exploring patterns in nature in light of existing theories helps to make sense of the patterns and to place them in the context of existing knowledge. Therefore, it is crucial to first be aware of what is currently known – and also what is not known – before starting to explore data and interpret findings in relation to existing knowledge.

2 · *Typical Data Collected by Community Ecologists*

Community ecologists apply theoretical, experimental and observational approaches to studying the processes that structure ecological communities (Morin 2011). Experimental approaches provide the most direct way of testing the effects of specific processes acting in particular communities. However, they typically focus on only a few axes of variation at a time, and their results can thus be hard to link to the full complexity of natural systems (Carpenter 1996; Schindler 1998). Likewise, while mathematical models can be used to clarify the links between underlying mechanisms to the dynamics of ecological communities, they typically rely on highly simplified assumptions, and can thus be difficult to relate to empirical data. Data acquired by non-manipulative observational approaches are shaped by the full complexity of assembly processes. However, as these processes can seldom be observed directly, it is difficult to causally relate the observed patterns to the underlying assembly processes when applying the observational approach. As all of these methods have their pros and cons, they are likely to provide the most comprehensive understanding when applied in combination. This book, however, mainly focuses on empirical research based on non-manipulative observational approaches.

Figure 2.1 describes the types of data that empirical community ecologists typically collect, and that HMSC can take as input. Understanding the basic features of these data and how they have been collected will be essential for properly setting up the HMSC model and for appropriately interpreting the results. In this chapter, we will briefly go through each of the data types: community data (Section 2.1), environmental data (Section 2.2), data describing the spatio-temporal context (Section 2.3), trait data (Section 2.4) and phylogenetic data (Section 2.5). Finally, we make some remarks about how to best organise the data prior to running the HMSC analyses and how to treat missing data (Section 2.6).

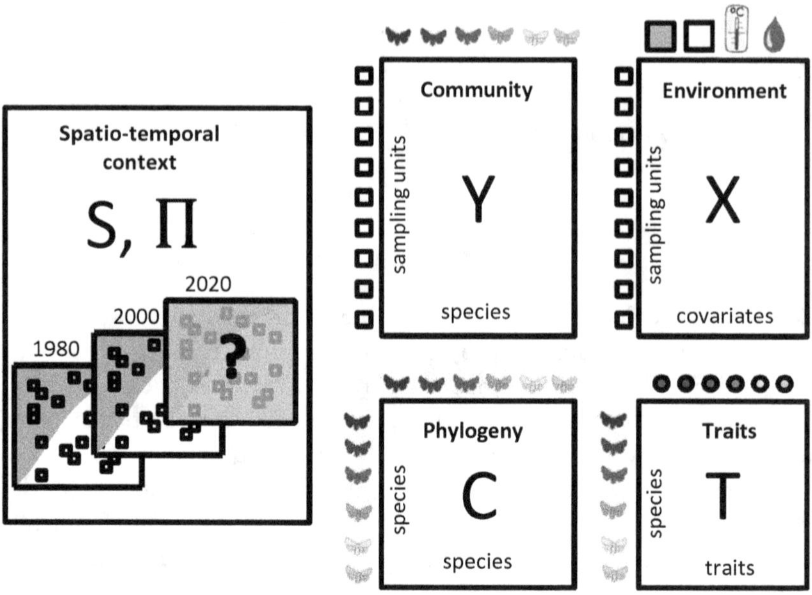

Figure 2.1 Data typically collected in community ecology. The community data (denoted as the **Y** matrix) include the occurrences or abundances of the species recorded in a set of temporal and/or spatial sampling units. The environmental data (denoted as the **X** matrix) consist of the environmental covariates measured over the sampling units. The trait data (denoted as the **T** matrix) consist of a set of traits measured for the species present in the **Y** matrix. The phylogenetic correlation matrix (denoted as the **C** matrix) quantifies phylogenetic relatedness among all species pairs, and can be derived e.g. from a phylogenetic tree. The spatio-temporal context includes location and time information about the samples, coded as hierarchical levels to which the sampling units belong (denoted as the **Π** matrix), and the spatial or temporal coordinates of the units included at each hierarchical level (denoted as the **S** matrix).

2.1 Community Data

Typical community data include field observations on the occurrence or abundances of species (**Y** matrix, Figure 2.1) in a set of temporal and/or spatial replicates. Throughout this book we refer to this matrix as 'community data' or, to emphasise the fact that these are the data to be modelled, we refer to them as 'response data'. Depending on the design of the observational or experimental study, the objectives of the study and the subject organisms, ecologists record the community data in various ways.

To start with, species can be observed either directly or indirectly. Most datasets based on traditional survey methods are constructed from direct visual encounters. For instance, vascular plant communities, insect communities and lichen communities are usually recorded by directly observing and counting the individuals of each species. Audial encounters, which are used to record e.g. bird and frog communities, also represent a direct way of recording species occurrence. In this book, we will present two case studies based on data collected via direct observation: vascular plant data obtained by surveying study plots (Section 6.7); and bird data recorded with audial encounters (Section 5.7 and Chapter 11).

Due to the difficulty in observing them directly, many mammalian species are often recorded using indirect cues such as tracks and droppings. Another way of indirectly recording species is through DNA-based molecular identification of environmental samples. With DNA-based methods, one can identify the occurrences of almost any target organism group from almost any environment or substrate, such as soil, wood, air or water (Bohmann et al. 2014). Most current applications of molecular species identification are based on amplifying and sequencing so-called barcoding genes. These markers have been selected as being especially suitable for discriminating between species, as they display a high degree of variation between species but only little variation within species (Hebert et al. 2003). In the molecular species identification workflow, the sequences are typically grouped based on their similarity into 'species hypotheses', referred to as operational taxonomic units (OTUs). The OTUs can then be connected to Linnaean taxonomies by comparing their sequences with those found in reference databases. In this book, we will present one case study based on DNA data: dead wood-inhabiting fungi surveyed by sequencing sawdust samples for the ITS region (the barcoding gene for fungi), followed by taxonomic placement of the sequences to the species level based on comparison with fungal reference databases (Section 7.9).

Community data on species can be measured in many kinds of units, including presence–absence, or abundance measured as e.g. percentage cover, biomass or counts of individuals or sequences. In studies of many larger animals, such as birds and mammals, it is often feasible and natural to measure abundance by counting the individuals. In vegetation studies, it can be difficult to distinguish among individuals, and thus species' abundances are often estimated as percentage cover. In studies focusing on energy flows, such as those on trophic interactions in arthropod communities, species' biomass may be the preferred abundance measure (Saint-Germain et al. 2007). In fungal studies based on fruit-body surveys, it is often difficult to reliably quantify the abundance of the study organisms, and hence the data may be collected simply as presence–absence. In this book, we will analyse abundance data based on counts of individuals of birds (Section 5.7), counts of occupied study quadrats of plants (Section 6.7), and counts of sequences of fungi (Section 7.9). Additionally, we will analyse presence–absence data in all three case study examples, by truncating the original data to presence–absence.

No data come without error. The most typical error in community data is that of 'false negatives', meaning that some individuals or species have remained unobserved because of a limited sampling effort, namely imperfect detection (Guillera-Arroita 2017). HMSC allows imperfect detection to be accounted for by including sampling effort-related covariates, such as variation in the total sequence count in the fungal case study (Section 7.9). However, separating the observation process more mechanistically from the biological processes remains to be implemented in HMSC, as discussed in the section dedicated to future development needs (Section 12.2.1). Another type of error can arise due to species misidentification, bringing both 'false negatives' (species that were actually present but misidentified) and 'false positives' (species that were actually not present but were recorded as present because they were misidentified as another species) into the community data. Species misidentification is especially common with DNA-based methods, partly because of the limited information contained in the short barcoding sequences, and partly because of the incomplete nature of the reference databases (Lou & Golding 2012). Species identification uncertainty can be quantified by using statistical methods that estimate the probabilities of the taxonomic assignments (Somervuo et al. 2017). While our case study on fungi will be based on species identified with probabilistic taxonomic placement (Section 7.9), the possibility of propagating species identification uncertainty in HMSC represents another challenge remaining for future developments (Section 12.2.1).

The last decade has experienced a rise of project initiatives aiming to compile already existing data, either from natural history collections, citizen data or scientific publications. Some of the highly used platforms include the Global Biodiversity Information Facility (GBIF 2018), national biodiversity atlas projects (e.g. Asher et al. 2001 for British and Irish butterflies; Saurola et al. 2013 for Finnish birds) and citizen science platforms such as iNaturalist (iNaturalist 2019). Scientific journals increasingly encourage data sharing, both by requiring the publication of primary data, and through the possibility of publishing specific data papers. While community ecologists originally tended to collect their own data, the increasing availability and large extent of already compiled data has greatly increased the reuse of data collected earlier. The use of data compiled from open sources is proving especially popular when analysing and predicting communities at a large scale (e.g. Brotons et al. 2007; Midgley et al. 2002; Pearman et al. 2008). Yet, such motley data come with added variation emanating from their mixed sources, adding new challenges to data modelling (e.g. Fithian et al. 2015; Pacifici et al. 2017). It is particularly problematic when the errors or biases in the data are correlated with the predictors of interest; for example, if using data from the GBIF without controlling for the variation in the observation effort, one might infer that biodiversity is highest in areas which are most accessible by humans, simply because most observations happen to be from such areas (Beck et al. 2014; Boakes et al. 2010). As another example, the species recorded in the global databases are not a random sample from all species, as they reflect the expertise and interests of the volunteers recording the observations (Troudet et al. 2017).

2.2 Environmental Data

The community data are usually accompanied by environmental data consisting of a set of covariates that the ecologist hypothesises to be important in explaining community composition (**X** matrix, Figure 2.1). Environmental covariates can be continuous (e.g. elevation, temperature, precipitation, pH) or categorical (e.g. vegetation type, whether the environment is pristine or disturbed, whether the sample originates from soil or water). In this section we only briefly discuss the types and sources of environmental data. Franklin (2009) and Guisan et al. (2017) provide much more extensive reviews on the types of environmental data that can be used in SDMs, and how they can be expected to constrain species distributions.

The environmental data can be directly measured by the community ecologist, or collected from an already existing dataset. Small-scale variables (e.g. microclimatic variables, soil properties) are typically measured directly when the community data are collected. As one example, we will use the decay class of the log as an environmental covariate in the fungal case study. This variable was measured in the field at the same time when the community data were sampled (Section 7.9). Large-scale variables (e.g. macroclimatic variables or vegetation types) are often collected afterwards from publicly available environmental datasets. Popular sources for climatic variables include the global climate Worldclim projections (Fick & Hijmans 2017; Hijmans et al. 2005) and the European climate maps constructed by Haylock et al. (2008). Other frequently used sources of large-scale environmental variation include habitat type classifications based on satellite remote sensing such as the CORINE land cover data (European Environment Agency 2016), vegetation map projections (e.g. Brus et al. 2012), and the Google Earth Engine platform (Gorelick et al. 2017). We will illustrate the use of such data in the bird case study (Section 5.7 and Chapter 11), where we utilise the CORINE land cover data (European Environment Agency 2016) for habitat type classification, and European climate maps (Haylock et al. 2008) for climatic variables.

Just as the community data, **Y**, may contain uncertainties, so do environmental predictors **X**. For example, global climate or vegetation maps have been produced from model predictions and thus contain their own uncertainties. Therefore, to minimise the degree of uncertainty, one should carefully select for the sources of environmental data. How much the remaining uncertainties matter for the ecological inference that is derived from SMDs very much depends on the nature of the uncertainties. For example, if one were to add independent random noise on top of perfectly clean data, one would expect the signal in the data to decrease but for no other biases to arise. Thus, with random data errors one would expect to overestimate the role of stochastic ecological drift in the data, in the same way that one would do if missing relevant predictors from the model.

2.3 Spatio-temporal Context

One unit of community data is one row of the matrix **Y** (Figure 2.1), i.e. the occurrences or abundances of the species as recorded for one sampling unit. Depending on the study, the general term 'sampling unit' may refer to a single visit that a researcher has made to a study plot or transect line, the catch of a trap over one day, a point count of birds based on fifteen minutes of listening or all of the butterfly observations that have been recorded in a particular 10 km × 10 km area over the last ten years. The sampling unit can thus represent widely different spatial and temporal scales – the choice of which will depend on the focal study organisms, the method used for acquiring the data, and the questions to be addressed by the study. In our bird case study (Section 5.7 and Chapter 11), each sampling unit will correspond to a transect line count conducted along a predefined survey route during one day. In our plant case study (Section 6.7), each sampling unit will correspond to one site on which twenty-five quadrats of 1 m × 1 m size were surveyed. In our fungal case study, each sampling unit will correspond to a single log from which several sawdust samples were pooled before sequencing (Section 7.9).

Concerning the spatial scale, in some cases there is a natural scale for defining local communities, such as the individual tree for saproxylic organisms, the host individual for microbiome studies, or the water body for planktonic communities. If the local community can be sampled exhaustively, as may be the case when surveying saproxylic organisms, the sampling unit of the empirical study can be selected to match the ecological scale of the local community. When the local community cannot be sampled exhaustively, for example in the case of planktonic communities, one may take several samples from the same local community. If these samples are pooled, the sampling unit corresponds to the local community. If the samples are not pooled, the sampling unit (in the sense of one row of the matrix **Y** in Figure 2.1) will be one sample, not the local community that the sample represents. In our fungal case study, the scale of sampling corresponds to the scale of local community, as interactions among dead-wood-inhabiting fungi take place within the well-defined unit of the log (Section 7.9). In our bird (Section 5.7 and Chapter 11) and plant case studies (Section 6.7), the spatial scale of sampling is arbitrarily decided rather than defined by the scale at which the ecological processes act.

Concerning the temporal scale, sometimes the sampling unit represents those species that were instantaneously present, such as the bacteria that are present in a soil sample at the time of sampling. In other cases, the sampling unit represents an accumulation of species over time, as is the case when animal tracks are counted in snow. Often the data are collected at higher temporal resolution but later pooled over time. For example, the time when a bird was heard by an audio recorder, or a mammal seen by a camera trap, may be known to the accuracy of a second, yet all observations made during a single day may be pooled into one sampling unit. Pooling into larger units over space and time is especially common in data compiled from multiple sources, such as the biodiversity atlas data mentioned above. In all of our case studies, the sampling only covers a short period of time, and thus represents the community that was present at the time of sampling.

In respect of HMSC, the spatio-temporal context of the study design is described by the two matrices **Π** and **S** (Figure 2.1). The matrix **Π** describes the units of the study. In addition to the sampling unit, which is the finest spatial and temporal unit, the units of the study may contain larger hierarchical scales. For example, in the case of saproxylic organisms, the sampling unit may be a single survey of a single tree, as in our fungal case study. The sampling-unit level can be nested within the level of the forest, meaning that other trees are surveyed within the same forest and other forests are surveyed in the same landscape. In addition to spatially nested hierarchical units, the sampling design may also contain cross-sectional structures. For example, the same trees may have been surveyed for saproxylic organisms over consecutive years. In this case, one tree-year pair would comprise one sampling unit, and the matrix **Π** would describe which tree, which forest and which year the sampling unit represents. While the plant data are collected at the quadrat level, we will use the plant abundances at the site level as the response variable (Section 6.7). Similarly, while the fungal data were collected from multiple samples within each log, we will not use the individual samples as the sampling unit, but rather the pooled sample representing the entire log (Section 7.9). However, HMSC can be used in a straightforward manner for hierarchical case studies, as we will illustrate with simulated data (see Section 5.6 for an example).

The matrix **S** describes the spatial and temporal coordinates of the units included in the matrix **Π**. In our case studies with fungal (Section 7.9) and plant data (Section 6.7), we will not include any spatial or temporal coordinates, and thus our models will be spatially and temporally implicit. In our case studies with bird data (Section 5.7 and Chapter 11), we will take a spatially explicit approach by including the spatial coordinates of the transect lines into the analyses. In Chapter 10 we will illustrate both spatially (Section 10.2) and temporally (Section 10.3) explicit HMSC analyses in the context of simulated data generated by an individual-based model.

2.4 Trait Data

If the aim is to understand how species traits influence community assembly processes, one needs to include data on species-specific traits (**T** matrix, Figure 2.1). These data may include morphological traits such as body size, physiological traits such as tolerance to climatic conditions, functional traits such as feeding type, or the position of the species within the surrounding food web. Much like environmental variables, traits can be continuous (e.g. body size in animals, seed size in plants or spore size in fungi), or categorical (e.g. feeding type in animals, life form in plants or trophic group in fungi). In our case study on birds, we will include the categorical trait of migratory behaviour, as well as the continuous trait of body mass (Chapter 11). In our case study on plants (Section 6.7), we will use the leaf tissue carbon-to-nitrogen ratio as a surrogate for competitive ability, representing a continuous trait.

Trait data can be collected simultaneously with community data, e.g. by measuring ten randomly selected individuals per species and then using the mean and standard deviation of these to represent the mean trait and its variability for each species. Alternatively, one may use trait data from published studies or already compiled trait database platforms. Frequently used global trait databases include the TRY database for plant traits (Kattge et al. 2011), and the EltonTraits database for foraging traits of birds and mammals (Wilman et al. 2014). As with environmental data, the researcher needs to be aware of potential errors and biases related to trait data. Clearly, global species trait databases are incomplete and contain errors, and may correspond more accurately with the actual traits of the species from one part of the world than from another part of the world.

2.5 Phylogenetic Data

For evaluating the extent to which species niches reflect phylogenetic relationships, one needs phylogenetic data. While we have represented such data by the phylogenetic correlation matrix **C** in Figure 2.1, these data typically come in the form of a phylogenetic tree. Phylogenetic trees are generally constructed by running genomic sequence data through phylogenetic analysis software (Nascimento et al. 2017). The availability of both the raw data as well as constructed phylogenetic trees is rapidly increasing in the literature, allowing quantitative phylogenetic trees to be integrated into analyses in many organisms. For example, our bird case study (Chapter 11) uses a phylogenetic tree constructed for the focal set of species with the BirdTree platform (https://birdtree.org/) (Jetz et al. 2012).

Where genetic data and quantitative phylogenies are lacking, data on taxonomic identity (at the level of genus, family, order, class, phylum, etc.) may be the best available proxy of phylogenetic relatedness. Taxonomical data can also be represented as a tree, but unlike quantitative phylogenetic trees, the branch lengths of a taxonomical tree are often assumed to be the same between all levels. Hence, branch length does not necessarily correlate with the times since the species have diverged from their common ancestors. We illustrate this approach in our plant case study, where we account for the relatedness among the species based on a taxonomical tree (Section 6.7).

2.6 Some Remarks about How to Organise Data

To run the statistical analyses of the data as smoothly as possible, we recommend organising the data as outlined in Figure 2.1. Thus, in the matrix **Y** of community data, each column represents a species and each row a sampling unit. As with the community matrix **Y**, the rows of the environmental matrix **X** and the study design matrix **Π** also correspond to the sampling units. Hence, since these three matrices share the same rows, they can all be compiled into a single data file. The species names in the trait data in the **T** matrix should be identical to the species names in the phylogenetic tree (from which the phylogenetic correlation matrix **C** is derived), which should be identical to those used in the community data **Y**. While this may sound trivial, small inconsistencies are often observed in practice, for example the bird species *Corvus monedula* could be named as 'Corvus monedula' in the **T** matrix, 'Corvus_monedula' in the phylogenetic tree and 'cor_mon' in the **Y** matrix. Unfortunately, while trivial for the bird researcher, it can be challenging for the computer to recognise that these three labels represent the same species.

Organising messy raw data into a clean format can sometimes take as much time as conducting the statistical analyses. Thus, it is important to learn how to keep the data clean directly from the beginning of the study. In simple studies, this can be done perhaps most straightforwardly by typing the data directly e.g. as csv files, whereas with more complex studies, utilising more sophisticated database structures may be worth the effort.

A common concern with data matrices is that they often contain missing data. For example, when the measurement of some environmental covariate is very time-consuming or expensive, it might be measured only for a subset of the sampling units. As another example, when the trait data are derived from existing databases, all species might not be found from the database. In the current implementation of HMSC, missing data are allowed for the community data matrix **Y** only, and hence not for the environmental data matrix **X**, the species trait data matrix **T**, the phylogenetic correlation matrix **C**, the study design matrix **Π**, nor the matrix **S** describing the spatio-temporal context of the study.

Concerning the community data matrix **Y**, it is important to keep in mind that 'missing data' (NA) is fundamentally different from zero or absence (0). A zero means that the species was searched for but not found, whereas 'missing data' means that the species was not even searched for. Thus, unlike zero or absence data, the missing observation is omitted when computing the likelihood of observing the data during model fitting. Concerning the other data matrices **X**, **T**, **C**, **Π** and **S**, there are basically two options if these have missing data. The first and simplest option is to exclude all the data that include missing elements. For example, if one or more environmental covariates have not been measured for a small number of sampling units, it may be wise to exclude those sampling units altogether from the statistical analyses. In contrast, if one or more environmental covariates are missing from most sampling units, a better option might be to exclude those environmental covariates, rather than restricting the analyses to a small subset of the sampling units. The other option is to apply the so-called data imputation; in other words, to make up the missing values. For example, if some particular trait of a particular species is missing, a trait value for a closely related species can be used as a proxy for the missing trait. More generally, data imputation refers to replacing missing data with an estimated value based on available information. Extensive imputation of missing data can introduce biases into the results, and thus should be applied with caution.

3 · *Typical Statistical Methods Applied by Community Ecologists*

In this chapter we introduce statistical methods commonly applied by ecologists working with community data. The aim is not to review every single statistical approach that can be applied to community data, but rather to give a general overview of the available tools. In this way, we will place JSDMs in general – and HMSC in particular – in the broader context of statistical community ecology.

We start by introducing ordination methods (Section 3.1), then discuss co-occurrence analysis approaches (Section 3.2), and finally methods based on generalised linear models of diversity metrics (Section 3.3). We will then move to species distribution modelling (Section 3.4), on top of which HMSC is built. While we will only verbally discuss the statistical methods in this chapter, we will apply some of the methods discussed to the bird community data in Section 11.5, where we illustrate how the outputs of these statistical methods compare to the outputs of HMSC.

3.1 Ordination Methods

In line with the historical development of community ecology theory, the analyses of observational data were largely descriptive until the 1950s. Conclusions were typically drawn from qualitative visualisation of the data e.g. in terms of diversity measures, such as distributions of species richness or species abundance. The first multivariate ordination methods emerged in the 1960s, as a response to the need for more quantitative methods (Bray & Curtis 1957; Gower 1966). The overall aim of ordination methods is to represent the multivariate nature of community data along a small number of axes representing the main trends. This is done by compressing information about the occurrences and co-occurrences of many species into a small number of axes that explain as much of the species variation as possible. Ordination methods provided community ecologists with new tools to link patterns in community composition with variation in environmental conditions, enabling more conclusive insights than were previously possible. Importantly, ordination methods allowed ecologists to relate their data to community processes by hypothesising the ecological and spatial processes that might correspond to the main axes of variation. The development and application of ordination methods has continued to expand since the 1950s, and they are currently by far the most widely used method for evaluating variation in species composition. A fundamental contribution to the development and application of multivariate ordination methods was set out in *Numerical Ecology*, published by Pierre and Louis Legendre in 1979, originally entitled *Écologie Numérique*. This book explained the wide array of multivariate ordination methods, both mathematically and ecologically, making them available for a large set of users (for the latest edition, see Legendre & Legendre 2012).

Ordination methods can be classified as unconstrained (also called 'indirect gradient analysis') or constrained (or 'direct gradient analysis'). Unconstrained ordination methods summarise the axes of variation in the community data without accounting for environmental data. Following Borcard et al. (2011), the basic idea of how unconstrained ordinations extract the axes of community variation is as follows: the community data matrix $\mathbf{Y}$, with dimensions of n sampling units and n_s species, can be illustrated graphically by placing the n sampling units in an n_s dimensional space. This will create a cloud of points, that will be elongated in some directions and flattened in others. The direction of the most elongated area corresponds to the greatest variation in the data, and this will be the first axis that an unconstrained ordination will extract. The next axis to be extracted will be the orthogonal axis (i.e. uncorrelated to the previous axis), that is the second most elongated area. This process is continued until all axes are extracted. Therefore, the first few axes contain most of the information about how communities are structured, and typically the first two axes are visualised in ordination plots. In ordination plots, both the sampling units (the locations of which are called site scores), and the species (the locations of which are called species scores) are plotted. Sites that are close to each other in the ordination space tend to have similar communities, whereas species that are close to each other tend to occur in the same sampling units.

Widely used unconstrained ordination methods include Principal Components Analysis (PCA), Principal Correspondence Analysis (PCoA) and non-metric multidimensional scaling (NMDS) (Legendre & Legendre 2012). Both PCA and PCoA utilise eigenanalysis, which yield an eigenvalue and an eigenvector for each axis of variation.

The eigenvalue measures the amount of variation, whereas the eigenvectors includes the site scores. Thus, the leading (largest) eigenvalue identifies the first axis of variation, and the corresponding eigenvector describes how the sites are ordered along this axis of variation. While PCA starts from the raw data and assumes the Euclidean distance, PCoA takes a distance matrix among the sites as the input. The choice of distance (or similarity) measure is an important part of any ordination analysis. For example, a commonly used similarity measure for abundance community data is Bray-Curtis, whereas the Sørensen distance is often used for presence–absence data. The eigenanalysis maximises the linear correlation between the original distances and those based on the main axes of variation. Since applying linear correlation is not a satisfactory approach for many kinds of data, NMDS relaxes the metric assumptions by maximising the rank order correlation rather than the linear correlation, and is thus not based on eigenanalysis.

A significant advance for statistical community ecology was the emergence of constrained ordination methods, that allow a more direct evaluation of the effects of environmental covariates. Unconstrained ordination methods are descriptive, in the sense that no statistical test is performed to assess the significance of the structures detected in relation to environmental or spatial constraints. However, in constrained ordination, one may test the hypotheses about whether and how environmental covariates influence the community composition, as the ordination axes only display the variation explained by the environmental (i.e. constraining) variables included in the analysis. This is done by combining the principles of unconstrained ordination, discussed above, with multiple regression.

The most widely used constrained methods in community ecology include Canonical Correspondence Analysis (CCA, ter Braak 1986) and distance-based Redundancy Analysis (db-RDA, Legendre & Anderson 1999). The axes of RDA are those linear combinations of the explanatory variables that best explain the variation in the community data. CCA is similar to RDA, but its implementation has been optimised to capture unimodal responses. Both methods make it possible to test the significance of the environmental covariates in explaining community variation, as well as to quantify their effects. An important development in constrained ordination has been the extensions that allow traits to be linked with environmental variation. These methods are the so-called RLQ (Dolédec et al. 1996) and fourth-corner analysis (Legendre et al. 1997). RLQ finds linear combinations that maximise covariance between the environmental data matrix and the trait data matrix, weighted by the community data. Fourth-corner analysis is similar, but focuses on the individual trait-environmental covariate relationships.

Accounting for spatial autocorrelation has been a long-standing challenge for ordination analysis. Ecological processes are generally spatially structured, as are ecological communities and the surrounding environmental variation. Therefore, sampling units that are close to each other tend to have more similar communities and environmental characteristics than those that are far from each other. If the spatial structure of the data is not accounted for, i.e. if sampling units are considered to be independent of each other regardless of their spatial distance, then the ecological signal can be confounded with the spatial structure (Legendre 1993; Legendre et al. 2002). This issue stimulated the development of a new generation of ordination approaches that enable assessing the spatial dependencies. For example, Principal Components of Neighbour Matrices (PCNM) uses the eigenvectors of the spatial distances among sampling units as explanatory variables of the RDA (Borcard & Legendre 2002). Another approach that has gained much popularity is the Permutational Multivariate Analysis of Variance (PERMANOVA, Anderson 2001). PERMANOVA can be used to statistically test the differences in community similarity among groups of sampling units that can represent larger spatial or temporal scales, along with assessing the effects of environmental covariates. Therefore, PERMANOVA is highly suitable for analysing community data collected from hierarchical study designs.

3.2 Co-occurrence Analysis

As discussed in Chapter 1, one of the most central topics in community ecology relates to species interactions. Sometimes interactions can be detected directly, for example by observing who feeds on whom. Such data are the starting point for network analysis, where species communities are represented by graphs, where the nodes are the species, and interacting species are connected by links. Various metrics describing the distribution of the links among nodes can then be derived (see Jordano 1987 for seminal work in this area). However, in many situations, interaction networks cannot be observed directly, and thus they need to be inferred indirectly. One way to hypothesise as to which species interact with each other is to assess species' co-occurrence patterns, based on the reasoning that interacting species should be non-randomly distributed with respect to each other. This idea dates back to Diamond

(1975), who found 'checkerboard distributions' among pairs of bird species that never co-occurred, driving the argument that this reflected negative interactions among species. The ordination methods discussed above are one way to study species' co-occurrence patterns: those pairs of species that have similar species scores (and are thus close to each other in the ordination space) will co-occur often, whereas those pairs of species that have dissimilar species scores (and are thus far away from each other in the ordination space) will co-occur infrequently.

While the analysis of co-occurrence patterns was implicitly included in ordination methods, these methods do not allow for direct testing of whether some sets of species do co-occur more or less often than expected by chance. Gotelli (2000) developed such tests by contrasting the realised co-occurrence patterns with those simulated by a null-model that assumed randomly distributed co-occurrences. However, non-random co-occurrence patterns can arise not only through species interactions, but also because of the species responding similarly or dissimilarly to environmental variation. To determine whether co-occurrence patterns deviated from those expected by the species responses to environmental variation, Peres-Neto et al. (2001) extended the pure null-model of Gotelli (2000) to an environmentally constrained null-model. In the approach of Peres-Neto et al. (2001), the responses of the species to environmental variation are first modelled one by one, after which co-occurrences are examined from the residual variation. In the early phase of joint species distribution modelling, the two steps of Peres-Neto et al. (2001) were combined in a single step with a model that included both the environmental responses of the species and the residual associations simultaneously (Ovaskainen et al. 2010).

3.3 Analyses of Diversity Metrics

Another way of analysing how community variation relates to environmental variation is to use univariate statistics, such as generalised linear mixed models (GLMM). Using GLMMs, community-level indices such species richness, Shannon evenness or Simpson similarity (Magurran 2004) are modelled as a function of environmental and spatial predictors. The flexibility of GLMMs allows one to incorporate many kinds of random effects that are needed to account for the nature of the study design, if it is for example hierarchical or spatial, or includes repeated visits (see e.g. Zuur et al. 2013). Analysing how broad-scale geographic species richness depends on environmental drivers has been termed macroecological modelling (MEM), which has been applied using both correlative (e.g. Hawkins et al. 2003) and process-based (e.g. Gotelli et al. 2009) statistical frameworks.

Univariate statistical models can also be applied to examine beta-diversity – in other words, how species composition varies over spatial or environmental gradients. In distance-based variance partitioning, a community distance matrix is considered as the response variable, and environmental and spatial distance matrices are considered as explanatory variables (Legendre et al. 2005; Smith & Lundholm 2010). A comparison of the explanatory powers of models with or without environmental and/or spatial predictors then identifies the unique and shared proportions of variance explained by the environmental and spatial predictors.

3.4 Species Distribution Modelling

Communities can also be approached through analysing each of the constituent species separately through species distribution models (SDM), also called habitat suitability models and niche models. SDMs are based on finding a statistical relationship between the abundance or occurrence of a species and its biotic and abiotic environment (Franklin 2009; Guisan et al. 2017; Peterson et al. 2011). Typical steps of SDM analyses involve: (i) defining one or more alternative SDM models and fitting them to the data; (ii) evaluating the abilities of the models to explain the fitted data and/or to predict held-out data not included for model fitting; (iii) using model evaluation to select the model that best captures the relationship between the environment and the species distribution; (iv) examining the parameter estimates of the best model to understand the drivers behind species distribution; and (v) using the best model to make predictions about species' abundances or occurrences in locations that have not been surveyed, or in scenarios representing different environmental conditions.

Since the introduction of the species distribution modelling package BIOCLIM in the 1980s (Booth et al. 2014; Nix 1986), SDMs have become highly popular, not only in community ecology but in ecology in general. They have also proved to be a powerful tool for testing (macro)ecological and biogeographical hypotheses on factors influencing species distribution ranges. Furthermore, SDMs are commonly used in applications in geography, paleoecology, phylogenetics, conservation biology and wildlife management (Araújo et al. 2019). Due to its wide applicability, there is a vast amount of literature on SDM approaches,

Table 3.1 *Summary of some popular and recently emerged SDM frameworks used to model community data. The SDM frameworks are classified either as single-species distribution models or joint species distribution models.*

Single-species distribution models	**Reference**
Boosted regression trees (BRT)	Hijmans et al. (2017); Ridgeway (2017)
Generalised additive model (GAM)	Wood (2011)
Generalised linear model (GLM)	R Development Core Team (2019)
Gradient nearest neighbour (GNN)	Crookston & Finley (2008)
Maximum-entropy approach (MaxEnt)	Phillips et al. (2006)
Multivariate adaptive regression spline (MARS-COMM)	Milborrow (2017)
Multivariate regression tree (MRTS)	De'ath et al. (2014)
Random forest (RF)	Liaw & Wiener (2002)
Support vector machine (SVM)	Meyer et al. (2017)
Gradient extreme boosting (XGB)	Chen et al. (2018)
Joint species distribution models	
Bayesian community ecology analysis (BC)	Golding & Harris (2015)
Bayesian ordination and regression analysis (BORAL)	Hui (2017)
Generalised joint attribute modelling (GJAM)	Clark et al. (2017)
Hierarchical modelling of species communities (HMSC)	Ovaskainen et al. (2017b)
Multivariate stochastic neural network (MISTN)	Harris (2015)
Species archetype model (SAM)	Hui et al. (2013)

which are covered extensively e.g. in Franklin (2009), Peterson et al. (2011) and Guisan et al. (2017). Below, we summarise this literature only briefly, with the aim of providing the background to how JSDM builds on single-species distribution modelling. In Chapter 5, we discuss how SDM relates to community ecology theory.

As illustrated in Table 3.1, there are many kinds of SDM frameworks to model data, either on individual or multiple species. The SDM frameworks differ in many aspects, including their structural assumptions (e.g. whether a generalised linear model or a random forest is assumed), statistical approaches (e.g. whether the model is fitted using maximum likelihood or Bayesian inference) and technical implementations (e.g. whether the method is available as an R-package or as self-standing software). The modelling frameworks of Table 3.1 can further be combined by applying so-called ensemble modelling, for example where predictions of several models are averaged (Breiner et al. 2015; Grenouillet et al. 2011). All SDMs listed in Table 3.1 are correlative, in the sense that they are based on finding statistical dependence between environmental data and species data. Thus, they do not directly model the assembly processes themselves, but the patterns generated by those processes (Elith & Leathwick 2009; Ovaskainen et al. 2017b). Assembly processes can be related more mechanistically to species distributions by the use of process-based SDMs (Boulangeat et al. 2012; Dormann et al. 2012; Morin et al. 2008; Talluto et al. 2016; Zurell et al. 2016). Process-based SDMs (also known as range dynamic models or hybrid SDMs) explicitly incorporate model structures and parameters describing some of the assembly processes (Zurell et al. 2016). Although this line of modelling can be linked more directly to ecological theory, in this book we focus on correlative SDMs. The motivation for doing so is that correlative SDMs are more generally applicable than process-based SDMs. This is because it is very difficult to derive a process-based framework that could readily be applied to the wide variety of data types typically available for community ecologists (see Chapter 2). For constructing process-based SDMs, one would ideally need more direct information on the underlying processes, such as data on fecundity, mortality and dispersal, and the dependency of these on the current population state of the focal and other species, as well as on their dependency on the prevailing environmental conditions. Such data are still rarely available. Even if most SDMs are correlative, one would like to link them to the underlying assembly processes as much as possible. To do so, it is important to understand the extent to which the results from correlative SMDs can or cannot be related to the underlying processes. In this context, the structural assumptions behind SDMs can be viewed as assumptions about how ecological communities are structured (D'Amen et al. 2017; Guisan & Thuiller 2005; Norberg et al. 2019).

The SDM frameworks in Table 3.1 are classified as single-species distribution models (SDMs) that model each species separately (Elith & Leathwick 2009; Ferrier & Guisan 2006; Guisan & Zimmermann 2000; Zimmermann et al. 2010), and joint species distribution models (JSDMs) that model all species at the same time (Clark et al. 2014; Ovaskainen et al. 2017b; Warton et al. 2015). Single-species distribution models were primarily developed for applications concerning only one species, whereas the focus of JSDMs is directly for communities comprising many species. Yet, single-species distribution models can also be applied to model the distributions of multiple species. This approach is called stacked species distribution modelling (SSDM), which is implemented by first fitting single-species distribution models separately for each species and then combining their predictions (Calabrese et al. 2014; Guisan & Rahbek 2011). In contrast, JSDMs combine species-level models into one model that is simultaneously fitted to all data. This allows JSDMs to seek community-level patterns in how species respond to their environment (e.g. Clark et al. 2014; Ovaskainen & Soininen 2011; Ovaskainen et al. 2017b), to relate such patterns to species traits and phylogenies (Abrego et al. 2017a; Brown et al. 2014; Pollock et al. 2012), and to quantify co-occurrence patterns among species (Ovaskainen et al. 2010; Ovaskainen et al. 2016a; Pollock et al. 2014). These two ways of analysing community data are also known as the 'predict first, assemble later' (SSDM) and the 'assemble and predict together' (JSDM) strategies (Ferrier & Guisan 2006). The Hierarchical Modelling of Species Communities framework – which this book is about – belongs to the latter JSDM strategy (Table 3.1).

4 · *An Overview of the Structure and Use of HMSC*

In this chapter, we will provide an overview of the HMSC framework. In Section 4.1, we first place HMSC in the context of the widely applied GLMM. In Section 4.2, we outline the mathematical structure of the HMSC model and introduce the notation that is used throughout the book. In Section 4.3, we introduce the conceptual theoretical framework under which HMSC has been developed. Finally, in Section 4.4, we illustrate the five steps of the HMSC workflow. As this chapter is aimed at providing a quick overview, it does not fully cover every mathematical feature nor the details about how different components of the HMSC model can be used for addressing specific study questions in community ecology. These will be covered in the remaining chapters of the book, where we build HMSC step by step, relate each component to ecological theory, and illustrate their use through examples.

4.1 HMSC Is a Multivariate Hierarchical Generalised Linear Mixed Model

Before describing what HMSC is all about, let us first clarify what HMSC is not about. Namely, HMSC is not a mechanistic, process-based model of community assembly, which would thus contain explicit descriptions of the assembly processes. In other words, HMSC does not include descriptions of how or why individuals interact with their environment and with other individuals of their own and other species. In particular, HMSC does not document who eats whom, or how energy and other resources flow through the network of interacting species. Furthermore, HMSC does not involve dispersal kernels or other quantifications of how individuals move around, nor does it contain descriptions about neutral processes such as ecological drift and demographic stochasticity. While it is feasible to incorporate mechanistic components to single-species distribution models in the so-called hybrid SDMs and process-based SDMs (e.g. Boulangeat et al. 2012; Dormann et al. 2012; Morin et al. 2008; Talluto et al. 2016; Zurell et al. 2016), such models are necessarily more case specific. The work leading to HMSC aimed to develop a model that would be generally applicable to many kinds of data collected by community ecologists (see Chapter 2). For this reason, even if we are fundamentally interested in the underlying processes, HMSC is not a mechanistic model but a correlative one. However, as illustrated in Section 4.3, it has been built in such a way that the model components can be conceptually related to community assembly processes. We will return to these links many times in Chapters 5–7, where the different HMSC components are developed in detail. Furthermore, we will focus particularly on the links between the underlying assembly processes and HMSC outputs in Chapter 9.5, where we apply HMSC to data generated by a mechanistic simulation model, with known assembly processes. By doing so, we will specifically ask how the results of HMSC analyses relate to the underlying community processes.

In the statistical terminology, HMSC is a *multivariate hierarchical generalised linear mixed model fitted with Bayesian inference.* This is a very general and widely applied statistical framework, and thus the novelty of HMSC is not in the statistical framework itself but in how the framework is applied to combine information from many types of data to infer community assembly processes. The above description of the statistical framework has much technical terminology. After this introductory chapter, we will gradually develop HMSC by addressing each of these terms individually. Thus, we will start Chapter 5 with the *linear model,* with which we expect most readers to be familiar. In Section 5.3, we will discuss link functions and error distributions to encompass *generalised* linear models. In Section 5.4, we will discuss random effects to extend to generalised linear *mixed* models. These components will form the core of Chapter 5 about single-species distribution modelling. HMSC can indeed be applied to single-species distribution modelling as well, as we illustrate using both simulated data (Section 5.6) as well as real data on the bird species *Corvus monedula* (Section 5.7). In Chapter 6, we will move to multi-species distribution modelling, thus extending to the *multivariate* case, and adding a *hierarchical* structure that builds a link from species traits and phylogenetic relatedness to their environmental niches. In Chapter 7, we discuss random effects in the context of multivariate models, where they are needed not only due to the nature of the study design but also to capture species-to-species associations. In Chapter 8, we cover the basics of *model fitting with Bayesian inference*, and in particular define the prior distribution of HMSC. In Chapter 9, we discuss topics related to model fit and model selection. Together, these chapters form Part II of the book, which builds the HMSC model step by step.

While the use of GLMMs is widespread in ecological literature, the use of the terminology is not always consistent (e.g. Bolker 2008; Fox et al. 2015; Gelfand et al. 2019; Zuur et al. 2009). For this reason, Table 4.1 defines the conventions that we follow in this book when referring to different kinds of statistical frameworks. All of the modelling frameworks presented in Table 4.1 can have one or more explanatory variables, or even none, in which case they are called intercept-only models. The explanatory variables can always be either continuous or categorical. In the context of species distribution modelling, the number of response variables is the same as the number of species. The types of the response variables relate to how species' occurrences or abundances are measured. An example of a normally distributed response variable might be log-transformed biomass, whereas presence–absence data or count data are by their nature not normally distributed. Random effects are generally needed to model dependency structures among the sampling units, for example due to hierarchical or spatial designs. In multivariate models, random effects can also model associations among the species.

4.2 The Overall Structure of HMSC

Bayesian models can be represented graphically with the help of a Directed Acyclic Graph (DAG), which describes how the parameters of the model are related to each other and to the data. The DAG of the core part of HMSC is shown in Figure 4.1. The boxes in Figure 4.1 refer to the data matrices that correspond to those shown in Figure 2.1; their precise meanings are described in Table 4.3. The ellipses in Figure 4.1 refer to the parameters of the HMSC model, described in greater detail in Table 4.4. The continuous arrows in Figure 4.1 correspond to statistical relationships that involve a random component, whereas the dashed arrows correspond to solely deterministic relationships. Both the statistical and deterministic relationships will be defined explicitly with the help of mathematical equations in Chapters 5–7. Those parameters that appear as the leaves of the DAG, i.e. those ellipses for which no arrows flow in, will require a prior distribution to be defined, as will be done in Chapter 8. The HMSC model is fitted to the data in the Bayesian inference using Markov chain Monte Carlo (MCMC) methods. While

Table 4.1 *Definitions of statistical frameworks as used in this book.*

Name of the statistical framework as used in this book	Number of explanatory variables	Types of explanatory variables	Number of response variables	Types of response variables	Random effects	Introduced in
Linear model	Zero or more	Continuous or categorical	One	Only normally distributed	Does not include	Section 5.2
Generalised linear model	Zero or more	Continuous or categorical	One	Can be non-normally distributed	Does not include	Section 5.3
Linear mixed model	Zero or more	Continuous or categorical	One	Only normally distributed	Can include	Section 5.4
Generalised linear mixed model	Zero or more	Continuous or categorical	One	Can be non-normally distributed	Can include	Section 5.4
Multivariate linear model	Zero or more	Continuous or categorical	One or more	Only normally distributed	Does not include	Section 6.1
Multivariate generalised linear model	Zero or more	Continuous or categorical	One or more	Can be non-normally distributed	Does not include	Section 6.1
Multivariate linear mixed model	Zero or more	Continuous or categorical	One or more	Only normally distributed	Can include	Section 7.3
Multivariate generalised linear mixed model	Zero or more	Continuous or categorical	One or more	Can be non-normally distributed	Can include	Section 7.3

Table 4.2 *Indices and their ranges in the core HMSC model.*

Index and its range	Refers to
$i = 1, \ldots, n$	Sampling unit
$j = 1, \ldots, n_s$	Species
$k = 1, \ldots, n_c$	Environmental covariate
$l = 1, \ldots, n_t$	Species trait
$h = 1, \ldots, n_f$	Latent factor
$u = 1, \ldots, n_u$	Hierarchical unit
$q = 1, \ldots, d$	Spatial coordinate in $\mathbb{R}^d$
$r = 1, \ldots, n_r$	Random effect

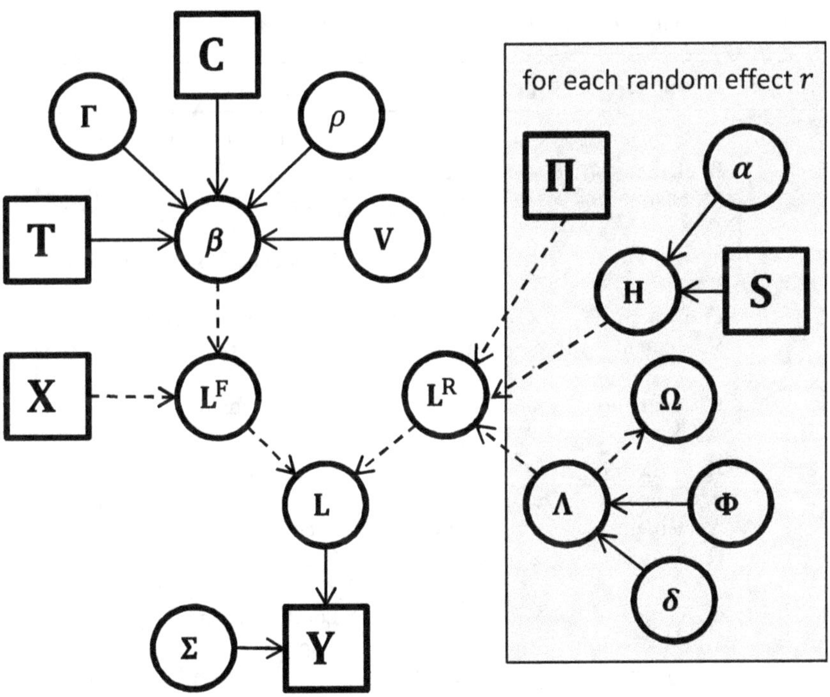

Figure 4.1 The Directed Acyclic Graph (DAG) of the core HMSC model. The input data are represented as squares and the estimated parameters as circles. The continuous arrows depict stochastic links modelled with the help of statistical relationships, and the dashed arrows depict deterministic links. The box illustrates a repeated structure, such that multiple random effects can be in the same model. The symbols are explained in Tables 4.2–4.4.

Table 4.3 *Data matrices and their dimensions in the core HMSC model. The spatial coordinates are defined separately for each random effect* r.

Data matrix	Dimension	Refers to
Y, element y_{ij}	$n \times n_s$	Community data
X, element x_{ik}	$n \times n_c$	Environmental data
T, element t_{jl}	$n_s \times n_t$	Species trait data
C, element c_{j_1j2}	$n_s \times n_s$	Phylogenetic data
Π, element π_{iu}	$n \times n_u$	Study design
S, element s_{uq}	$n_u \times d$	Spatial coordinates

Table 4.4 *Parameters and their interpretations in the core HMSC model. The 'Category' column indicates whether the parameter is related to the fixed effect (F), random effect (R), or data model (D) part of HMSC. The parameters of the random effect part are defined separately for each random effect* r.

Category	Parameter	Type	Interpretation
F	$\mathbf{L}^{\mathrm{F}}$, element L_{ij}^{F}	$n \times n_s$ matrix	Linear predictor of fixed effects
F	**B**, element β_{kj}	$n_c \times n_s$ matrix	Species niches
F	**M**, element μ_{kj}	$n_c \times n_s$ matrix	Expected species niches based on traits
F	ρ	scalar	Phylogenetic signal in species niches
F	**Γ**, element γ_{kl}	$n_c \times n_t$ matrix	Influence of traits on niches
F	**V**, element V_{k_1k2}	$n_c \times n_c$ matrix	Residual covariance of species niches
R	$\mathbf{L}^{\mathrm{R}}$, element L_{ij}^{R}	$n \times n_s$ matrix	Linear predictor of random effects
R	**H**, element η_{uh}	$n_u \times n_f$ matrix	Site loadings
R	$\boldsymbol{\alpha}$, element α_h	vector of length n_f	Spatial scale of site loadings
R	**Λ**, element λ_{hj}	$n_f \times n_s$ matrix	Species loadings
R	**Ω**, element Ω_{j_1j2}	$n_s \times n_s$ matrix	Species associations
R	**Φ**, element ϕ_{hj}	$n_f \times n_s$ matrix	Local shrinkage of species loadings
R	$\boldsymbol{\delta}$, element δ_l	vector of length n_f	Global shrinkage of species loadings
D	**L**, element L_{ij}	$n \times n_s$ matrix	Linear predictor
D	**Σ**, element σ_j^2	$n_s \times n_s$ diagonal matrix	Residual variance

we illustrate the use of Bayesian inference in the Chapters 5–7, we postpone a more thorough discussion of this topic until Chapter 8. HMSC includes several tools for evaluating model fit and comparing different models. Such cross-validation approaches are illustrated throughout Chapters 5–7, and a more detailed discussion of model selection can be found in Chapter 9.

The DAG of Figure 4.1 and Tables 4.2–4.4 contain a large amount of notation, which at this point is not expected to make full sense to the reader. However, while reading Chapters 5–8 that build the HMSC model in more detail, we expect that this information will be useful to return to, as it compactly summarises the model structure and thus helps to keep the big picture in mind.

4.3 Linking HMSC to Community Ecology Theory

Even if HMSC is based on a correlative GLMM framework, it is conceptually linked to the assembly rules framework (Ovaskainen et al. 2017b). This is illustrated in Figure 4.2, which links the illustration on community assembly processes (Figure 1.1) with the statistical structure of HMSC (Figure 4.1). Here we provide a brief overview of these links, which we will discuss more deeply in Chapters 5–7.

The fixed effects part of HMSC (shown on the left-hand side of Figure 4.2) models environmental filtering, i.e. how the interplay between species niches and environmental heterogeneity influences species occurrence and abundance. In HMSC, the parameters denoted by β are interpreted as species niches. These address environmental filtering from the species-specific viewpoint, measuring how environmental variation influences the occurrence of each species. Each species has its own niche, and thus its own β parameters. Species niches depend on response traits, the dependency on which is captured by the parameters denoted by γ. However, not all response traits may be known or measured. If the missing response traits are phylogenetically correlated, they can be expected to leave a phylogenetic signal on species niches. In HMSC, the presence and strength of the phylogenetic signal is captured by a parameter denoted by ρ.

The random effects part of HMSC (shown on the right-hand side of Figure 4.2) models biotic filtering, namely how the ecological interactions among species influence their occurrences, particularly their co-occurrences. A key parameter of this part of the model is the species-to-species association matrix $\mathbf{\Omega}$, which describes those species pairs that are

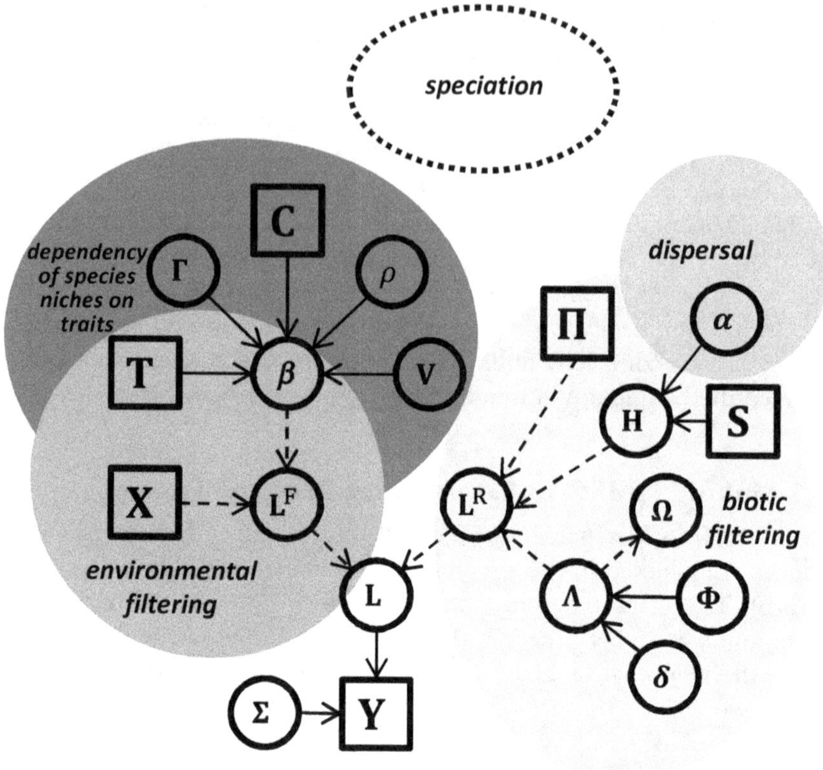

Figure 4.2 Links between the theory of community ecology and the statistical structure of HMSC. Differently shaded areas depict community assembly processes, whereas the boxes, circles and arrows depict the statistical structure of HMSC. Because speciation is not explicitly modelled in HMSC, it is represented by a dashed shape.

found together either more or less often than expected by chance. For adequate interpretation of the results presented by the species association matrix, it is important to understand exactly what 'by chance' means in this context, as co-occurrence patterns are generated both by environmental filtering and by biotic filtering. As the fixed effect part of HMSC controls for environmental filtering, two species that are found together 'more often than by chance' means that the two species are found together more often than can be expected by their niche similarity. Thus, the matrix $\mathbf{\Omega}$ describes residual (or environmentally constrained, see Section 3.2) co-occurrence rather than raw co-occurrences. While residual co-occurrences are causally more closely linked to biotic interactions than raw co-occurrences, the fixed effect part of the HMSC model will seldom be able to account for all other factors aside from biotic interactions. For this reason, interpreting residual associations as biotic interactions should be done with great caution, as we will emphasise in Section 7.6.

Dispersal limitation also causes variation in species' occurrences that cannot be explained by environmental filtering, resulting for example in a species being absent from an area in which the environmental conditions are suitable. Thus, the influences of dispersal limitation will also be captured by the random effects part of the HMSC model. In particular, if including spatially explicit random effects, a parameter denoted by α will capture the spatial scale of the unexplained random variation; this parameter can thus be expected to be linked to the spatial scale of dispersal. However, this interpretation comes with the same caveat as interpreting the random effects as biotic interactions: namely, if the fixed effect part of the HMSC model will miss some relevant environmental covariates, then the α parameters may capture the spatial scale at which those missing covariates vary, rather than the spatial scale at which dispersal takes place.

4.4 The Overall Workflow for Applying HMSC

Following Tikhonov et al. (2019), we summarise the typical workflow of an HMSC analysis into the five steps, illustrated in Figure 4.3.

The first step of the HMSC workflow (Figure 4.3) consists of setting up the model and fitting it to the data. Setting up the model involves making many kinds of choices. Some concern the structure of the model, such as whether spatial random effects are included and what kind of error distribution is assumed. Other choices relate to the selection of the predictors, e.g. which environmental covariates and species traits are included, and whether higher order effects or interactions among predictors are included. As HMSC is a rather complex model (Figure 4.1), describing its structure and discussing the relevant options will take a large part of this book, namely Chapters 5–7. Furthermore, many choices relate to the prior distribution. As setting up the prior distribution in an informed way can be challenging, the default prior distribution has been selected to be as generally applicable as possible, as described in Chapter 8.

While the first step of setting up the model requires an ecological understanding of the system and a clear view of the hypotheses to be addressed, the second step of the HMSC workflow (Figure 4.3) is purely technical. Specifically, it consists of checking whether MCMC

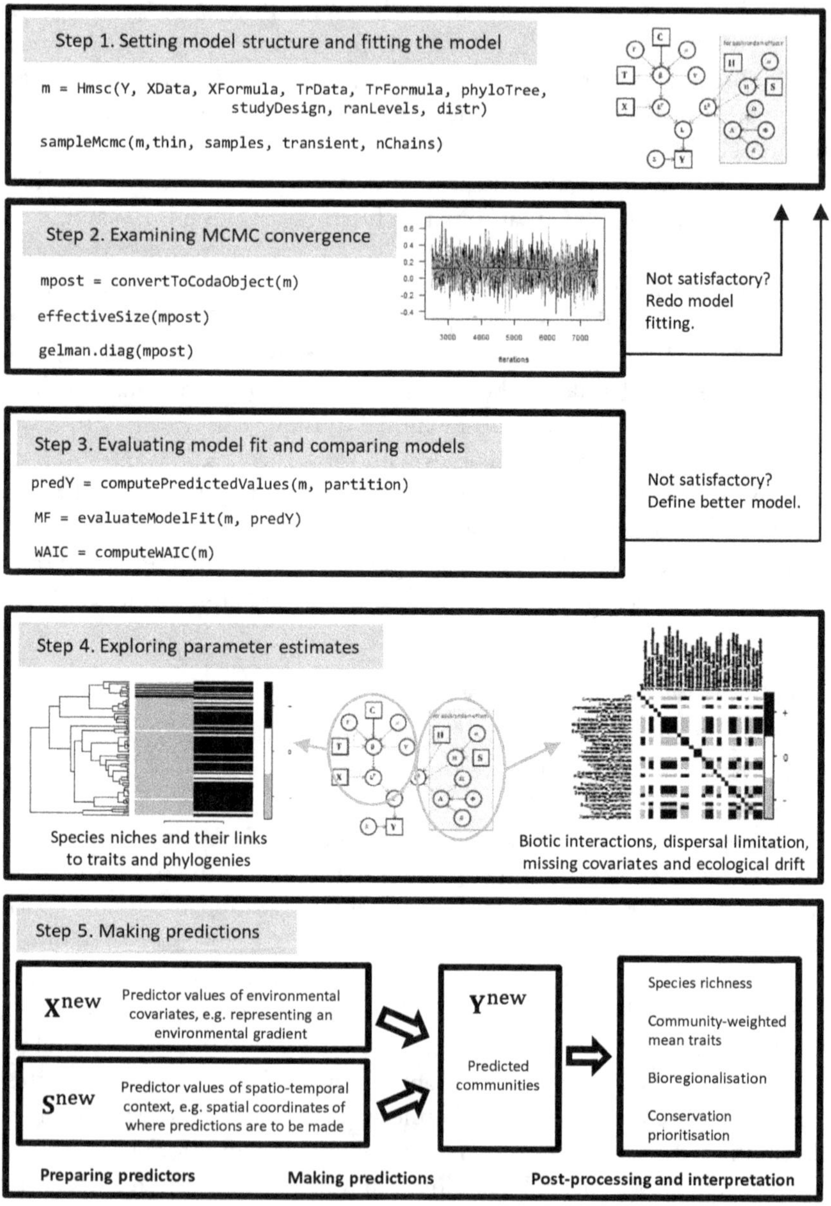

Figure 4.3 The five steps of a typical workflow of HMSC analyses. The computer code in Steps 1–3 illustrates the syntax of the R-package Hmsc. The graph in Step 2 shows an MCMC trace plot, and the graphs in Step 4 illustrate the estimates of some key model parameters.

convergence is satisfactory. If not, the inference derived from the fitted model cannot be fully trusted, and thus the model should be refitted by running more MCMC iterations. MCMC convergence can be examined with the help of visual inspection of MCMC trace plots, or using more formal diagnostics, such as the effective sample size and the potential scale reduction factor. We will illustrate these in Chapters 5–7, and discuss them more specifically in Chapter 8.

The third step of the HMSC workflow (Figure 4.3) consists of examining the model fit, and possibly comparing multiple models and selecting which will be used as the final model(s). The explanatory power of the model is determined by examining how well the model predicts the data used to fit it, whereas the predictive power is determined by examining how well the model predicts independent test data not used for model fitting. HMSC includes functionality to perform cross-validation in various ways, including partitioning either the sampling units or some higher-level units into different folds. HMSC also includes functionality for performing what we will call conditional cross-validation. Conditional cross-validation is based on partitioning not only the sampling units but also the species into different folds, and examining whether and how much accounting for species interactions (or more precisely, species associations) improves predictive power. Another approach to model selection is to use information criteria, out of which HMSC implements the so-called Widely Applicable Information Criterion (WAIC). While we illustrate cross-validation in Chapters 5–7, we postpone a more comprehensive discussion about model selection until Chapter 9.

The fourth step of the HMSC workflow (Figure 4.3) consists of exploring the parameter estimates. This is where most of the ecological inference takes place, and thus where the results of the analyses are related to ecological theory and knowledge on the study system. Key parameters relate to the species niches (measuring how species occurrences or abundances relate to abiotic variation), the dependency of species niches on species traits and phylogenetic relationships (identifying response traits and signals of niche conservation), and residual species-to-species associations (measuring a combination of biotic interactions, dispersal limitation, ecological drift, and missing covariates). We discuss the interpretation of these parameters extensively in Chapters 5–7, and illustrate them further with the help of a simulated case study in Chapter 10 and a real data case study of Finnish birds in Chapter 11.

The fifth step of the HMSC workflow (Figure 4.3) consists of using the model for making predictions. Predictions can be made for locations for which environmental covariates are known but where the species have not been surveyed. This can be done for creating maps of species distributions, species richness or trait distributions. In a comparison of a large number of species and joint species distribution modelling methods, HMSC generally performed the best in these kinds of prediction tasks (Norberg et al. 2019). Predictions can also be done to examine how the community is expected to respond to environmental perturbations, such as land-use change or climatic change in future scenarios. Furthermore, predictions can be used to illustrate how some aspect of community structure (for example, species richness or community-weighted mean trait) depends on environmental or spatial variation, as sometimes model predictions clarify such dependencies more clearly than the raw parameter estimates. We will illustrate many kinds of model predictions while building the HMSC framework in Chapters 5–7. In the context of the case study on Finnish birds in Chapter 11, we further illustrate how HMSC predictions can be post-processed for the purposes of bioregionalisation and conservation prioritisation.

Part II

Building a Joint Species Distribution Model Step by Step

5 · *Single-Species Distribution Modelling*

In this chapter we cover the basics of GLMMs in the univariate context of single-species distribution modelling. We start this chapter by discussing in Section 5.1 how SDMs relate to the theory on environmental species niches introduced in Chapter 1. We next introduce the linear model (Section 5.2), then move to generalised models (Section 5.3) and mixed models with both fixed and random effects (Section 5.4), and finally describe how the explained variance can be partitioned among the explanatory variables (Section 5.5). Although we assume that much of this material may already be familiar to a reader experienced in GLMMs, we still recommend that the reader at least skim through these sections, as they will provide the basic framework and notation for the rest of the book. In the applied part of this chapter, we will use both simulated data (Section 5.6) and real data (Section 5.7) to illustrate how the R-package Hmsc can be used to analyse GLMMs. While these analyses are rather standard and could also be conducted with many other packages (e.g. Zuur et al. 2013), we encourage the reader to go through them, as they provide the simplest way of getting familiar with the syntax of Hmsc. We will build on these examples in the more advanced parts of the book.

5.1 How Do Species Distribution Models Link to Species Niches?

SDMs assume that environmental filtering results in an association between environmental conditions and species occurrences. Since such an association is closely related to the niche concept, SDMs are sometimes called niche models (Peterson et al. 2011). However, the use of the term 'niche' in the context of correlative SDMs has been criticised, with the argument that a more mechanistic understanding of how environmental conditions affect fitness is required for modelling a niche (see Kearney 2006). To highlight the descriptive rather than mechanistic way in which the association between environmental conditions and species occurrence is built in SDMs, some researchers prefer to refer to SDMs with the terms 'environmental model' or "habitat suitability model" (Guisan et al. 2017). The term "envelope model" is also used, especially in studies that aim at predicting future distributions of species based on the species-climate association (Araújo & Peterson 2012; Hijmans & Graham 2006). Through this book, we use the term 'species distribution model', and although we acknowledge the static and correlative nature of SDMs, we conceptually consider SDMs to model the realised niche. We note that SDMs model realised rather than fundamental niche simply because they utilise data on actual species occurrence that are simultaneously influenced by the full range of assembly processes, not just the match between the environmental conditions and the fundamental niche of the species.

In SDMs, the relevant axes of the niche are defined *a priori* by the user through the choices of the environmental predictors included in the model. Thus, in the predictions, the model will return the environmental hypervolume (*sensu* Hutchinson 1959, see Chapter 1) only in terms of those environmental predictors. Therefore, when fitting an SDM, deciding which environmental predictors should be included is one of the most important choices to be made. There is no general theory telling which and how many environmental predictors the user should include in order to represent species niches. It is thus the job of the user to first investigate the ecology of the focal species/community, and then decide about potentially important environmental filters. To ensure that the important environmental filters are included, some users might be tempted to include as many predictors as possible. However, that is actually unwise, as increasing the number of predictors increases model complexity and thus the risk of overfitting, which will decrease the model's predictive power rather than increase it. Thus, even if applying variable selection tools (the discussion of which we will postpone until Chapter 9), it is important to make prior choices on the predictors based on their ecological relevance.

Furthermore, the user will need to decide whether to include only linear and additive effects of the different environmental predictors, or also their non-linear or interactive effects, implemented through polynomials of multiple predictors or by applying e.g. generalised additive models. Related to the structural relationship between environmental predictors and species occurrences, there is great variation among different SDM frameworks (Guisan & Thuiller 2005; Merow et al. 2014; Norberg et al. 2019). There is a gradient among SDMs allowing for very flexible predictor functions (e.g. random forest and generalised additive models) to more rigid ones (e.g. generalised linear models) (Guisan & Thuiller 2005; Merow et al. 2014) (Table 3.1). Similar to the choice of environmental predictors, the choice of SDM framework, in the sense of assumed relationship between species occurrence and environmental predictors, is not a matter of which one fits better to general theory, but which fits best to the focal study system (Guisan et al. 2002).

5.2 The Linear Model

The core starting point of HMSC is the linear model. We assume most readers to be familiar with the most basic linear model, which can be written as:

$$y_i = \alpha + \beta x_i + \varepsilon_i \tag{5.1}$$

In Equation 5.1, the subscript i is an index for the data point, which represents the sampling unit. If there are data from 100 sampling units, then i takes the values from 1 to 100; more generally, with n sampling units we write $i = 1, 2, \ldots, n$. The variable x is the explanatory variable, also called independent variable, or simply the x-variable, and its value for sampling unit i is denoted by x_i. The variable y is the response variable, also called the dependent variable, or simply the y-variable, and its value for sampling unit i is denoted by y_i. In the context of species distribution modelling, the response variable y is typically the occurrence or the abundance of a species, whereas the explanatory variable x is at this point a single environmental variable (e.g. climatic condition) that is hypothesised to influence the distribution of the species.

The basic question that is addressed when applying the linear model is how the response variable y depends on the explanatory variable x, for example how species occurrence or abundance depends on climatic conditions. This question is answered by estimating the model parameters, called the intercept (α) and slope (β). If the slope is positive ($\beta > 0$), then the response variable increases with the explanatory variable, i.e. the species is more abundant under warm than under cool climatic conditions. If the slope is negative ($\beta < 0$), the response variable decreases with the explanatory variable, i.e. the species is more abundant under cool than under warm climatic conditions. If the slope is zero ($\beta = 0$), the response variable does not depend on the explanatory variable, i.e. the species is equally abundant under cool and warm climatic conditions. If temperature is measured in the units of Celsius degrees, then $\beta = 0.1$ means that the response variable y is predicted to increase by one unit if the temperature increases by ten Celsius degrees. The intercept α is typically of less ecological interest, but nevertheless important, as it models the expected value of the response variable when the explanatory value equals zero. In the context of species distribution modelling, the intercept can relate to the mean occurrence probability or the mean abundance of the species, and hence it can be of interest as well. This can especially be the case if the model includes multiple species, in which case we may wish to ask how the rare species differ from the common ones. We will return to this issue in the next chapter, where we consider multiple species simultaneously.

In Equation 5.1, the term ε_i is the residual related to the sampling unit i. Rearranging the equation, we can write $\varepsilon_i = y_i - (\alpha + \beta x_i)$, which shows that the residual is the difference between the observed (y_i) and predicted ($\alpha + \beta x_i$) values of the response variable. The residual is needed because the response variable cannot be expected to be fully predicted by the explanatory variable, and thus some of the variation in the response variable will remain unexplained. This unexplained variation is called residual variation. In the linear model, the residual variation is assumed to be normally distributed, an assumption that can be written mathematically as $\varepsilon_i \sim N(0, \sigma^2)$. Here $N(\mu, \sigma^2)$ stands for the normal distribution with mean μ and variance σ^2, or equivalently, with standard deviation σ. The mean, or expected value, of the residuals can be set to zero without loss of generality, because the overall mean of the response variable is already captured by the intercept α. The variance σ^2 is a parameter of the model, and hence it is estimated with α and β when fitting the linear model to the data.

5.2.1 Continuous and Categorical Explanatory Variables

Equation 5.1 is a special case of the full linear model, as it contains only a single explanatory variable x. The linear model can contain any number of explanatory variables, which may be continuous (covariates) or categorical (factors). Let us assume that there are two covariates, denoted x_1 and x_2, such that their values for sampling unit i are denoted by x_{i1} and x_{i2}. The linear model can then be written as $y_i = \alpha + \beta_1 x_{i1} + \beta_2 x_{i2} + \varepsilon_i$, where β_1 and β_2 are the slopes of the covariates x_1 and x_2. More generally, if there are n_c covariates, we may write the linear model as:

$$y_i = \sum_{k=1}^{n_c} x_{ik}\,\beta_k + \varepsilon_i \tag{5.2}$$

Here the summation incorporates the covariates $k = 1, \ldots, n_c$ included in the model. To simplify the notation, we have dropped the intercept α from the model. This can be done by deciding that the first explanatory variable x_1 models the intercept, so that $x_{i1} = 1$ for all sampling units i. Thus, the actual number of explanatory variables in the notation of Equation 5.2 is $n_c - 1$. We note that Equation 5.2 can be written equivalently as $y_i \sim N(L_i, \sigma^2)$, where L_i is called the linear predictor:

$$L_i = \sum_{k=1}^{n_c} x_{ik}\,\beta_k \tag{5.3}$$

While Equation 5.2 was motivated in terms of continuous covariates, it applies equally well to factors, as long as we code them with the help of so-called dummy variables. For example, assume that we wish to use the habitat type in which the sampling unit is located as an explanatory variable, and that there are three habitat types, for instance coniferous forests, broadleaved forests and mixed forests. We can do this by setting $x_{i1} = 1$ as the intercept, x_{i2} as an indicator variable for broadleaved forest ($x_{i2} = 1$ if sampling unit i is in a broadleaved forest, otherwise $x_{i2} = 0$), and x_{i3} as an indicator variable for mixed forest ($x_{i3} = 1$ if sampling unit i is in a mixed forest, otherwise $x_{i3} = 0$).

Note that we did not include an indicator variable for coniferous forests, as it was considered as the reference level. With factors, one of the levels is always considered as the reference level, otherwise the model would be over-parameterised. In the model that we have just described, there are three parameters to estimate, the influences β_1, β_2 and β_3 of the explanatory variables x_1, x_2 and x_3, respectively. These three parameters are sufficient to model the expected value of the response variable for all three habitat types: the predicted value is β_1 for coniferous forests, $\beta_1 + \beta_2$ for broadleaved forests, and $\beta_1 + \beta_3$ for mixed forests. This is because the response to the intercept β_1 is always included in the prediction, and the dummy variables x_2 and x_3 determine whether β_2 or β_3 is to be included. For this reason, β_2 for example, does not model the effect of the sampling unit being in a broadleaved forest, but rather the difference that it makes if the sampling unit is in a broadleaved forest (the focal level) compared to being in a coniferous forest (the reference level). This somewhat complicates the interpretation of the parameter estimates related to categorical variables. For this reason, it is often more informative to assess their effect sizes by translating them into model predictions.

The explanatory power of the linear model is measured by R^2, called the coefficient of determination, which is defined as the proportion of variance that the model prediction explains out of the total variance in the response variable. In the linear model, R^2 equals the square of the Pearson correlation coefficient between the response variable and the model prediction. If $R^2 = 0$, the explanatory variables fail to explain any variation in the response variable, whereas if $R^2 = 1$, the response variable can be fully predicted based on the explanatory variables.

To clarify the terminology, even if Equation 5.2 can contain any number of explanatory variables, it is a univariate – not multivariate – model. This is because the terms 'univariate' and 'multivariate' refer to the number of response variables, not the number of explanatory variables. If we wish to emphasise that there are several explanatory variables, we may call Equation 5.2 a multiple linear regression. We will return to multivariate models in the next chapter, where we extend our analysis to multiple species.

5.3 Generalised Linear Models

For abundance data such as log-transformed biomass, the assumption of normal distribution in the linear model might be appropriate. However, survey data on species occurrences or abundances are typically not normally distributed. By occurrence we mean specifically the presence–absence of the species, which is typically coded as $y = 1$ for presence and $y = 0$ for absence. Abundance is often measured by counting the number of individuals, hence y is a non-negative integer. In both of these cases, the assumptions made by the linear model are not valid, and hence applying the linear model may lead to wrong conclusions. This is why generalised linear models are needed.

We recall that the linear model can be written as $y_i \sim N(L_i, \sigma^2)$, which means that the response variable y is assumed to be normally distributed. Here, the mean of the normal distribution is given by the linear predictor L_i, which is specific to each sampling unit i, and the variance is given by σ^2, which is common to all sampling units. Generalised linear models also build on the linear predictor, and thus Equation 5.3 remains unchanged for them.

5.3.1 Probit Model for Presence–Absence Data

We first consider the case of a binary response variable such as species occurrence, i.e. the presence ($y_i = 1$) or absence ($y_i = 0$) of the species.

A natural distribution for binary variables is given by the Bernoulli distribution, and thus we write $y_i \sim \text{Bernoulli}(\mu_i)$. Here μ_i is the expected value, which in the case of the Bernoulli distribution equals the probability of the response variable being one. Thus, $y_i = 1$ with a probability of μ_i and consequently $y_i = 0$ with a probability of $1 - \mu_i$.

The reason why generalised linear models based on the Bernoulli distribution will need a non-linear link function is that simply setting $\mu_i = L_i$ would not lead to a good model, as the linear predictor L_i can obtain any real numbers, whereas the expected value μ_i needs to be a probability, i.e. a number between zero and one. We expect many readers to be familiar with logistic regression, for which the link function is the logit function $\text{logit}(p) = \log(p/(1-p))$. Setting $L = \text{logit}(\mu)$ makes the probability μ range appropriately from zero to one as the linear predictor L ranges from minus infinity to infinity. Equivalently, we may write $\mu = \text{logit}^{-1}(L)$, where the inverse logit (also called the logistic function) is defined as $\text{logit}^{-1}(L) = \exp(L)/(1 + exp(L))$. When $L = 0$ it holds that $\mu = 0.5$, and negative values of the linear predictor correspond to probabilities smaller than 0.5, whereas positive values of the linear predictor correspond to probabilities greater than 0.5.

Even if the logistic regression is often used in statistical ecology, there is no ecological reason why the choice of this link function would be especially natural for presence–absence data. HMSC does not apply logistic link function for presence–absence data, but instead it applies the so-called probit link function. The probit link function is based on the cumulative distribution of the standard normal distribution $N(0, 1)$, i.e. the normal distribution with zero mean and unit variance. The cumulative distribution of the standard normal distribution is denoted by $\Phi(x)$. By the definition of a cumulative distribution, $\Phi(x)$ is the probability by which a random deviate from the standard normal distribution is at most x. The function $\Phi(x)$ behaves qualitatively similarly to the logistic function: when x increases from minus infinity to infinity, $\Phi(x)$ increases from zero to one. Thus, equally well as the logistic function, we may also apply $\Phi(x)$ to map the linear predictor into a probability. Thereby, the probit model is defined as:

$$y_i \sim \text{Bernoulli}(\Phi(L_i)) \tag{5.4}$$

where the linear predictor is given by Equation 5.3.

The reason why HMSC specifically applies the probit model instead of the logistic model is not a philosophical or an ecological reason, but simply a pragmatic one: in the full multivariate and hierarchical context of the HMSC, the probit model is easier to fit to data than the logistic model. This is because the probit model of Equation 5.4 can be written equivalently as $y_i = 1_{z_i \geq 0}$, where $z_i \sim N(L_i, 1)$, and the indicator function $1_{z_i \geq 0}$ takes the value of one if $z_i \geq 0$ and the value of zero if $z_i < 0$. Writing the probit model in terms of a normally distributed auxiliary variable z_i makes it possible to parameterise it using the broad array of methods available for the linear model. We note that even if the auxiliary variable z_i follows a normal distribution, the probit model does not have a residual term $\varepsilon_i \sim N(0, \sigma^2)$ where the variance σ^2 would be estimated. Instead, in the probit model the variance of z_i is fixed as one. The normal distribution behind z_i models Bernoulli randomness, i.e. the random variable obtains the value of one with probability $\Phi(L_i)$ and the value of zero with the probability of $1 - \Phi(L_i)$. While a typical user of HMSC may not need to worry about how parameters are estimated in practice (see Section 8.6), we wanted to include this technical note to clarify why HMSC applies probit regression instead of logistic regression.

5.3.2 Poisson and Lognormal Poisson Models for Count Data

We next consider the case of count data. In the context of species distribution modelling, species abundances are often measured as the number of individuals observed, in which case y_i is a count that can obtain any non-negative integer value $y_i = 0, 1, 2, \ldots$ The linear model does not apply naturally to such data because it can also predict negative values and any real numbers, not just integers. To restrict the predictions to non-negative integers, we need a suitable link function and error distribution. One simple option is to apply the Poisson distribution with the log-link function. The Poisson model is defined by:

$$y_i \sim \text{Poisson}(\exp(L_i)) \tag{5.5}$$

where the linear predictor is again given by Equation 5.3. The exponential function ensures that the mean of the Poisson distribution $\mu_i = \exp(L_i)$ is positive. Note that we can write equivalently $L_i = \log(\mu_i)$, showing that Equation 5.5 indeed has the log-link function hidden in it. The Poisson distribution is a discrete probability distribution restricted to non-negative integers, and thus the outcomes y_i of Equation 5.5 are appropriate for count data.

One feature of the Poisson distribution is that it does not have a variance parameter to be estimated. This is because the variance of the

Poisson distribution equals its mean, so it cannot be estimated separately from the mean. In other words, if $y_i \sim \text{Poisson}(\mu_i)$, then $\text{E}[y_i] = \text{Var}[y_i] = \mu_i$, where E[·] and Var[·] denote the expected value and variance of a random variable. This feature of the Poisson distribution often makes it unrealistic for ecological count data, because ecological count data often show more variation than it allows for. For example, consider the data points $y_1 = 3$, $y_2 = 0$, $y_3 = 0$, $y_4 = 2$, and $y_5 = 30$. These could represent e.g. the counts of some bird species, where in sampling units 1–4 we would have encountered just a few or not any individuals, whereas in sampling unit 5 we would have encountered a flock consisting of thirty individuals. These data could not plausibly arise from the standard Poisson distribution, if assuming the same mean for all sampling units. To illustrate this, we note that the mean of these data is seven, so Poisson (7) seems like a reasonable choice. But if the data were distributed according to the Poisson distribution with a mean of seven, the probability of seeing no individuals would be less than 0.1 per cent. Yet, in our example, there are no observed individuals in two sampling units. Even worse, the probability of seeing thirty or more individuals would be only $2 \cdot 10^{-11}$, so we would expect to see such a high number far less often than once per million. Thus, to model data of these characteristics, some distribution other than Poisson is needed.

One common choice for an error distribution that allows for more variance than the standard Poisson distribution is the Negative Binomial distribution. However, this is not implemented in HMSC, for the reason that it is highly challenging to technically implement in the multivariate and hierarchical context. Instead, HSMC implements another choice that essentially achieves the same goal, the so-called lognormal Poisson model. The lognormal Poisson model is given by:

$$y_i \sim \text{Poisson}(\exp(L_i + \varepsilon_i)),\ \varepsilon_i \sim N(0, \sigma^2) \tag{5.6}$$

Thus, the lognormal Poisson model differs from the baseline Poisson model by adding normally distributed noise to the linear predictor. Now, increased variance σ^2 in the normal distribution leads to increased variation in the outcomes y_i. To illustrate this, we randomised five random deviates from the lognormal Poisson distribution with $L_i = \log(7)$ and $\sigma^2 = 1$; the outcome was 0, 14, 43, 0 and 9. Thus, unlike the standard Poisson distribution, this distribution simultaneously allows for both small and large values. By letting σ^2 be a free parameter to be estimated, the lognormal Poisson distribution can adjust the amount of variance to that observed in the data.

As a technical point, we note that the expected value of y_i in Equation 5.6 is not $\exp(L_i)$, but instead $\exp(L_i + \sigma^2/2)$. The reason for this is the presence of the non-linear exponential function that maps $L_i + \varepsilon_i$ to the expectation of the Poisson distribution. We made this point to avoid confusion as to why the expected prediction under this model is not $\exp(L_i)$, as it would be under the standard Poisson model.

5.3.3 Hurdle Models for Zero-Inflated Data

Ecological data often have excess zeros – more than expected based on the distribution of the non-zeros – making them zero-inflated data. In the example above, we randomised the five random deviates 0, 14, 43, 0 and 9 from the lognormal Poisson distribution. With these data, the species would be absent from two sites, and rather abundant in the other three sites. To analyse these data, we would actually not need a zero-inflated model, as the lognormal Poisson distribution would explain perfectly both the distribution of zeros and non-zeros. But assume that the data would look instead like 0, 42, 39, 0 and 43. Now the non-zeros have much less variation, and it might be less plausible that the zeros originate from the same distribution as the non-zeros. For these data, a zero-inflated Poisson distribution model could be a perfect choice, as it explicitly models the excess zeros (Zuur et al. 2012).

Zero-inflated models are not currently implemented in HMSC, simply because doing so is technically challenging in the full hierarchical and multivariate context of HMSC. Furthermore, when implementing a zero-inflated model, one needs to decide about the distribution of the non-zeros, such as the use of the Poisson distribution in the context of the zero-inflated Poisson distribution. However, it is always possible to use the closely related hurdle modelling approach (Barry & Welsh 2002). A hurdle model contains two components, one of which models presence–absence, and the other models abundance conditional on presence. While a hurdle model can be viewed as a single model, it can be fitted as two separate models. This is an advantage in the context of HMSC, as it means that one can fit a hurdle model even if there is no specific implementation of hurdle models there. To fit a hurdle model, one first fits a presence–absence model (e.g. a probit model) for data truncated to presence–absence, i.e. for data where the zeros have been kept as zeros and all non-zero values have been set to one. To fit the second model on abundance conditional on presence, all data points that are zero are declared as missing data points, as "conditional on presence" means that they should be ignored. The model that is chosen for the abundance conditional on presence depends on the nature of the data, i.e. on the distribution of the non-zeros.

Once the two components of the hurdle model have been fitted, they can be combined to produce predictions that predict the full distribution of counts in the original response variable, including the zeros and non-zeros. To do so, first the presence–absence model is used to predict whether the species is present or not. If it is predicted to be absent, then the prediction is zero, and the model for abundance (conditional on presence) is not needed at all. If the species is predicted to be present, then the model for abundance (conditional on presence) is used to predict how many individuals there are. We do not illustrate hurdle models in this chapter, but we will return to them in the context of sequence count data in Section 7.9.

Finally, let us clarify the difference between zero-inflated models and hurdle models. The sole difference is that in a zero-inflated model, the zeros occur for two reasons. First, it may be that the binary part of the model predicts that there is an excess zero. Second, it may be that the binary part of the model does not predict an excess zero, but the count part of the model still predicts a zero. In contrast, in a hurdle model, all zeros are excess zeros. This difference implies that the two parts of a zero-inflated model need to be fitted to data simultaneously, whereas the two parts of a hurdle model can be fitted separately.

5.4 Mixed Models

In correlative species distribution models, spatial and temporal patterns that cannot be explained by environmental predictors can be accounted for by including spatial or temporal predictors or spatial or temporal covariance structures. These capture the amount and type of residual variation in the data that cannot be attributed to the variation in observed abiotic or biotic environmental conditions (Dormann et al. 2007).

One of the 'hidden' assumptions of the basic linear model of Equation 5.1 is that it ignores any spatial or temporal structure, in the sense that the residuals ε_i are assumed to be independent of each other. This assumption is implicitly present in the equation $\varepsilon_i \sim N(0, \sigma^2)$, but to make it explicit, we should specify that the residuals for different sampling units are assumed to be independent of each other.

What does it mean in practice that the residual ε_1 for sampling unit 1 is independent of the residual ε_2 for sampling unit 2? This question is probably most easily answered by discussing when the residuals are *not* independent. For example, consider a hierarchical study design that includes several plots, each of which includes several sampling units. Let us assume for simplicity that we apply the linear model with normally distributed residuals, and that there are no covariates in the model, neither at the plot level nor the sampling unit level. Thus, we would fit an intercept-only model, defined simply as $y_i = \alpha + \varepsilon_i$, where α is the common mean for all sampling units. The residual ε_i describes how the species abundance in the sampling unit i differs from the mean abundance of the species over all sampling units. Now, if sampling units 1 and 2 belong to the same plot, one could expect species' abundances to be simultaneously either higher or lower than average in both of the sampling units. This is because the environmental conditions in the focal plot could be either especially favourable or unsuitable for the species, making the species generally common or rare in the entire plot. In this case, the residuals ε_1 and ε_2 would not be independent of each other, but instead positively correlated. Dependency among residuals can be expressed mathematically by writing $\mathrm{Cov}[\varepsilon_1, \varepsilon_2] = \rho\sigma^2$. Here, $\mathrm{Cov}[\cdot]$ stands for covariance, ρ is the correlation and σ^2 is the residual variance. While in the above example the correlation would be positive, generally the correlation ρ can have any value in the range of $-1 \leq \rho \leq 1$.

Random effects are generally needed to account for dependency structures in the data, for example those generated by the hierarchical study design consisting of sampling units within higher levels such as plots. If data that includes a dependency structure is analysed with a model that fails to account for it, the conclusions derived from the model may be wrong (see e.g. Beale et al. 2007). HMSC implements two kinds of random effects: those needed for hierarchical study designs, and those needed for spatial or temporal study designs. It further implements any combination of these, so that it is possible to include in a single HMSC model both a temporal and a spatial random effect; the latter can be further defined either at the level of plots or sampling units, or even at both of these levels. To set up the basics, we will next discuss hierarchical and spatial study designs separately. The case of the temporal study design will be treated as a special case of the spatial study design, as in this context time can be considered as one-dimensional space.

5.4.1 Hierarchical Study Designs

Perhaps the most common situation with ecological data that calls for the inclusion of a random effect is that of the hierarchical study design discussed above, when multiple sampling units have been surveyed within higher levels such as plots. As addressed above, it can be expected that data points originating from the same plot are more similar than data points originating from different plots, and thus that the residuals can be expected to be positively correlated within plots. Therefore, we might expect that $\mathrm{Cov}[\varepsilon_i, \varepsilon_j] = \rho\sigma^2$ with $0 < \rho \leq 1$ if the sampling units i and j belong to the same plot, whereas $\mathrm{Cov}[\varepsilon_i, \varepsilon_j] = 0$ if the sampling units i and j belong to different plots.

A standard way of implementing this dependency structure into the linear model is to add a plot-level random effect. To do so, we use a_p to denote the effect of the plot p. The random effects are assumed to be normally distributed, so that $a_p \sim N(0, \sigma_P^2)$, where σ_P^2 is the variance in the plot-level effects. We denote by $p(i)$ the plot behind sampling unit i, so that e.g. $p(2) = 1$ if sampling unit $i = 2$ belongs to the plot $p = 1$. We can now extend the basic linear model into the mixed model:

$$y_i = L_i + a_{p(i)} + \varepsilon_i, a_p \sim N(0, \sigma_P^2), \varepsilon_i \sim N(0, \sigma^2) \qquad (5.7)$$

Equation 5.7 is called a mixed model because it includes both fixed effects, i.e. the regression parameters included in the linear predictor L_i, as well as random effects, i.e. the effect of the plot.

To understand how Equation 5.7 is linked to the above discussion about correlated residuals within a plot, let us denote the sum of the plot effect and the residual by $\epsilon_i = a_{p(i)} + \varepsilon_i$, so that the model reads $y_i = L_i + \epsilon_i$. As ϵ_i is the difference between the observed data and the linear predictor, it corresponds to the residual of such a model that would fail to include the random effect. It is now easy to see that $\mathrm{E}[\epsilon_i] = 0$, $\mathrm{Var}[\epsilon_i] = \sigma_P^2 + \sigma^2$, that $\mathrm{Cov}[\epsilon_i, \epsilon_i] = \sigma_P^2$ if the sampling units i and j belong to the same plot, and that $\mathrm{Cov}[\epsilon_i, \epsilon_j] = 0$ if the sampling units i and j belong to different plots. Thus, the correlation that the random effect generates between two sampling units from the same plot is $\rho = \sigma_P^2 / (\sigma_P^2 + \sigma^2)$, i.e. the proportion of the residual variance that can be attributed to the plot level.

Another way of writing Equation 5.7 is to use the multivariate normal distribution for describing the distribution of the response variables y_i for all sampling units i simultaneously. To do so, we denote by $\boldsymbol{y}$ the vector of length n (number of sampling units) containing the values of all of the y_i, by $\boldsymbol{L}$ the vector of all linear predictors L_i, and by $\boldsymbol{\epsilon}$ the vector of all residuals ϵ_i. We may then write Equation 5.7 as:

$$y = \mathbf{L} + \boldsymbol{\epsilon},\ \boldsymbol{\epsilon} \sim N(\mathbf{0}, \boldsymbol{\Sigma}) \tag{5.8}$$

where $N(\boldsymbol{\mu}, \boldsymbol{\Sigma})$ stands for the multivariate normal distribution with mean vector $\boldsymbol{\mu}$ and variance-covariance matrix $\boldsymbol{\Sigma}$. In Equation 5.8, the mean vector $\boldsymbol{\mu}$ is set to zero because the expectation of each ϵ_i is zero. The diagonal elements of the variance-covariance matrix $\boldsymbol{\Sigma}$ model the variances, and thus based on our discussion above they are set to $\Sigma_{ii} = \sigma_P^2 + \sigma^2$. The off-diagonal elements of the variance-covariance matrix $\boldsymbol{\Sigma}$ model the covariances, and thus based on our discussion above they are set to $\Sigma_{ij} = \sigma_P^2$ if the sampling units i and j belong to the same plot, and $\Sigma_{ij} = 0$ if the sampling units i and j belong to different plots.

It is often wondered what difference is made if a factor is declared as a random or fixed effect. Indeed, it would be technically possible to set a plot as a fixed effect, so that the effect of each plot (except the one used as a reference level) would be modelled through a dummy variable that indicates whether the sampling unit belongs to that particular plot. The sole difference in declaring plot as a random effect rather than a fixed effect is that, as a random effect, the effects of the individual plots are assumed to follow a common distribution, $a_p \sim N(0, \sigma_P^2)$. This assumption is natural to make if there are many plots and only a few sampling units per plot. In such a case, making the assumption $a_p \sim N(0, \sigma_P^2)$ can help estimate the plot-specific effects. Further, it is natural to consider a factor as a random effect if one is not particularly interested in the effects of the specific levels of the factor, but merely on the overall amount of variation among the levels of the factor. This is usually the case for the effect of a plot in a hierarchical sampling design, as the sampled plots are considered to represent a much larger set of all possible plots that could have been sampled. In contrast, if the factor is the habitat type, for example classified as coniferous, broadleaved and mixed forests, one is likely to be interested in the specific effects of each of these levels, and thus in this case it is natural to set the factor modelling habitat type as a fixed effect, as discussed in Section 5.2.1.

5.4.2 Spatial and Temporal Study Designs

As another example of where a random effect is needed to account for dependent residuals, we will discuss spatial and temporal study designs. In the case of a spatial study design, the sampling units are associated with spatial coordinates, and dependency among residuals is created by a phenomenon called spatial autocorrelation. Positive spatial autocorrelation means that observations are likely to be more similar for two sampling units located near each other than for two sampling units located far away from each other. This can be the case for exactly the same reasons as to why observations are likely to be similar for two sampling units belonging to the same plot, as discussed earlier. If not accounted for, spatial autocorrelation can lead to biased model inference and prediction (Dormann 2007a; Segurado et al. 2006).

In the case of spatial data, it is natural to assume that the level of residual correlation decreases with increasing distance between the sampling units. When modelling spatial data with Equation 5.8, we define the elements of the variance-covariance matrix $\mathbf{\Sigma}$ as:

$$\Sigma_{ij} = f(d_{ij}) + \delta_{ij}\sigma^2 \tag{5.9}$$

where d_{ij} is the distance between the sampling units i and j, and $f(d)$ is called a spatial covariance function. The symbol δ_{ij} is the Kronecker delta, defined as $\delta_{ij} = 1$ for $i = j$, and $\delta_{ij} = 0$ for $i \neq j$, so that the term $\delta_{ij}\sigma^2$ models the independent residual variation on top of the spatial variation.

Out of the many possible choices available for a spatial covariance function, HMSC implements the exponential covariance function defined as $f(d) = \sigma_S^2 \exp\left(-d_{ij}/\alpha\right)$. Here, σ_S^2 is the spatial variance, and α is the characteristic spatial scale of spatial autocorrelation. With these assumptions, Equation 5.9 together with Equation 5.8 will explicitly describe how similarity among residuals decreases with increasing distance between them. The amount (σ_S^2) and spatial scale (α) of the spatial random effect are positive parameters that are both estimated as the model is fitted.

To avoid confusion arising from the same symbol used for different meanings, we recall that in the beginning of this chapter we used the symbol α for the intercept, but then we included the intercept within the β parameters. From now on, we will use α solely for the spatial scale, thus it should not be confused with the intercept.

In the discussion above, we have not addressed whether the space is one-, two- or three-dimensional. This is because the discussion thus far is independent of the dimensionality of the space. In fact, it extends beyond the usual Euclidian distances over space to more general measures of distance. As one example, we may think about time as a one-dimensional coordinate. In this way, the above discussion also holds for time-series data, in which the species abundances would be repeatedly sampled from the same locations over different years. In such a case, two observations from consecutive time points are likely to be more similar to each other than two observations separated by a long time lag; this phenomenon is called temporal autocorrelation. One way to account for temporal autocorrelation is to apply the model described above for spatial autocorrelation, but replacing the spatial distance d_{ij} by the amount of time that elapsed between the observations were made for sampling units i and j.

5.4.3 How Do Spatial Structures Link to Ecological Theory?

As dependency structures of residual variation may be generated by a myriad of processes, an appropriate interpretation of the fitted random effects requires a good ecological understanding of the study system.

The impact of stochastic processes such as dispersal and ecological drift on species distributions has received relatively little attention in the SDM literature, partly because it is challenging to derive straightforward hypotheses about these processes from non-manipulative observational data (Araújo & Guisan 2006). Indeed, the most appropriate way to account for such processes in the context of SDMs is to construct process-based SDMs (Thuiller et al. 2013). Stochastic processes, historical contingencies and dispersal processes (and generally missing covariates) generate spatial or temporal variation in species communities that cannot be attributed to environmental predictors, and thus they produce distributions with unexplained residual spatial or temporal autocorrelation (Bokma et al. 2001; Kessler 2009; Rangel et al. 2007).

Dispersal within the established range of a species is also likely to generate spatial variation in species distributions. For example, in wind-dispersing sessile organisms, such as many plants and fungi, the deposition probability – and thus the probability of colonisation success of propagules – will be higher near the parental individuals. Mobile organisms, such as most animals, also show distance-dependent dispersal. For actively moving organisms, the landscape structure (for instance, presence of movement barriers) can also be expected to have a great impact on the movement rates. In either case, variation in the species' occurrence and abundance can be expected to show more aggregated patterns in space than expected solely from environmental variation.

Concerning phylogeographic assembly processes over large spatial and temporal scales, species may be found from specific continents simply because they originated on their respective continents and have not dispersed to others. An SDM that accounts only for the climatic niche of the species would however predict that a species is equally likely to be found on both continents, assuming there are areas that are within its climatic niche. Thus, the fact that the species is completely absent from one continent due to historical contingencies is one example of spatial variation that cannot be attributed to environmental predictors.

5.5 Partitioning Explained Variation Among Groups of Explanatory Variables

In addition to measuring the overall model fit, it is of interest to know which of the explanatory variables are most important in contributing to the model predictions. This question can be addressed by partitioning the explained variance among the individual explanatory variables, or among groups of explanatory variables. In HMSC, the variance partitioning is conducted at the level of the linear predictor. Thus, while variance partitioning can be applied to any model, one should be cautious about the interpretation of the probit and Poisson models, because they involve non-linear link functions that convert the linear predictor to the scale of the data.

We recall that if X and Y are two random variables, then the variance of their weighted sum can be expanded as $\mathrm{Var}[aX+bY] = a^2\mathrm{Var}[X] + b^2\mathrm{Var}[Y] + 2ab\mathrm{Cov}[X, Y]$, where $\mathrm{Cov}[X, Y]$ denotes the covariance between X and Y. Following this equation, we may expand the variance in the linear predictor as:

$$\mathrm{Var}[L] = \sum_{k=1}^{n_c} \beta_k^2 \mathrm{Var}[x_{\cdot k}] + 2\sum_{k_1=1}^{n_c-1} \sum_{k_2=k_1+1}^{n_c} \beta_{k_1}\beta_{k_2}\mathrm{Cov}[x_{\cdot k_1}, x_{\cdot k_2}] \tag{5.10}$$

where the dot notation in $x_{\cdot k}$ means that $x_{\cdot k}$ is the vector consisting of the x_{ik} values for all sampling units i. If the explanatory variables were not correlated among each other, it would hold that $\mathrm{Cov}[x_{\cdot k_1}, x_{\cdot k_2}] = 0$ for $k_2 \neq k_1$, and thus Equation 5.10 would simplify as $\mathrm{Var}[L] = \sum_{k=1}^{n_c} \beta_k^2 \mathrm{Var}[x_{\cdot k}]$. In this case, the proportion of the explained variance attributed to covariate k would simply be given as $\beta_k^2 \mathrm{Var}[x_{\cdot k}]/\mathrm{Var}[L]$.

If the explanatory variables are correlated, their covariances complicate the picture, as the terms $\beta_{k_1}\beta_{k_2}\mathrm{Cov}[x_{\cdot k_1}, x_{\cdot k_2}]$ in Equation 5.10 may contribute either positively or negatively to the variance of the linear predictor. To account for the covariances, it is thus beneficial to group correlated explanatory variables, especially if they belong to the same theme, such as ‘climatic variables’ or ‘habitat variables’. We index by

$g = 1, \ldots, n_g$ the groups of explanatory variables, we denote by G_g those explanatory variables k that belong to group g, and we denote by $L_i^{(g)} = \sum_{k \in G_g} x_{ik}\beta_k$ the contribution of group g to the linear predictor, so that $L_i = \sum_{g=1}^{n_g} L_i^{(g)}$. Then the variation in the linear predictor can be partitioned among the predictor groups as:

$$\mathrm{Var}[L] = \sum_{g=1}^{n_g} \mathrm{Var}\left[L^{(g)}\right] + \sum_{g_1=1}^{n_g-1} \sum_{g_2=g_1+1}^{n_g} 2\mathrm{Cov}\left[L^{(g_1)}, L^{(g_2)}\right] \quad (5.11)$$

Variance partitioning in HMSC is based on ignoring the covariances among the groups of explanatory variables, i.e. approximating Equation 5.11 by $\mathrm{Var}[L] \approx \sum_{g=1}^{n_g} \mathrm{Var}\left[L^{(g)}\right]$. The covariances among explanatory variables within groups are included by applying the full Equation 5.10 for each group of explanatory variables.

5.6 Simulated Case Studies with HMSC

After setting up the statistical foundations of univariate generalised linear mixed models, it is time to apply the R-package Hmsc! We will start with simulated case studies, where we generate simulated data according to the models described in Sections 5.2–5.4, and then use Hmsc to fit the same models back to the data. Note that while we refer by HMSC generally to the modelling framework, we refer by Hmsc specifically to the R-package. While this may not be as exciting as working with real data, we find the use of simulated data helpful for several reasons. First, simulating data from models that we have described mathematically makes the link between equations and computer code explicit. This hopefully helps in understanding equations for those who are more familiar with computer code, and vice versa, helps to understand computer code for those who are more familiar with equations. Second, with simulated data, the true parameter values are known, thus allowing us to evaluate the accuracy of the match between parameter estimates from the fitted models and the true parameter values. This helps to understand the relationship between the type and amount of data and parameter uncertainty, as well as how to interpret parameter uncertainty. Third, simulated data provide clean case studies that allow quick demonstrations on how to set up the model structures in HMSC that correspond to different link functions and random effect structures. Thus, after walking through the simulated case studies in this section, we will be well-equipped to model some real data in the next section!

5.6.1 Generating Simulated Data

We first generate data where we simulate variation in a single covariate x on $n = 50$ sampling units. We then construct the linear predictor assuming the intercept $\beta_1 = 0$ and slope $\beta_2 = 1$. Finally, we use the same linear predictor to construct three response variables, which conform to the assumptions of the normal model (y_1), the probit model (y_2), and the lognormal Poisson model (y_3).

```
set.seed(1)
n = 50
x = rnorm(n)
beta1 = 0
beta2 = 1
L = beta1 + beta2*x
y1 = L + rnorm(n, sd = 1)
y2 = 1*((L + rnorm(n, sd = 1)) > 0)
y3 = rpois(n = n, lambda = exp(L + rnorm(n, sd = 1)))
par(mfrow = c(1,3))
plot(x,y1, main = "Normal")
plot(x,y2, main = "Probit")
plot(x,y3, main = "Lognormal Poisson")
```

Note that in the script above, we have first set the random number seed to make the results reproducible. We further note that while the above script can be used to essentially reproduce Figure 5.1, we

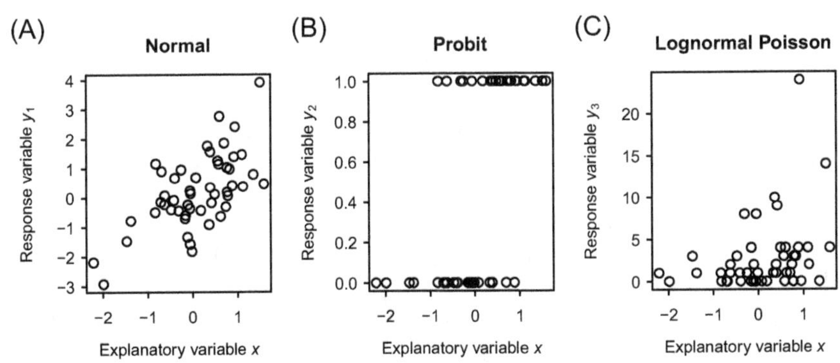

Figure 5.1 Scatterplots of the simulated data. The three panels correspond to response variables, distributed according to the normal distribution (A), the Bernoulli distribution with the probit link function (B) and the lognormal Poisson distribution (C).

have applied some additional formatting to make this (and other) figures look visually nicer. In order to keep the focus on the essential elements, we have suppressed the code that was used to improve the visualizations.

5.6.2 Fitting Models and Examining Parameter Estimates

The standard way of analysing normally distributed data (i.e. the response variable y_1) with maximum-likelihood inference is to use the lm function. As we expect to be familiar for most readers, this can be done as follows:

```
df = data.frame(x, y1)
m.lm = lm(y1 ~ x, data = df)
summary(m.lm)

##
## Call:
## lm(formula = y1 ~ x, data = df)
##
## Residuals:
##      Min       1Q   Median       3Q      Max
## -1.92760 -0.66898 -0.00225 0.48768 2.34858
##
## Coefficients:
##             Estimate Std. Error t value Pr(>|t|)
## (Intercept)   0.1219     0.1394   0.875    0.386
## x             0.9545     0.1681   5.679 7.73e-07 ***
## ---
## Signif. codes: 0 '***' 0.001 '**' 0.01 '*' 0.05 '.' 0.1 ' ' 1
##
## Residual standard error: 0.9781 on 48 degrees of freedom
## Multiple R-squared: 0.4019, Adjusted R-squared: 0.3894
## F-statistic: 32.25 on 1 and 48 DF, p-value: 7.726e-07
```

As expected, the parameter estimates roughly correspond to the values we assumed for the intercept ($\beta_1 = 0$) and slope ($\beta_2 = 1$) when generating the data.

We next conduct the analogous analyses with Hmsc. The first step in using Hmsc is installing it from the CRAN repository (R Development Core Team 2019) with install.packages("Hmsc"), and then loading it:

```
library(Hmsc)
```

We can now construct the model as:

```
Y = as.matrix(y1)
XData = data.frame(x = x)
m.normal = Hmsc(Y = Y, XData = XData, XFormula = ~x)
```

While the lm function both constructs and fits the model at the same time, the Hmsc function only constructs the model object. We have called the model m.normal to distinguish it from the model m.lm fitted to the same data by the standard lm function, and to distinguish it from the m.probit and m.lognormal.poisson models that we will construct later for the response variables y_2 and y_3.

Fitting an HMSC model with Bayesian inference is done with the sampleMcmc function. When calling sampleMcmc, one needs to decide how many chains to sample (nChains), how many samples to obtain per chain (samples), how much thinning to apply (thin), what length of a transient (also called burn-in) to include (transient), and how frequently we wish to see the progress of the MCMC sampling (verbose).

```
nChains = 2
thin = 5
samples = 1000
transient = 500*thin
verbose = 500*thin
```

After choosing these parameters (the meaning of which we will return to later), we are ready to call sampleMcmc and to sample the posterior distribution with MCMC methods. In the context of Bayesian inference, sampling the posterior distribution is the jargon used for estimating model parameters. The exact meaning of this will be discussed in Chapter 8.

```
m.normal = sampleMcmc(m.normal, thin = thin, samples = samples,
                 transient = transient, nChains = nChains,
                    verbose = verbose)
## [1] "Computing chain 1"
## [1] "Chain 1, iteration 2500 of 7500, (transient)"
## [1] "Chain 1, iteration 5000 of 7500, (sampling)"
## [1] "Chain 1, iteration 7500 of 7500, (sampling)"
## [1] "Computing chain 2"
## [1] "Chain 2, iteration 2500 of 7500, (transient)"
## [1] "Chain 2, iteration 5000 of 7500, (sampling)"
## [1] "Chain 2, iteration 7500 of 7500, (sampling)"
```

After sampling, the model object m.normal includes the estimated parameters, in the same way as the object m.lm includes the parameters estimated by the lm function.

We next apply the function convertToCodaObject to extract the posterior distribution from the model object and to convert it into a format that is understood by the coda package (Plummer et al. 2018).

```
mpost = convertToCodaObject(m.normal)
```

The parameter estimates may then be viewed in numerical format using the summary function. We use this to inspect the estimates of the intercept and the slope, i.e. the β parameters, referred to as Beta in Hmsc.

```
summary(mpost$Beta)
##
## Iterations = 2505:7500
## Thinning interval = 5
## Number of chains = 2
## Sample size per chain = 1000
##
## 1. Empirical mean and standard deviation for each variable,
##    plus standard error of the mean:
##
##                              Mean     SD Naive SE Time-series SE
## B[(Intercept)(C1), sp1(S1)] 0.1167 0.1534 0.003430       0.003430
## B[x (C2), sp1 (S1)]         0.9411 0.1810 0.004047       0.004141
##
## 2. Quantiles for each variable:
##
##                                 2.5%     25%    50%   75%  97.5%
## B[(Intercept) (C1), sp1 (S1)] -0.1843 0.01175 0.1171 0.219 0.4181
## B[x (C2), sp1 (S1)]            0.6142 0.81562 0.9405 1.057 1.3098
```

The explanatory power of the model R^2 can be assessed with the function evaluateModelFit. Before doing so, the posterior distribution of the predicted values needs to be computed with the function computePredictedValues.

```
preds = computePredictedValues(m.normal)
MF = evaluateModelFit(hM = m.normal, predY = preds)
MF$R2

## [1] 0.4018712
```

We note that the parameter estimates and the R^2 given by HMSC are highly consistent with those given by the lm function. However, the two approaches are not identical – lm applies the maximum-likelihood framework (and thus yields confidence intervals), whereas HMSC applies the Bayesian framework (and thus yields credible intervals).

5.6.3 Checking MCMC Convergence Diagnostics

As we mentioned when presenting the overall HMSC workflow (Section 4.4), and as we will discuss more formally in Section 8.6, the application of HMSC (and more generally, the use of any MCMC sampling algorithm) requires confirmation that the MCMC chain has converged. Upon convergence, the samples provided by the MCMC chain will provide an accurate approximation of the posterior distribution. But if the chain has not converged, the samples provided by the MCMC chain can yield biased parameter estimates and a biased view on the amount of uncertainty in the parameter estimates. For this reason, it is important to know how to check if MCMC convergence has been achieved, and what to do if it has not. We next illustrate how this can be done in practice, at the same time explaining the meaning of the parameters that guide how posterior sampling is done in the MCMC algorithm. We start by showing in Figure 5.2 trace plots of the β-parameters, generated by:

```
plot(mpost$Beta)
```

The black and grey lines show the two independent MCMC chains (nChains=2). The chains do not start from iteration 1 because by setting transient=2,500 we have chosen not to store the values before 2,500 iterations. The chains have run in total 7,500 iterations each – we selected to obtain 1,000 samples, and we have recorded every fifth step (thin=5) of the iterations.

The posterior trace plots look as good as possible. First, the two chains yield essentially identical results. Second, the chains mix very well, i.e. they rapidly rise and fall without any apparent autocorrelation. Third, they seem to have reached a stationary distribution, as the first half of the recorded iterations looks essentially identical to the second half.

We may also evaluate MCMC convergence more quantitatively in terms of effective sample sizes and potential scale reduction factors (Brooks & Gelman 1998; Gelman & Rubin 1992).

```
effectiveSize(mpost$Beta)

## B[(Intercept) (C1), sp1 (S1)]          B[x (C2), sp1 (S1)]
##                       2000.000                     1913.439

gelman.diag(mpost$Beta,multivariate = FALSE)$psrf

##                                 Point est. Upper C.I.
## B[(Intercept) (C1), sp1 (S1)]    0.9998872 1.000708
## B[x (C2), sp1 (S1)]              1.0003806 1.002444
```

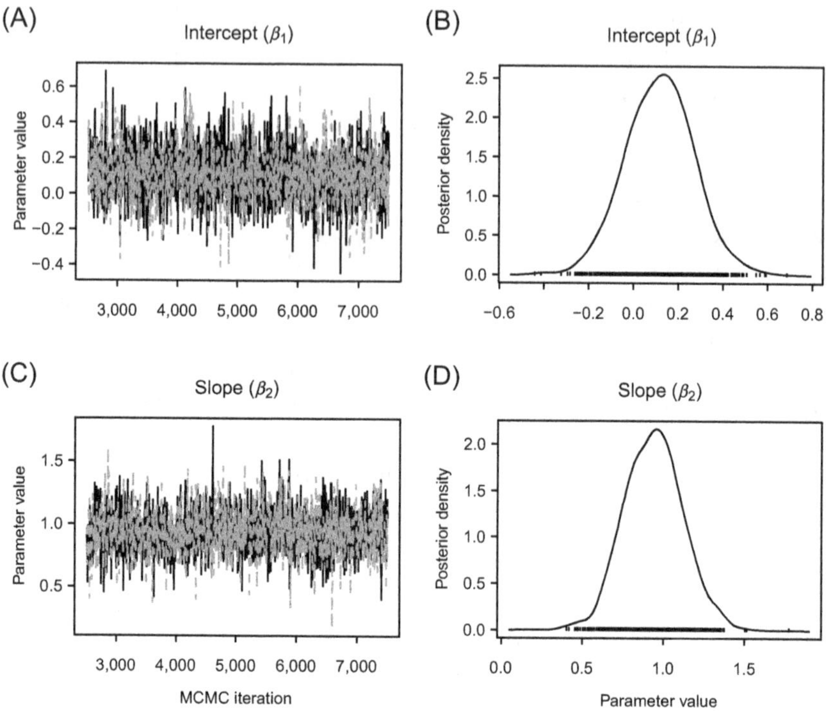

Figure 5.2 Posterior trace plots of the β parameters (panels A and C) and the posterior densities estimated based on the posterior samples (panels B and D). The upper A and B panels show the intercept β_1 and the lower panels C and D show the slope β_2.

We observe that for all parameters the effective sample sizes are very close to the actual sample sizes, which are 2,000 (1,000 per chain). This indicates that there is very little autocorrelation among consecutive samples. The potential scale reduction factors are very close to one, which indicates that the two chains gave consistent results, as was also suggested by visual inspection of the trace plots.

In summary, the MCMC diagnostics did not indicate any problems with MCMC convergence. This means that the posterior sample is likely to be representative of the true posterior distribution, and thus the inference from the model can be trusted. If the MCMC convergence had indicated problems, the model would need to be refitted using different sampling parameters. If the posterior trace plots would have suggested the presence of a transient (the early iterations would have looked different from the later iterations), the amount of transient

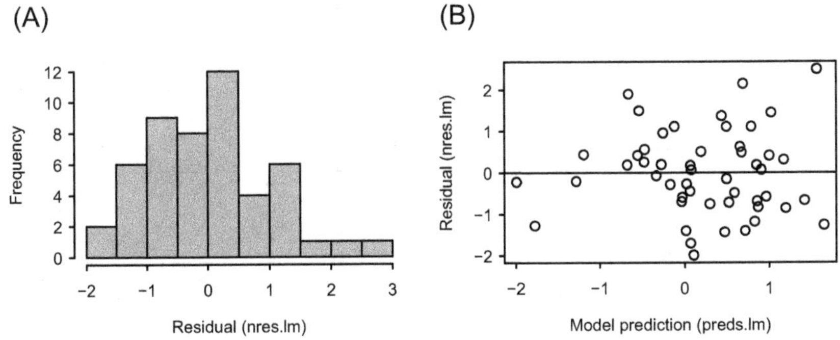

Figure 5.3 Residuals of the linear model fitted with the lm function, shown as a histogram (A) and as a function of the fitted values (B).

iterations to be discarded should be increased. If the sampled chains would have shown autocorrelation and/or if the different chains would have yielded dissimilar estimates, the number of iterations should be increased. This can be done by increasing either the number of samples, or by keeping the number of samples fixed but increasing the thinning interval. We recommend the latter, as increasing the number of samples can make the model objects quite large and lead to computationally expensive post-processing of the results. If one needs to run one million iterations, we recommend doing so by choosing samples = 1,000 and thin = 1,000 rather than samples = 10,00,000 and thin = 1. Based on our experience, 1,000 samples are typically sufficient to evaluate posterior means and credible intervals with high accuracy, as well as to account for parameter uncertainty when using the model to make predictions.

5.6.4 Checking the Assumptions of the Linear Model

When applying the linear model, it is a good practice to examine if the assumptions of that model are upheld. In the context of the lm function, this can be done by constructing the following diagnostic plots, shown in Figure 5.3:

```
nres.lm = rstandard(m.lm)
preds.lm = fitted.values(m.lm)
par(mfrow = c(1,2))
hist(nres.lm, las = 1)
plot(preds.lm, nres.lm, las = 1)
abline(a = 0, b = 0)
```

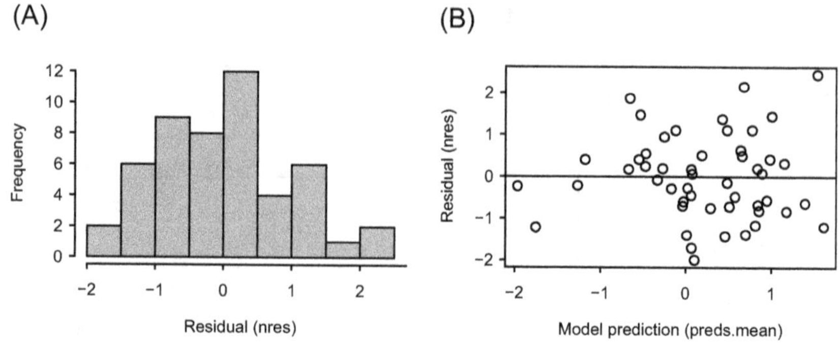

Figure 5.4 Residuals of the linear model fitted with HMSC, shown as a histogram (A) and as a function of the fitted values (B).

The first plot (Figure 5.3A) shows that the residuals conform well to the assumption of normality, and the second plot (Figure 5.3B) shows that the residual variation is homoscedastic. This is not surprising, since the data were simulated from the linear model that makes these assumptions.

Structural model assumptions can be checked with basically any model-based approach, and thus also with HMSC. However, while simply typing plot(m.lm) would provide many diagnostic plots 'automatically' for models fitted with the lm function, there is no such built-in functionality with HMSC. This is because typical applications of HMSC relate to much more complex models than the univariate linear model, and with that greater complexity it is not straightforward to decide what a 'standard' diagnostic plot would be.

To generate the diagnostic plots for the linear model we just fitted with HMSC, we first summarise the posterior distribution of predicted values into the posterior mean, and then extract and standardise the residuals.

```
preds.mean = apply(preds, FUN = mean, MARGIN = 1)
nres = scale(y1-preds.mean)
par(mfrow = c(1,2))
hist(nres)
plot(preds.mean, nres)
abline(a = 0, b = 0)
```

These diagnostic plots, shown in Figure 5.4, are essentially identical to those obtained for the linear model fitted with the lm function. This is to be expected, as the parameter estimates – and hence the fitted values and residuals – were consistent between the two approaches.

From now on, the structural assumptions of the model will not be checked, since this would take up a lot of space and possibly distract the reader from the core message of each of the examples. Thus, the take-home message of this example was that structural model validation can (and should) be done exactly as with any model-based approach.

5.6.5 Fitting Generalised Linear Models

Above, we have fitted a linear model with normally distributed residuals. Since this has been implemented as the default option in Hmsc, we did not need to specify it explicitly. To specify explicitly that one assumes the normal model, one can set the distr argument as:

```
m.normal = Hmsc(Y = Y, XData = XData, XFormula = ~x,
     distr = "normal")
```

As discussed in Section 5.3, the other models implemented in Hmsc are the probit model for presence–absence data (distr="probit"), and Poisson (distr="poisson") and lognormal Poisson (distr="lognormal poisson") models for count data.

To fit a probit model for the response variable y_2, the model is defined as:

```
Y = as.matrix(y2)
m.probit = Hmsc(Y = Y, XData = XData, XFormula = ~x,
     distr = "probit")
```

Posterior sampling can be done as usual.

```
verbose = 0
m.probit = sampleMcmc(m.probit, thin = thin, samples = samples,
    transient = transient, nChains = nChains, verbose = verbose)
```

We note that we have set above verbose = 0 to suppress the output on how the MCMC sampling is progressing. Generally, this is not recommended; instead, we encourage the user to check if the MCMC sampling is moving as expected, e.g. to be able to estimate the remaining run time. In this and forthcoming examples, we have suppressed the output simply to compact it.

We next evaluate MCMC convergence exactly as we did for the linear model.

```
mpost = convertToCodaObject(m.probit)
effectiveSize(mpost$Beta)
```

```
## B[(Intercept) (C1), sp1 (S1)]           B[ x (C2), sp1 (S1)]
##                       1462.083                      1165.303

gelman.diag(mpost$Beta,multivariate = FALSE)$psrf

##                                 Point est. Upper C.I.
## B[(Intercept) (C1), sp1 (S1)]    1.002103   1.009992
## B[x (C2), sp1 (S1)]              1.003507   1.013312
```

Compared to the case of the linear model, we observe that the effective sample size is somewhat smaller. This is because achieving MCMC convergence is generally more challenging for non-normal models than for normal models.

We then look at the parameter estimates and evaluate the model's explanatory power.

```
round(summary(mpost$Beta)$quantiles, 2)

##                                 2.5%   25%   50%  75% 97.5%
## B[(Intercept) (C1), sp1 (S1)] -0.54 -0.25 -0.11 0.03 0.30
## B[x (C2), sp1 (S1)]             0.71  1.08  1.32 1.56 2.05

preds = computePredictedValues(m.probit)
evaluateModelFit(hM = m.probit, predY = preds)

## $RMSE
## [1] 0.3972632
##
## $AUC
## [1] 0.8461538
##
## $TjurR2
## [1] 0.3501425
```

The estimated parameters are compatible with their true values, but there is now more parameter uncertainty than observed in the normally distributed data. This is because presence–absence data are less informative than normally distributed data. We note that to keep the output more compact, we have outputted only the posterior quantiles of the β parameters, whereas for the linear model we also outputted the posterior means and standard deviations. We further note that instead of the R^2 computed for the linear model, model fit is now evaluated in terms of Area Under the Curve (AUC) and Tjur R^2 (Pearce & Ferrier 2000; Tjur 2009). We will discuss the different measures of model fit more extensively in Section 9.2.

Fitting a lognormal Poisson model to the response variable y_3 and evaluating MCMC convergence can be done as usual.

```
Y = as.matrix(y3)
m.lognormal.poisson = Hmsc(Y = Y, XData = XData, XFormula = ~x,
                          distr = "lognormal poisson")
m.lognormal.poisson = sampleMcmc(m.lognormal.poisson,
                                 thin = thin, samples = samples,
                                 transient = transient,
                                 nChains = nChains,
                                 verbose = verbose)
mpost = convertToCodaObject(m.lognormal.poisson)
effectiveSize(mpost$Beta)

## B[(Intercept)    (C1), sp1 (S1)]      B[x (C2), sp1 (S1)]
##                         140.8157                  263.0327

gelman.diag(mpost$Beta,multivariate = FALSE)$psrf

##                                    Point est.  Upper C.I.
## B[(Intercept) (C1), sp1 (S1)]      1.142399    1.508606
## B[x (C2), sp1 (S1)]                1.009958    1.016613
```

The MCMC diagnostics are again worse than for the normal model. This is to be expected, as the normal model is the easiest case from the point of MCMC sampling.

We next look at parameter estimates and evaluate model fit.

```
round(summary(mpost$Beta)$quantiles,2)

##                                  2.5%   25%  50%  75% 97.5%
## B[(Intercept) (C1), sp1 (S1)]  -0.46 -0.06 0.11 0.27 0.56
## B[x (C2), sp1 (S1)]              0.05  0.42 0.61 0.78 1.13

preds = computePredictedValues(m.lognormal.poisson,
                               expected = FALSE)
evaluateModelFit(hM = m.lognormal.poisson, predY = preds)

## $RMSE
## [1] 4.319722
##
## $SR2
## [1] 0.04952168
##
## $O.AUC
## [1] 0.5828877
##
## $O.TjurR2
```

```
## [1] 0.02401248
##
## $O.RMSE
## [1] 0.474574
##
## $C.SR2
## [1] 0.02109143
##
## $C.RMSE
## [1] 4.957869
```

As illustrated by the output above, there are many kinds of model fit that can be computed for count data. The measure SR2 can be viewed as a pseudo-R^2. Whereas Hmsc computes the R^2 of a linear model as the squared Pearson correlation between predicted and true values, it computes the pseudo-R^2 for count data as the squared Spearman correlation between observed and predicted values. Counts can be used to separate whether a species is present (count > 0) or absent (count = 0), and thus a model fitted to count data can also be evaluated from the perspective of how well it predicts occurrences. This is indicated by O. in the measures of model fit, and thus the measures O.AUC and O.TjurR2 evaluate model fit in terms of AUC and Tjur R^2. It is also of interest to examine how well the model is able to predict abundance variation in sampling units where the species is present, ignoring absences. This is done with the measures indicated by C., where C refers to 'conditional on presence'. Note that here we have generated the predictions with the option expected = FALSE. In this case, the predictions are not expected values (e.g. on average we might expect to see 2.3 individuals), but rather a posterior predictive distribution of data (i.e. integer-valued counts 0,1,2,3, ...). We have selected expected = FALSE because it allows presence–absences to be inferred from the predictions, enabling Hmsc to also compute the O. and C. variants of measures of model fit.

5.6.6 Predicting New Sampling Units

Next, we illustrate how Hmsc can be used to generate predictions of new sampling units.

While generalised linear models can be used to predict the response variable for any values of the explanatory variables, often one simply wishes to visualise how the response value changes within the range of the explanatory variables (i.e. from smallest to largest values in the data). This can be done using the constructGradient, predict and plotGradient

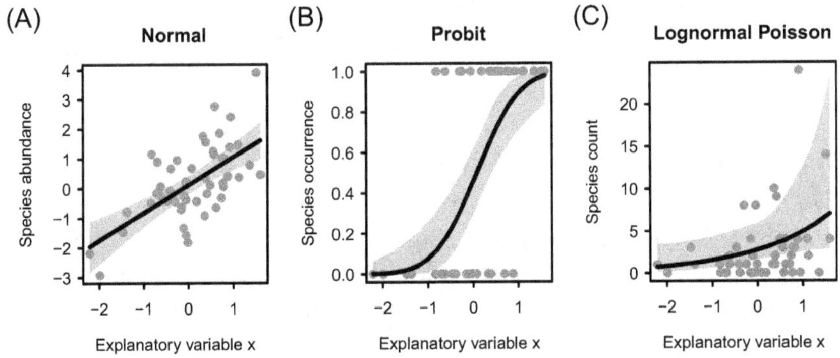

Figure 5.5 Model predictions as a function of the explanatory variable x. The panels correspond to models fitted to normally distributed data y_1 (A), Bernoulli distributed data y_2 assuming the probit link function (B), and lognormal Poisson distributed data y_3 (C).

functions. In Figure 5.5 we plot model predictions over a gradient of the sole explanatory variable x included in the model.

```
par(mfrow = c(1,3))
for (i in 1:3){
  m = switch (i, m.normal,m.probit, m.lognormal.poisson)
  Gradient = constructGradient(m, focalVariable = "x")
  predY = predict(m, Gradient = Gradient, expected = TRUE)
  plotGradient(m, Gradient, pred = predY, measure = "Y",
        index = 1, showData = TRUE, main = c("Normal",
        "Probit","Lognormal Poisson")[ i] )
}
```

The code above loops over the three models we have fitted (normal, probit and lognormal Poisson), and plots the model prediction together with the data used to fit the models. All predictions reproduce what we assumed when generating the data, i.e. that the response variable y increases with the explanatory variable x. In more detail, the predictions made by the three models differ quite fundamentally from each other. The data y_1 as well as the predictions of the normal model can obtain any real values. The data y_2 are zeros or ones, and the predictions of the probit model are probabilities. The data y_3 are counts (non-negative integers), and the predictions of the lognormal Poisson model are expected counts (non-negative real numbers).

In the plots of Figure 5.5, the lines show the posterior mean and the shaded area the 95 per cent credible interval of the model prediction. We note that many of the data points fall outside the 95 per cent credible interval. This is because the shaded areas illustrate the credible intervals for the model predictions rather than the credible intervals for the data.

5.6.7 Hierarchical Random Effects

We next consider a hierarchical study design in which $n = 120$ sampling units have been sampled in $n_p = 12$ plots (on average 10 sampling units per plot). We assume that the plots have an additive effect to the response variable, for which we include a random intercept as introduced in Equation 5.7.

```
n = 120
x = rnorm(n)
beta1 = 0
beta2 = 1
sigma = 1
np = 12
sigma.plot = 1
L = beta1 + beta2*x
plot.id = sample(1:np, n, replace = TRUE)
ap = rnorm(np, sd = sigma.plot)
a = ap[plot.id]
y = L + a + rnorm(n, sd = sigma)
```

In this script, the variable *ap* includes the random effects of the twelve plots, and the variable *a* has copies of each of these, so the effects of the plots are assigned to the sampling units. Let us plot these data so that we colour the sampling units by the plot from which they originate (Figure 5.6).

```
cols = rainbow(np)
plot(x, y, col = cols[ plot.id] , las = 1)
for (p in 1:np){
  abline(beta1+ap[p] , beta2, col = cols[p] )
}
```

In the maximum-likelihood framework, one option for fitting a mixed model to these data would be to apply the lmer function (Bates et al. 2015) as lmer(y ~ x + (1 | plot.id)). Fitting a similar model with HMSC can be done as follows:

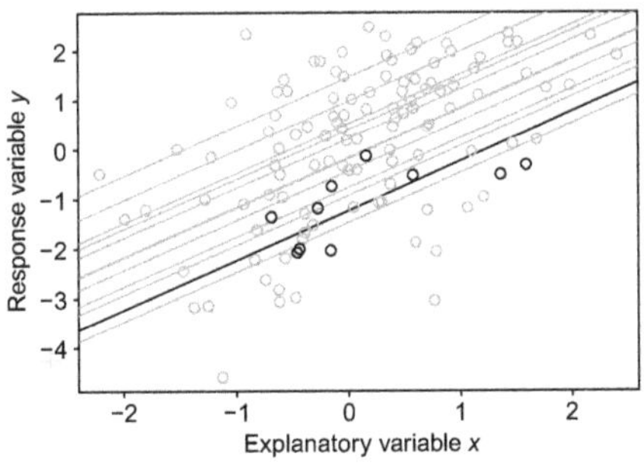

Figure 5.6 Scatterplot of the simulated data originating from twelve plots. The lines show the plot-specific models that were used to generate the data. In the version of the figure shown in the Colour Plate, each plot is indicated by one colour. Here one of the plots is highlighted black whereas the remaining plots are shown grey.

```
plot.id = as.factor(plot.id)
sample.id = as.factor(1:n)
XData = data.frame(x = x)
Y = as.matrix(y)
studyDesign = data.frame(sample = sample.id, plot = plot.id)
rL = HmscRandomLevel(units = studyDesign$plot)
m = Hmsc(Y = Y, XData = XData, XFormula = ~x,
     studyDesign = studyDesign, ranLevels = list("plot" = rL))
```

Here we have included two new input parameters. First, studyDesign defines the nature of the study design, which in this case are the individual sampling units and the plots that the sampling units belong to. Second, ranLevels includes a list of random effects to be included in the model, which in this case is the random effect of the plot created with the function HmscRandomLevel. We note that we have also included the sampling unit in studyDesign. The purpose of this is to illustrate that it is possible to include levels that are redundant, from the model fitting point of view, but may be needed later when post-processing the results.

We next fit the model as we have done before.

```
m = sampleMcmc(m, thin = thin, samples = samples, transient =
     transient, nChains = nChains, verbose = verbose)
```

To avoid repetition, we do not show the results of MCMC convergence, but we recall that it is always important to perform those checks. Unless stated otherwise, we have chosen the MCMC sampling parameters that achieve sufficient convergence for all examples. We next measure the explanatory power of the model.

```
preds = computePredictedValues(m)
MF = evaluateModelFit(hM = m, predY = preds)
MF$R2

## [1] 0.5712603
```

Here, as well as in the previous examples, we have computed R^2 based on predicting the same data that were used for model fitting. This is why we have called R^2 as a measure of explanatory power rather than a measure of predictive power. If we would add more predictors, we could make R^2 approach one, even if the predictors would represent just random noise – a phenomenon known as overfitting. For this reason, the predictive power of a model should always be evaluated more critically by cross-validation.

5.6.8 Evaluation of Model Fit Through Cross-validation

To apply cross-validation, we first assign the samples randomly into a number of groups ('folds'). For example, we may apply two-fold cross-validation across the samples by making the following partition:

```
partition = createPartition(m, nfolds = 2, column = "sample")
```

This is how the partition looks for the first few sampling units:

```
##        sample.id plot.id partition
## [1,]           1       4         2
## [2,]           2       7         2
## [3,]           3      11         2
## [4,]           4       4         1
## [5,]           5       5         1
## [6,]           6       3         1
```

The idea behind cross-validation is that when making predictions for a particular fold, data from the focal fold is not used for parameter estimation. This is to avoid the possibility of obtaining an overly optimistic assessment of the model's predictive power because of overfitting. Model fitting and predictions are first made separately for each fold, after which the predictions are combined to provide one matrix of predictions for all sampling units.

We next make model predictions according to this partition, and use the model predictions to compute a predictive R^2.

```
preds = computePredictedValues(m, partition = partition)

## [1] "Cross-validation, fold 1 out of 2"
## [1] "Computing chain 1"
## [1] "Computing chain 2"
## [1] "Cross-validation, fold 2 out of 2"
## [1] "Computing chain 1"
## [1] "Computing chain 2"
```

Performing the two-fold cross-validation required fitting the model twice. In general, performing *k*-fold cross-validation requires fitting the model *k* times, which can become computationally intensive. For this reason, we have applied here two-fold cross-validation rather than e.g. leave-one-out cross-validation. Increasing the number of folds means that more data are available for fitting the model, which can be expected to lead to greater predictive performance. For this reason, the predictive power estimated by two-fold cross-validation is likely to underestimate the true predictive power of the full model fitted to all data.

```
MF = evaluateModelFit(hM = m, predY = preds)
MF$R2

## [1] 0.3326187
```

As expected, the cross-validation-based predictive R^2 is lower than the explanatory R^2.

Cross-validation can be performed in many ways, and how exactly it should be performed depends on which aspect of the predictive power one wishes to measure. To illustrate this point, let us partition the plots (rather than the sampling units) into different folds.

```
partition = createPartition(m, nfolds = 2, column = "plot")
```

This is how the partition looks for the first fifteen sampling units:

```
##          sample.id plot.id partition
## [1,]             1       4         1
## [2,]             2       7         2
## [3,]             3      11         1
## [4,]             4       4         1
## [5,]             5       5         2
## [6,]             6       3         2
## [7,]             7       6         1
## [8,]             8       2         2
## [9,]             9       6         1
```

```
## [10,]          10        12        1
## [11,]          11        10        1
## [12,]          12        12        1
## [13,]          13         6        1
## [14,]          14         8        2
## [15,]          15         6        1
```

As seen by examining the correspondence between the plots and the partition, all sampling units belonging to a particular plot have now been included into the same fold.

We can calculate cross-validation-based predictive power for this partition just as we did above.

```
preds = computePredictedValues(m, partition = partition)
MF = evaluateModelFit(hM = m, predY = preds)

MF$R2

## [1] 0.08268645
```

We observe that the predictive power of the model is smaller when assigning entire plots rather than individual sampling units into different cross-validation folds. This is because the prediction task is now more difficult: when making a prediction for a particular sampling unit, the model was fitted without training data from any other sampling units in the same plot. Thus, the model does not have the possibility to estimate the actual random effect for the focal plot, and its predictive power is based solely on the fixed effects.

5.6.9 Spatial Random Effects

We next consider spatial data by assuming that each sampling unit is associated with two-dimensional spatial coordinates. We generate the data by the model described in Equation 5.9, so that we assume spatially structured residuals with an exponentially decreasing spatial covariance function. When generating the data, we randomise the sampling units into the unit square, set variance of the spatial random effect as $\sigma_S^2 = 4$, and set its spatial scale to $\alpha = 0.5$. We thus assume a strong spatial effect (there is four times as much spatial variation as residual variation, given that we assume $\sigma = 1$), and that the characteristic spatial scale of spatial variation is 0.5 spatial units.

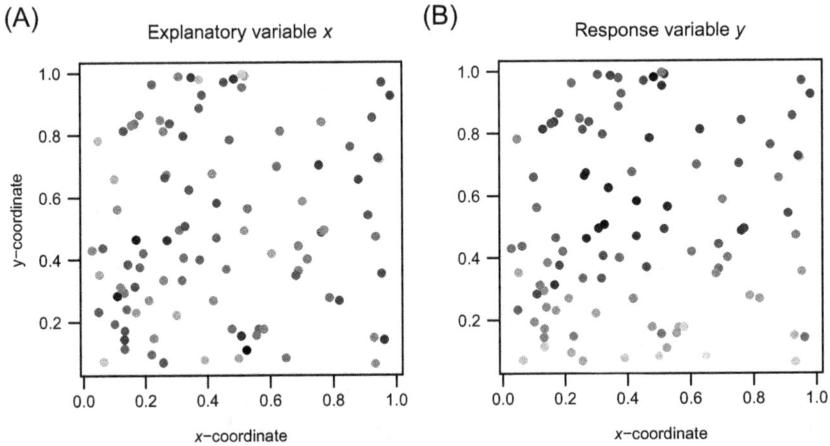

Figure 5.7 Plots of simulated spatially structured data. The panels show the spatial variation in the explanatory variable x (A) and in the response variable y (B). Lowest values are shown in grey, highest in black. In the version of the figure shown in the Colour Plate, lowest values are shown in blue, highest in red.

```
n = 100
beta1 = 0
beta2 = 1
sigma = 1
sigma.spatial = 2
alpha.spatial = 0.5
x = rnorm(n)
L = beta1 + beta2*x
xycoords = matrix(runif(2*n), ncol = 2)
Sigma = sigma.spatial^2*
  exp(-as.matrix(dist(xycoords))/alpha.spatial)
a = mvrnorm(mu=rep(0,n), Sigma = Sigma)
y = L + a + rnorm(n, sd = sigma)
```

The spatial structure is visible in the data (Figure 5.7), as nearby sampling units have similar responses y even if the predictor x is not spatially autocorrelated (due to the assumptions we made when generating the data).

A spatially explicit random effect can be included in a HMSC model by bringing in the locations of the sampling units with the sData input argument of the function HmscRandomLevel.

```
sample.id = as.factor(1:n)
studyDesign = data.frame(sample = sample.id)
rownames(xycoords) = sample.id
rL = HmscRandomLevel(sData = xycoords)
XData = data.frame(x)
Y = as.matrix(y)
m = Hmsc(Y = Y, XData = XData, XFormula = ~x,
    studyDesign = studyDesign, ranLevels = list("sample" = rL))
```

Model fitting and evaluation of explanatory power can be done as before.

```
m = sampleMcmc(m, thin = thin, samples = samples, transient
     = transient, nChains = nChains, verbose = verbose)
preds = computePredictedValues(m)
MF = evaluateModelFit(hM = m, predY = preds)
MF$R2

## [1] 0.9089852
```

We next convert the posterior to a coda object to examine the estimate that we obtained for the spatial scale α of the random effect. We first evaluate MCMC convergence in terms of this parameter in Figure 5.8, and then examine its posterior quantiles.

```
mpost = convertToCodaObject(m)
plot(mpost$Alpha[[1]])
```

```
summary(mpost$Alpha[[1]])$quantiles
##      2.5%       25%       50%       75%     97.5%
## 0.3064236 0.5468278 0.7602240 0.9736202 1.2803772
```

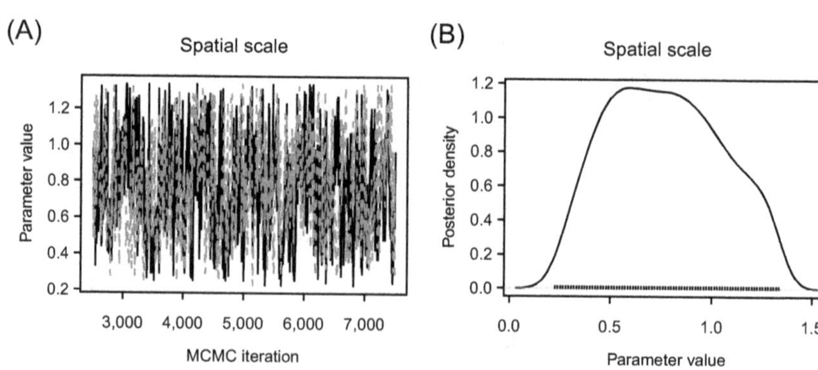

Figure 5.8 Posterior trace plot (A) and posterior density (B) of the spatial scale parameter.

The spatial scale α of the estimated random effect is given by the parameter Alpha[[1]]. Because there can be multiple random effects in a HMSC model, in this example we specify by [[1]] that the spatial random effect at the sampling unit level is the first (and only) random effect included in the model. Satisfactorily, the MCMC convergence is good (Figure 5.8), and the estimated α is consistent with the assumed value of 0.5, in the sense that this value belongs to the core part of the posterior distribution.

We next evaluate the predictive power of the model through two-fold cross-validation.

```
partition = createPartition(m, nfolds = 2, column = "sample")
preds = computePredictedValues(m, partition = partition)
MF = evaluateModelFit(hM = m, predY = preds)

MF$R2

## [ 1] 0.7173674
```

As the model includes both the environmental covariate x and the spatially structured random effect, its predictive power builds on both of these. Regarding space, the model can utilise spatial interpolation, i.e. its predicted values are influenced by the observed data in the nearby sampling units.

For the purpose of illustrating the benefit of using a spatial model for prediction, we also evaluate the explanatory and predictive powers of the corresponding non-spatial model, and compare them to the spatial model.

```
m = Hmsc(Y = Y, XData = XData, XFormula = ~x)
m = sampleMcmc(m, thin = thin, samples = samples, transient
     = transient, nChains = nChains, verbose = verbose)
preds = computePredictedValues(m)

MF = evaluateModelFit(hM = m, predY = preds)
MF$R2

## [1] 0.1013627
preds = computePredictedValues(m, partition = partition)
MF = evaluateModelFit(hM = m, predY = preds)

MF$R2

## [1] 0.1006852
```

We observe that both the explanatory and the predictive powers of the non-spatial model are much lower than for the spatial model. The superiority of the spatial model is expected because the data y have a strong spatial signal that cannot be attributed to the explanatory variable x.

5.7 Real Data Case Study with HMSC: The Distribution of *Corvus Monedula* in Finland

Now that we are familiar with how to conduct univariate analyses with Hmsc, we are ready to handle the first real empirical case study of this book! We start by loading the data on Finnish birds.

```
da = read.csv(file.path(data.directory, "bird data\\data.csv"))
da = droplevels(subset(da, Year==2014))
XData = data.frame(hab = da$Habitat, clim = da$AprMay)
Y = as.matrix(da$Corvus_monedula)
colnames(Y) = "Corvus monedula"
xy = as.matrix(cbind(da$x, da$y))
```

The full bird data include multiple survey years and multiple species. To keep our example simple, the script above selects only a small subset of the data. First, to restrict the analyses to the univariate case, the script selects data for only one species, namely the jackdaw *Corvus monedula.* Second, in this example we only use data from a single year (2014). Third, we select two environmental covariates, one describing the habitat type (classification to either broadleaved forest, coniferous forest, open habitat, urban habitat, or wetlands), and the other describing climatic conditions (mean spring temperature in April–May). As illustrated in Figure 5.9, the selected data consist of environmental covariates and counts of *C. monedula* in 137 spatially explicit sampling units.

5.7.1 Setting up and Fitting HMSC Models

As the response variable is a count, a lognormal Poisson model will be fitted to these data. As fixed effects, we include the linear and squared effects of the covariate of climate (which is continuous) and the effect of habitat type (which is a categorical factor). We include the squared effect of climate to allow the species' abundances to peak at intermediate climatic conditions. We further include a spatial random effect to account for the spatial nature of the study design. This is the full model,

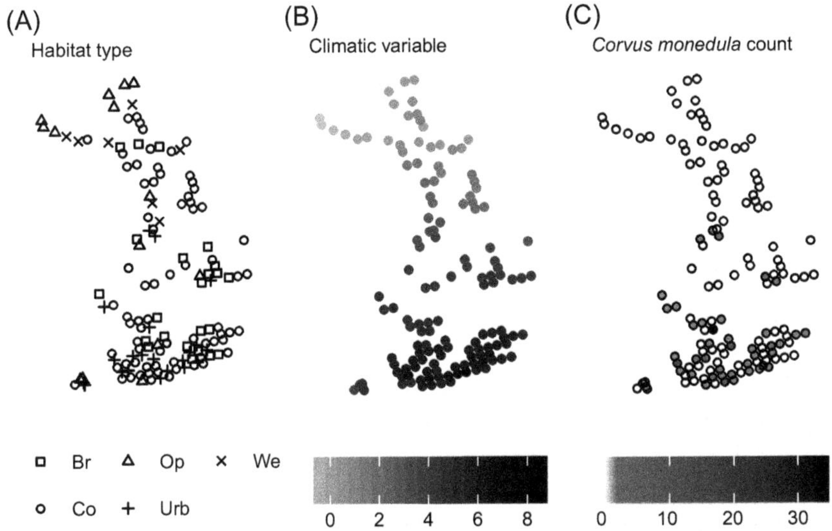

Figure 5.9 An illustration of environmental and species data used in this example. The panels show spatial variation in habitat type (A), climatic conditions (B), and the counts of the target species across Finland (C). For a colour version of the figure, see the Colour Plate.

named mFULL in the script. To disentangle the effects of the environmental and spatial predictors, we also fit two additional alternative models. In the script these are named mENV (model including environmental predictors but no spatial random effect), and mSPACE (model including spatial random effect but no environmental predictors).

```
studyDesign = data.frame(route = as.factor(da$Route))
rownames(xy) = studyDesign[ ,1]
rL = HmscRandomLevel(sData = xy)
XFormula = ~ hab + poly(clim, degree = 2, raw = TRUE)
mFULL = Hmsc(Y = Y, XData = XData, XFormula = XFormula,
        distr = "lognormal poisson", studyDesign = studyDesign,
        ranLevels = list(route = rL))
mENV = Hmsc(Y = Y, XData = XData, XFormula = XFormula,
        distr = "lognormal poisson")
mSPACE = Hmsc(Y = Y, XData = XData, XFormula = ~1,
        distr = "lognormal poisson",
        studyDesign = studyDesign,
        ranLevels = list(route = rL))
```

After defining the models, we fit them to the data. To do model fitting as well, as all forthcoming steps, in a compact notation, we include all three models into a single object (the list models). Now we can fit all three models by simply looping over them.

```
models = list(mFULL, mENV, mSPACE)
for (i in 1:3){
  models[[i]] = sampleMcmc(models[[i]], thin = thin,
     samples = samples, transient = transient,
     nChains = nChains, verbose = verbose,
     initPar = "fixed effects")
}
```

When fitting the models, we use the option initPar = "fixed effects". With this option, the initial condition for the MCMC chain is selected by first fitting species-specific models in the maximum-likelihood framework instead of randomising them from the prior. While it would be better to use independent initial conditions for different MCMC chains, it is generally difficult to achieve MCMC convergence in non-normal models. For this reason, it is often a more practical alternative to start from more informed initial conditions, which can shorten the transient period.

Let us evaluate MCMC convergence for the β parameters, as well as for the spatial scale parameter α in the full model mFULL.

```
mpost = convertToCodaObject(models[[1]], spNamesNumbers =
  c(T,F), covNamesNumbers = c(T,F))
ess.beta = effectiveSize(mpost$Beta)
psrf.beta = gelman.diag(mpost$Beta,multivariate = FALSE)$psrf
ess.alpha = effectiveSize(mpost$Alpha[[1]])
psrf.alpha = gelman.diag(mpost$Alpha[[1]],multivariate =
  FALSE)$psrf
```

In the script above, we use spNamesNumbers = c(T,F) to refer to the species by their names instead of by their numbers in the coda object. Similarly, covNamesNumbers = c(T,F) is used to refer to the explanatory variables by their names instead of by their numbers.

```
## Alpha1[factor1]
##         137.9184
##                  Point est.  Upper C.I.
## Alpha1[factor1]  1.001457     1.009525
```

The effective sample size is relatively high and the potential scale reduction factor is relatively close to one (Figure 5.10), meaning that

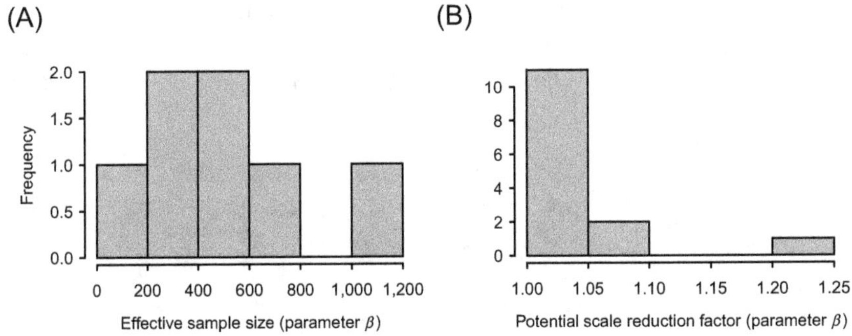

Figure 5.10 MCMC Convergence diagnostics for the β parameters measured in terms of the effective sample size (A) and the potential scale reduction (B).

MCMC convergence is fairly good, even if it is not as ideal as with the simulated data. It is thus meaningful to move on.

5.7.2 Examining what Influences the Distribution of *C. Monedula*

We next evaluate the explanatory powers of the models.

```
MF = list()
for (i in 1:3){
  preds = computePredictedValues(models[[i]], expected = FALSE)
  MF[[i]] = evaluateModelFit(hM = models[[i]], predY = preds)
}
```

Among the many measures of model fit that can be computed for count data (Section 9.2), we look at AUC for discriminating presences and absences (O.AUC), and the pseudo-R^2 for discriminating abundances conditional on presence (C.SR2).

```
##                                    O.AUC     C.SR2
## Model FULL (covariates + space) 0.8570381 0.207254291
## Model ENV (covariates only)     0.8526393 0.127769067
## Model SPACE (space only)        0.8448192 0.001894612
```

All three models have a relatively high AUC and hence relatively high power to discriminate presences from absences. We note, however, that this does not necessarily mean that the models are great, since high explanatory power can also arise from overfitting. We will return to this issue later when evaluating the predictive power of the models. In contrast, the models have a rather low ability to discriminate among abundance variation, even for the same data that were used to fit the models.

We will perform a variance partitioning among the fixed effects related to habitat and climate and the random effects. To do so, we first need to look at the so-called design matrix **X** that Hmsc has constructed from XData and XFormula.

```
round(head(models[[1]]$X),2)
## (Intercept) habCo habOp habUrb habWe poly(clim)1 poly(clim)2
## 1          1     0     1      0     0        7.57       57.31
## 2          1     1     0      0     0        7.74       59.97
## 3          1     0     0      1     0        7.71       59.40
## 4          1     1     0      0     0        7.63       58.27
## 5          1     0     0      1     0        6.90       47.56
## 6          1     1     0      0     0        7.11       50.59
```

We next group the explanatory variables. The columns 2–5 of the design matrix **X** relate to habitat type, and the columns 6 and 7 to climate (the linear and squared effects of the mean spring temperature in April–May). The reason why the categorical explanatory variable of habitat type with five levels has been expanded into four dummy variables is explained in Section 5.2.1. The first column models the intercept, which does not explain any variance so we can group this column arbitrarily e.g. with the habitat variable. Thus, we follow Equation 5.11 with the following grouping of the columns of **X**.

```
groupnames = c("habitat", "climate")
group = c(1,1,1,1,1,2,2)
```

We are now ready to call computeVariancePartitioning, which implements Equation 5.11.

```
VP = list()
for (i in 1:2){
  VP[[i]] = computeVariancePartitioning(models[[i]],
     group = group, groupnames = groupnames)
}
```

In the script above, we loop only over the models mFULL and mENV, because model mSPACE includes only the spatial random effect and thus all variance is trivially explained by the random part. In model mENV we have included only fixed effect, and thus by definition the spatial random effect does not explain anything in that model. After making these trivial modifications manually, we can plot the table of variance partitioning for all the three models.

```
##                                     habitat  climate  random
## Model FULL (covariates + space)      0.228     0.763   0.007
## Model ENV (covariates only)          0.224     0.775   0.000
## Model SPACE (space only)             0.000     0.000   1.000
```

In both Models FULL and ENV, the climatic variables explain the most of the variation, but the habitat variable is important as well. In the Model FULL, the random part explains very little, suggesting that we have included the relevant predictors. We note that these variance proportions are measured at the scale of the underlying linear predictor (Equation 5.11) rather than at the scale of species occurrence, which should be kept in mind when interpreting the results for non-normally distributed models. For example, for the lognormal Poisson model the linear predictor corresponds to the log-scale of the original counts, and thus the variance partitioning relates more closely to the relative than the absolute variation in abundance.

Let us next look at the parameter estimates from Model FULL.

```
round(summary(mpost$Beta, quantiles = c(0.025, 0.5, 0.975))
 [[2]],2)

##                                                    2.5%   50% 97.5%
## B[(Intercept), Corvus monedula]                   -9.52 -6.21 -3.76
## B[habCo, Corvus monedula]                         -1.45 -0.40  0.74
## B[habOp, Corvus monedula]                         -0.98  0.60  2.15
## B[habUrb, Corvus monedula]                         1.00  2.19  3.50
## B[habWe, Corvus monedula]                         -3.74 -0.67  1.70
## B[poly(clim, degree = 2, raw = TRUE)1, Corvus monedula]
                                                  -0.03  0.67  1.61
## B[poly(clim, degree = 2, raw = TRUE)2, Corvus monedula]
                                                  -0.08  0.01  0.08
```

While the parameter estimate table describes how environmental conditions translate to the expected species counts, these effects are somewhat difficult to interpret from the table. This is because the categorical variable of habitat type includes the reference level (here, broadleaved forests) modelled by the intercept, and because we have included both the first and second order effects of the climatic variable. To gain more insight into what these parameter estimates mean, we next construct prediction plots.

5.7.3 Predicting the Distribution of C. *Monedula* over Environmental and Spatial Gradients

SDMs are widely used to make predictions, and the types of predictions can be roughly classified as interpolation and extrapolation. Interpolation refers to making predictions for environmental conditions or spatio-temporal ranges that are within (in the case of extrapolation, outside) those in the data used for model fitting. One example of extrapolation is when an SDM is applied to predict the future distribution of species due to climate change, which classifies as extrapolation because the predictions is be done by transferring the fitted models to the novel environmental conditions (e.g. Guisan & Thuiller 2005; Pearson & Dawson 2003). As thoroughly discussed in the literature, making reliable predictions outside the environmental and spatio-temporal ranges seen in the training data is highly challenging, because it makes the assumption that the used model correctly captures all the factors that influence the relationships between environmental predictors and species occurrence (Dormann 2007b; Randin et al. 2006; Yates et al. 2018). The fact that an environmental predictor significantly influences the occurrence of a given species does not necessarily mean that the species is present *because of* the environmental variable (Grafen & Hails 2002). The link between the species' presence and the environmental variable might, for example, arise if the environmental predictor included in the model correlates with another environmental variable to which the species is actually responding.

To see how the β-parameters translate into model predictions, we utilise the functionality of Hmsc to make and visualise predictions over environmental gradients. Let us first consider the climatic gradient.

```
m = models[[1]]
par(mfrow = c(1,2))
Gradient = constructGradient(m, focalVariable = "clim",
                             non.focalVariables = list(hab = 1))
predY = predict(m, Gradient = Gradient, expected = TRUE)
plotGradient(m, Gradient, pred = predY, measure = "Y",
             index = 1,showData = TRUE)

Gradient = constructGradient(m, focalVariable = "clim",
                            non.focalVariables = list(hab = 2))
## # weights: 15 (8 variable)
## initial value 220.492994
## iter 10 value 160.509875
## iter 20 value 158.882636
## final value 158.882627
## converged

predY = predict(m, Gradient = Gradient, expected = TRUE)
plotGradient(m, Gradient, pred = predY, measure = "Y",
             index = 1,  showData = TRUE)
```

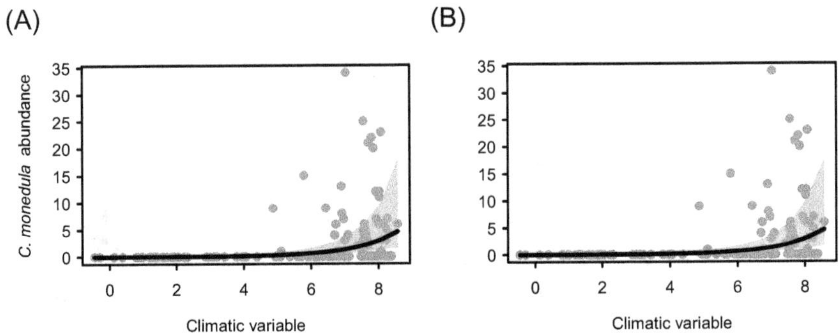

Figure 5.11 Model predictions of the expected count of the target species over the climatic range in the data. In panel A the predictions have been made for the most common habitat type in the entire data, whereas in panel B it has been made for the most common habitat type conditional on the climatic conditions.

When making predictions for over the gradient of one variable, we also need to decide about the values of the other variables. In the script above, we have made these choices in two different ways with the help of the option non.focalVariables. First, we have selected hab = 1 to make predictions for the most common habitat type in the data, which in this case is coniferous forest. Thus, when moving along the climatic gradient in Figure 5.11A, we assume that the species survey has always been conducted in a coniferous forest. Second, we have selected hab = 2 to make predictions for the most common habitat type under each climatic condition. Selecting hab = 2 means that the values of habitat are modelled as a function of climate, using generalised linear models. This is why the output above includes information about the fitting procedure. With hab = 2, when moving along the climatic gradient in Figure 5.11B, we assume that under the most cold climatic conditions the species survey has been conducted in wetlands, whereas under the other climatic conditions it has been conducted in a coniferous forest. To visualise this, let us look at the environmental variables at the coolest end of the climatic gradient.

```
head(Gradient$XDataNew)

##             clim hab
## 1 -0.45885000  We
## 2  0.01604474  Co
## 3  0.49093947  Co
## 4  0.96583421  Co
## 5  1.44072895  Co
## 6  1.91562368  Co
```

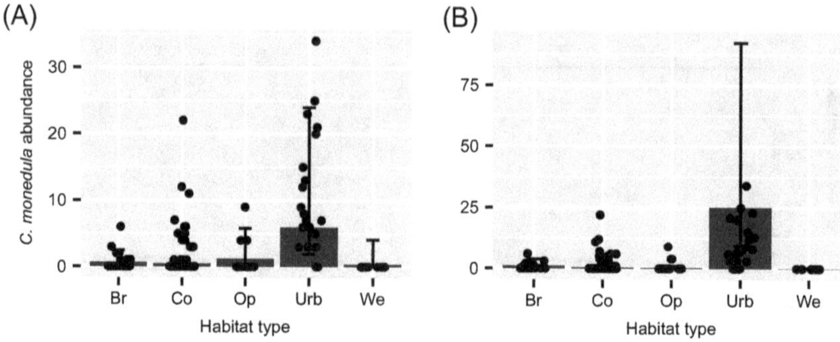

Figure 5.12 Model predictions of the expected count of the target species for each habitat type in the data. In panel A, the predictions have been made for the average climatic conditions for all the data, whereas in panel B they have been made for the average climatic conditions specific to each habitat type.

Whichever prediction we make, we observe that the expected count of *C. monedula* increases with temperature, with hardly any difference between the two approaches: the two panels in Figure 5.11 look essentially identical. This corresponds to the pattern in the raw data that indicates that the species is absent from northern Finland, where the climate is coldest (Figure 5.9).

Let us then consider the habitat gradient. In this case, we also make the predictions in two ways, setting the climatic conditions either to their overall mean value or to the local mean conditional on the habitat type.

We observe that the species is most common in urban habitats under both types of predictions (Figure 5.12). However, now the predictions vary somewhat between the two panels. For example, the prediction for urban environments is higher in panel B than in panel A. This is because the urban habitats are primarily located in southern Finland, where climatic conditions are warmer. Panel A thus predicts the count for urban habitats located under cooler climatic conditions than panel B, which explains the difference between the two predictions.

Let us then look at the estimated spatial scale of the random effect. As the random effect played only a minor role in the Model FULL, we evaluate its spatial scale in Model SPACE, which include only the random effect.

```
mpost = convertToCodaObject(models[[3]])
round(summary(mpost$Alpha[[1]], quantiles = c(0.025, 0.5,
  0.975))[[2]], 2)
```

```
##    2.5%     50%    97.5%
## 252.79 796.28 1238.66
```

While the spatial scale parameter α includes much posterior uncertainty, there is clear evidence of a spatial signal, as the 95 per cent credible interval does not include zero. The estimated spatial scale is in the order of several hundreds of kilometres, which corresponds to the scale of the leading variation in the raw data: the species is present in the southern half of Finland and absent in the northern half.

We next evaluate the predictive powers of the models through two-fold cross-validation.

```
partition = createPartition(models[[1]], nfolds = 2,
  column = "route")
MF = list()
for (i in 1:3){
 preds = computePredictedValues(models[[i]],
     partition = partition)
 MF[[i]] = evaluateModelFit(hM = models[[i]], predY = preds)
}
##                                      O.AUC     C.SR2
## Model FULL (covariates + space) 0.8406647 0.10529266
## Model ENV (covariates only)     0.8311339 0.07464396
## Model SPACE (space only)        0.7559873 0.09482771
```

The cross-validated measures of model fit are not much lower than the explanatory measures, indicating that the models are not over-parameterised. We will next use the full model to predict the distribution of the species across Finland.

```
m = models[[1]]
grid = read.csv(file.path(data.directory,
  "bird data\\grid_10000.csv"))
grid = droplevels(subset(grid, !(Habitat=="Ma")))

xy.grid = as.matrix(cbind(grid$x, grid$y))
XData.grid = data.frame(hab = grid$Habitat,
  clim = grid$AprMay)
Gradient = prepareGradient(m, XDataNew = XData.grid,
                        sDataNew = list(route = xy.grid))

predY = predict(m, Gradient = Gradient)
EpredY = apply(abind(predY,along = 3), c(1,2), mean)
EpredO = apply(abind(predY,along = 3), c(1,2), FUN =
  function(a) {mean(a > 0)})
```

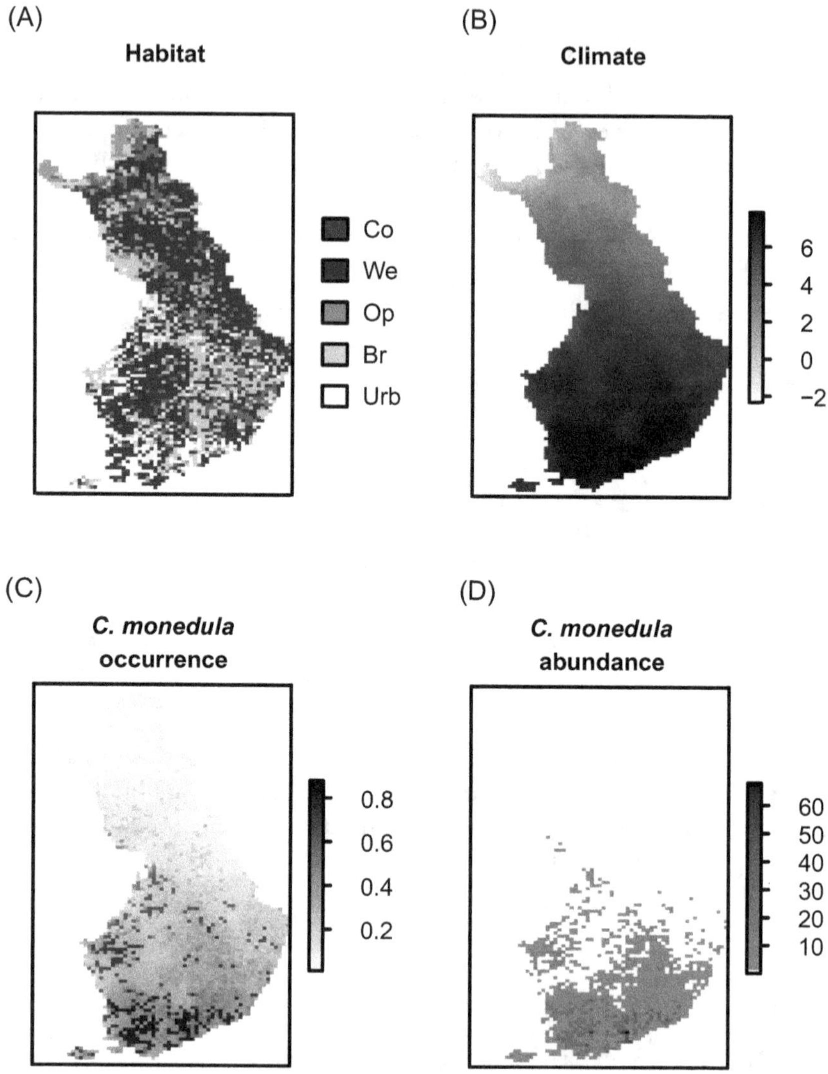

Figure 5.13 Environmental data and predicted species distribution over all of Finland. The panels show habitat (A) and climatic (B) variables for 10,000 prediction locations, and the predicted occurrence (C) and abundance (D) of the target species. For a colour version of the figure, see the Colour Plate.

What we have done in the script above is to read in a grid of 10,000 Finnish locations, which brings along information about the habitat type, climatic conditions and spatial coordinates. These are the input data needed to predict the distribution of our focal species. We first need to drop the grid locations for which the habitat type is Ma (marine waters or beaches), because our training data do not include this habitat type and thus the model cannot make predictions for it.

We next use the prepareGradient function to convert the environmental (XData.grid) and spatial (xy.grid) predictors into a format that can be used as input for the predict function. This function creates objects related to the study design and random effects, which are needed by the predict function. The posterior predictive distribution is then generated using the predict function, as was done above when constructing the plots over environmental gradients. We then summarise the posterior predictive distribution by its mean value (expected count) over the posterior distribution, denoted EpredY. We also summarise the posterior predictive distribution by computing the posterior probability that the count is non-zero, denoted EpredO, as this is the predicted occurrence probability.

The model predicts that *C. monedula* occurs mainly in southern Finland (Figure 5.13). On top of this general trend, there is small-scale spatial variation in predicted abundance, which relates to the fact that the species prefers urban habitats. We note that the predicted distribution seems largely consistent with the raw data shown in Figure 5.9.

6 · *Joint Species Distribution Modelling*

Variation in Species Niches

In the previous chapter, we introduced single-species distribution modelling with HMSC. However, HMSC is not really meant for this. For single-species modelling there are many other approaches available that have many features lacking in HSMC. Single-species distribution modelling is covered extensively in Franklin (2009), Guisan et al. (2017) and Peterson et al. (2011). In this chapter, we move to the area for which HMSC is really meant, namely multi-species modelling. In practice this means that the response variable $\boldsymbol{y}$ from Chapter 5 now becomes a matrix of $n \times n_s$ dimensions, denoted the $\mathbf{Y}$ matrix. The columns of this matrix correspond to the species and the rows to the sampling units. Thus, the element y_{ij} of matrix $\mathbf{Y}$ is the occurrence or abundance of species j in sampling unit i.

We start this chapter by discussing the difference between stacked species distribution modelling and joint species distribution modelling (Section 6.1). We then start building HMSC as a joint species distribution model, first discussing how to model variation among species niches in general (Section 6.2), and then adding hierarchical levels to model species niches in particular as a function of species traits (Section 6.3), phylogenetic relationships (Section 6.4), or a combination of the two (Section 6.5). We then move to case studies, where we apply HMSC first to simulated data (Section 6.6) and then to real data on a plant community (Section 6.7).

6.1 Stacked versus Joint Species Distribution Models

As mentioned in Section 3.4, one important difference between stacked species distribution models (SSDMs) and joint species distribution models (JSDMs) is that the former assume species-specific responses to the environment, whereas the latter assume shared responses. This is because SSDMs are first fitted separately for each species and thus assume that species respond individualistically to variation in environmental conditions (D'Amen et al. 2017; Ferrier & Guisan 2006). In contrast, JSDMs combine the species-specific models with a hierarchical layer that models shared responses to the environment (Hui et al. 2013; Ovaskainen & Soininen 2011). In JSDMs, the shared responses may represent, for example, that species with similar traits have similar responses. In complex communities it is difficult to predict *a priori* the joint structure of species' responses to environmental variation, and thus one might assume that treating each species individually is more in line with the current understanding of community assembly. However, if there are different functional groups such as trophic groups in the focal community, then one may hypothesise that species belonging to the same group do respond similarly to variation in resource types, and thus the underlying assumptions of JSDMs are more in line with theory. More generally, all species have evolved from the same common ancestors and under the general environmental conditions that characterise the planet Earth. Thus, while species have diverged from each other in a wide variety of traits, their variation is ultimately structured by common underlying mechanisms. How species vary in their responses to environmental variation can thus be considered as a collection of related stories rather than a collection of idiosyncratic stories, suggesting some joint structure among species niches.

The assumption of joint responses has been proven to be advantageous when modelling communities with large numbers of rare species (Hui et al. 2013; Madon et al. 2013; Ovaskainen & Soininen 2011; Ovaskainen et al. 2016a). This is because the assumption of joint responses allows the models for rare species to 'borrow' information from the more common species, which facilitates model parameterisation. Since most ecological communities consist of a few common and many rare species (Magurran & Henderson 2003) – the latter of which are of much interest in many community-wide studies (e.g. McCune 2016; Zhang & Vincent 2018) – the assumption of joint responses might be generally beneficial in community ecology studies.

Some JSDMs include model structures that allow use of data on species-specific traits and phylogenetic trees (Abrego et al. 2017a; Brown et al. 2014; Pollock et al. 2012), and hence enable direct testing of hypotheses related to response traits and niche conservatism (see Chapter 1). With such JSDMs, it is possible to test e.g. whether closely related species have more similar environmental niches than distantly related ones, as well as to quantify the amount of variation in species niches or species occurrences that can be attributed to traits (Abrego et al.

2017a; Ovaskainen et al. 2017b). In such analyses, it is desirable to use a quantitative phylogenetic tree where the branch lengths represent the duration of time that the species have belonged to the same ancestral state during their evolutionary history. If such information is not available, a taxonomical tree can be used as a proxy for a phylogenetic tree. However, using a taxonomical tree with equal branch lengths may fail to capture how phylogenetic relatedness connects with environmental niches, because equal branch lengths tends to overestimate the phylogenetic distance between close relatives and underestimate the divergence between distantly related species (e.g. Whitfeld et al. 2012).

SDMs vary in their assumptions regarding not only environmental filtering but also biotic filtering – that is, whether and how biotic interactions influence species occurrences (Kissling et al. 2012; Wisz et al. 2013). As SSDMs fit models for each species independently of other species, they assume that species distributions are statistically independent of each other, beyond the dependencies generated by the environmental covariates. In contrast, JSDMs are fitted to all data at once, and hence can estimate statistical dependence among species even beyond those generated by species' responses to environmental covariates.

To define SSDMs and JSDMs in mathematical notation, we expand the single-species notation of Chapter 5 to the multi-species context. While the environmental covariates x_{ik} are the same for all species, each species responds to them independently. Thus, the β parameters that model species niches need to be species-specific. This can be done by adding the extra index j, so that β_{kj} is the response of species j to the covariate k. The linear predictor is also species-specific, and thus written as $L_{ij} = \sum_{k=1}^{n_c} x_{ik}\beta_{kj}$. Note that while in the single-species case the linear predictor $\boldsymbol{L}$ and the species niches $\boldsymbol{\beta}$ were vectors, they are now matrices, just as the response data $\mathbf{Y}$. The linear predictor is denoted by the $n \times n_s$ matrix $\mathbf{L}$, and the species niches by the $n_c \times n_s$ matrix $\mathbf{B}$. Generally, we denote vectors by bold italic font and matrices by bold non-italic font. In the matrix product notation, we can write compactly $\mathbf{L} = \mathbf{XB}$.

In the multi-species case, the data model (i.e. the link from the linear predictor to the observations) also needs to be specified for each of the species. For normally distributed data, we write $y_{ij} \sim N(L_{ij}, \sigma_j^2)$, where the subscript j in the variance parameter σ_j^2 indicates that the amount of residual variation can be species-specific. In the same way, we could apply other error distributions and link functions to define probit or lognormal Poisson models for each species.

What we described above can be viewed as a collection of single-species distribution models. As we did not connect the models for different species in any way, we could fit them separately for each species. This approach is called stacked species distribution modelling (Calabrese et al. 2014; Guisan & Zimmermann 2000; Guisan & Rahbek 2011). In the stacked species distribution model, single-species models are fitted separately for each species, and then the predictions are combined to obtain inferences at the community level (Ferrier & Guisan 2006). The stacked species distribution approach belongs to the category of 'predict first, assemble later' of Ferrier and Guisan (2006).

Because HMSC can be used for single-species distribution modelling, it can also be used for stacked species distribution modelling; however, it is primarily meant for joint species distribution modelling. Joint species distribution modelling differs from stacked species distribution modelling in that the species-specific models are statistically connected to each other, and hence they are fitted to data together. In HMSC, the species-specific models are connected by two model components. First, in this chapter we will build a hierarchical structure that allows information to be shared among the species when modelling their responses to environmental covariates. This model component allows modelling the species niches β as a function of species traits and phylogenetic relationships. Second, as we will discuss in Chapter 7, HMSC involves a latent variable structure that can be used to estimate the species-to-species associations. This and the next chapter will thus cover the core ideas of how JSDM is implemented in HMSC.

6.2 Modelling Variation in Species Niches in a Community

The β parameters of HMSC model the responses of the species to environmental covariates – in other words, the species environmental niches. Let us begin modelling species niches by considering a probit model with a single explanatory variable x. Following the convention of always including the intercept in the predictor matrix, a single-species model would have the two parameters of β_1 (intercept) and β_2 (slope). Of these, the slope β_2 would model how the species occurrence probability depends on the variable x. While we are usually not so interested in the intercept, now it is actually important to understand its meaning. In the present case, the intercept β_1 would model the baseline occurrence probability of the species; more precisely, the probability that the species

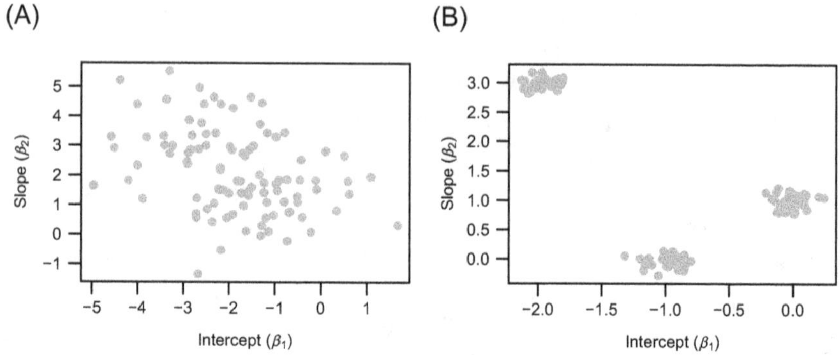

Figure 6.1 Illustration of variation in species niches. In both panels, each dot corresponds to one species in a community of 100 species. In panel A, there is continuous variation among species niches, whereas in panel B the species niches form three clusters.

occurs when $x = 0$. The baseline occurrence probability, by which we mean occurrence probability in a sampling unit with $x = 0$, of the species is $\Phi(\beta_1)$. Assuming that we have scaled the variable x so that its mean is 0, the baseline probability corresponds to the occurrence probability at mean environmental conditions. The value $\beta_1 = 0$ corresponds to the baseline occurrence probability of 0.5, whereas $\beta_1 < 0$ makes the baseline occurrence probability smaller than 0.5 and $\beta_1 > 0$ makes it greater than 0.5. To make this more quantitative, we note that in the probit model for example, $\beta_1 = 2$ corresponds to the baseline occurrence probability of 0.98 and $\beta_1 = -2$ to the baseline occurrence probability of 0.02. In a multi-species model, each species would have its own intercept and slope, denoted by β_{1j} and β_{2j} for species j. If we would know these parameters for each species, we could visualise the species parameters in a two-dimensional graph, as illustrated in Figure 6.1.

In Figure 6.1, each of the 100 dots corresponds to one species, and the location of the dot in the graph specifies its intercept (β_1) and slope (β_2). The two panels in Figure 6.1 illustrate two communities differing in how the environmental niches are structured among the species. Concerning the community represented by Figure 6.1A, we can make three observations based on visual examination. First, most species are rare, as the intercept β_1 is mostly negative. Hence for these species, the baseline occurrence probability is less than 0.5. Second, the slope β_2 is positive for most species, meaning that the occurrence probability of these species increases with an increasing value of the covariate x. As a direct consequence of this, species richness is also expected to increase with an increasing value of the covariate x. Third, there appears to be a negative association between the intercept β_1 and the slope β_2. Therefore, species with the smallest intercepts have the highest slopes, meaning that the occurrence probabilities of rare species are especially influenced by the environmental covariate x. This could be the case, for example, if the rare species would be specialised to a resource represented by x, whereas the common species would be generalists using many kinds of resources.

As the slope β_2 measures how the occurrence of the species depends on the environmental condition x, it is one niche axis of the species. It is arguable whether it makes sense to consider the intercept β_1 as a niche axis. As it is part of the β parameters, we consider it as a niche parameter as well. To justify this, we recall that the intercept β_1 measures the baseline occurrence probability and thus the commonness of the species. Why some species are rare and others are common ultimately depends on how they interact with their environment and with other species. For example, let us assume that the community from Figure 6.1 is a community of wood-decaying fungal species in Europe, and that individual logs (fallen dead trees) are the sampling units. The environmental condition x could represent, for example, the diameter of the log. In this case, a positive response to x would mean that species are primarily found from large logs. In addition to the community patterns related to resource use, wood-decaying fungi also respond to variation in macroclimatic conditions (Abrego et al. 2017b). Some of the species depicted in Figure 6.1 may reach their climatic optimum in a forest patch from southern Finland, whereas for others this may be their range margin. While the resource availability x varies locally among the sampling units, the macroclimatic conditions might be uniform across the forest patch. Thus, it is not possible to include variation along this niche axis in the model, but the responses of the species to macroclimate would be reflected by their baseline occurrence probability. In other words, the species that reach their climatic optimum in the focal forest patch would be the common species and thus have the largest intercepts, whereas the species that are at their range margin would be rare and have the smallest intercepts.

The idea of continuous variation in species niches (Figure 6.1A) was introduced by Ovaskainen and Soininen (2011), who modelled species niches with the multivariate normal distribution

$$\boldsymbol{\beta}_{\cdot j} \sim N(\boldsymbol{\mu}, \mathbf{V}) \tag{6.1}$$

Here $\boldsymbol{\beta}_{\cdot j}$ denotes the vector of all regression coefficients for species j, representing its entire multivariate niche, including its responses to all covariates in the model. In Equation 6.1, $\boldsymbol{\mu}$ is the mean and $\mathbf{V}$ is the variance-covariance matrix of the multivariate normal distribution. If there are n_c covariates (including the intercept), then $\boldsymbol{\beta}_{j\cdot}$ and $\boldsymbol{\mu}$ are vectors of length n_c, whereas $\mathbf{V}$ is a $n_c \times n_c$ matrix. In fact, we have assumed Equation 6.1 when simulating the community of Figure 6.1A with the parameters $\boldsymbol{\mu} = \boldsymbol{\mu}_1$, and $\mathbf{V} = V_1$, where $\boldsymbol{\mu}_1 = (-2, 2)$ and $V_1 = ((2, -1), (-1, 2))$. We have named these parameters $\boldsymbol{\mu}_1$ and V_1 as we will compare them later to other values of these same parameters.

While in the community of Figure 6.1A variation in species niches is of continuous nature, the species niches in the community of Figure 6.1B form three clusters. These clusters could arise, for example, from the evolutionary histories of the species, so that species from each cluster would share a common ancestor. Assuming a high degree of niche conservatism, the present species would still resemble their ancestors, and hence also each other. The idea of clustered variation in species niches (Figure 6.1B) was introduced by the SAM of Hui et al. (2013). In this model, the species are assumed to be structured in groups, and the niches of the species belonging to each group are assumed to be similar. This corresponds to the idea that each group is represented by one species archetype.

6.3 Explaining Variation in Species Niches by Their Traits

In real communities, species are not likely to exhibit as simple variation in their niche space as the hypothetical communities depicted in Figure 6.1. Yet, the idea that species niches have some kind of joint structure is sound from the ecological theory perspective (see Chapter 1); it is also a powerful starting point for statistical developments. HMSC builds on Equation 6.1 by assuming continuous (multivariate normal) variation in species niches. However, instead of assuming a common expected value μ for each species, it utilises trait information to model the expected value of the species niche in a species-specific way. To explain how this is done, we extend the example of Figure 6.1A by incorporating a species trait in Figure 6.2. In this figure, the size of each dot shows the value of some measured trait for each species. We have assumed that the measured trait is correlated with the specialisation levels of the species to the covariate x.

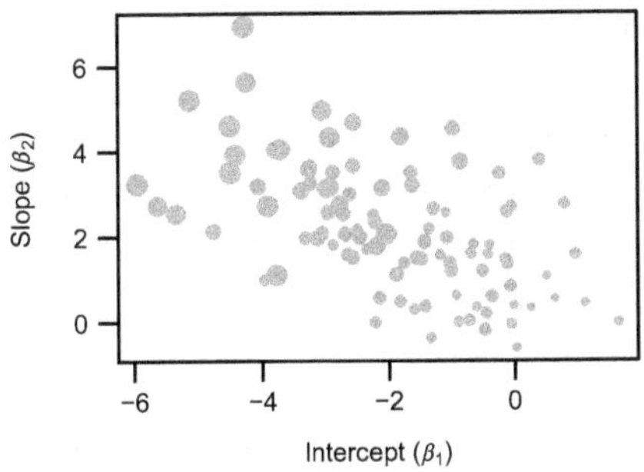

Figure 6.2 Illustration of variation in species niches structured by their traits. In the figure, the size of the dot represents a measured trait of the species.

The trait data could arise for example from laboratory experiments, where the species are observed to preferentially use resource x instead of z when given the choice of the two types. In Figure 6.2, the large size of the dot represents a stronger preference for the resource type x, whereas a small dot size represents the rare use of the resource type x.

We recall that in Figure 6.2 the locations of the dots represent species niches, as measured by the β parameters of the underlying probit regression model. Specifically, the β parameters measure how species' occurrences depend on variation in environmental covariates. Thus, while the trait value measured in the laboratory could be interpreted as the species fundamental niche, the β parameters relate to the realised niche. In Figure 6.2, these two are positively correlated, meaning the species that showed a strong preference for the resource type x in the laboratory (i.e. species shown by large dots) also tended to primarily occur in sampling units with high covariate x values (i.e. species with a high slope parameter β_2). On top of this overall trend, there is variation generated by the factors that make the fundamental niches different from the realised ones (see Chapter 1).

With data on real communities, laboratory-based measurements of fundamental niches are seldom available. However, many other kinds of trait measurements might be available (see Section 2.4), and some of these may turn out to be good predictors of species niches. With the help of trait data, we may thus sharpen Equation 6.1 by replacing the common expected value $\boldsymbol{\mu}$ by the expected value for species j, denoted by $\boldsymbol{\mu}_{\cdot j}$. The expected value $\boldsymbol{\mu}_{\cdot j}$ is a vector, with the element μ_{kj} measuring the expected response of the species j to the covariate k. To illustrate this, let us denote by t_{j2} a trait for species j, e.g. the trait value measuring the preference for the resource type x, indicated by the size of the dot in Figure 6.2. We have added the second subscript 2 into t_{j2} because we will use t_{j2} as a predictor in a linear regression model, and thus we wish to reserve the first predictor t_{j1} for the intercept, so that $t_{j1} = 1$ for all species j. We now model the expected intercept of species j as $\mu_{1j} = t_{j1}\gamma_{11} + t_{j2}\gamma_{12}$, and the expected slope of species j as $\mu_{2j} = t_{j1}\gamma_{21} + t_{j2}\gamma_{22}$. Here γ_{kl} measures the effect of trait l on the response to the covariate k.

In fact, the data depicted in Figure 6.2 was generated with this model. In this figure, we assume that the species traits t_{j2} are distributed according to the normal standard distribution $N(0, 1)$. We then assume the parameters $\gamma_{11} = -2$, $\gamma_{12} = -1$, $\gamma_{21} = 2$, and $\gamma_{22} = 1$. Next, we assume some residual variation in species niches, on top of the variation related to the variation in the species trait t_{j2}. To do so, we model the realised niche $\boldsymbol{\beta}_{\cdot j}$ with the multivariate normal distribution

$$\boldsymbol{\beta}_{\cdot j} \sim N(\boldsymbol{\mu}_{\cdot j}, \mathbf{V}) \tag{6.2}$$

where we have set $\mathbf{V}$ equal to the diagonal matrix $\mathbf{V}_2 = ((1,0),(0,1))$.

The fact that the distribution of species niches in Figure 6.2 closely resembles that of Figure 6.1A is not a coincidence, since we have set up the parameters so that the theoretical distribution of the species niches in these two figures is identical. We can thus consider them to represent the same community; the only difference is that in Figure 6.1A we did not know what caused variation in the realised species niches $\mathbf{B}$, whereas in Figure 6.2 we have been able to explain some of that variation with the help of the species trait t_{j2}.

While we have considered the case of a single-species trait (excluding the intercept), any number of species traits can be used as predictors. Denoting the number of traits (including the intercept) by n_t, we model the expected response of the species j to covariate k as:

$$\mu_{kj} = \sum\nolimits_{l=1}^{n_t} t_{jl}\gamma_{kl} \tag{6.3}$$

where γ_{kl} measures the effect of trait l on the response to the covariate k. The traits can be continuous covariates or categorical factors, where in the latter case they are expanded to dummy variables in the same way as we did for categorical environmental variables in Section 5.2.1. Equation 6.3 can be compactly written in the matrix notation as $\mathbf{M} = \boldsymbol{\Gamma}\mathbf{T}^{\mathrm{T}}$, where $\mathbf{M}$ is the $n_c \times n_s$ matrix with elements μ_{kj}, $\boldsymbol{\Gamma}$ is the $n_c \times n_t$ matrix with elements γ_{kl}, $\mathbf{T}$ is the $n_s \times n_t$ matrix with elements t_{jl}, and the superscript T in $\mathbf{T}^{\mathrm{T}}$ denotes the transposition of the matrix $\mathbf{T}$.

We note that the clustered distribution of species niches in Figure 6.1B could also result from Equation 6.3 if we assume a categorical trait with three levels, each level corresponding to one cluster. However, applying Equation 6.3 with a categorical trait is not the same as applying a species archetype model, because with Equation 6.3 the species clusters would be predicted based on the trait value, whereas with the species archetype model the species niches are clustered without the use of any trait information (Hui et al. 2013).

Above, we have explained that the communities depicted in Figure 6.1A and Figure 6.2 could be the same, with the only difference that in Figure 6.2 we have been able to explain some of the variation by the species traits. More precisely, with the parameters we chose, the traits explain half of the variation among species for both the intercept $\beta_{1\cdot}$ and the slope $\beta_{2\cdot}$. To measure how much of species niches the traits explain, HMSC reports the explanatory power $R^2_{T\beta}$ of the multivariate linear model mapping traits into species niches (Equations 6.2 and 6.3). For our toy example, $R^2_{T\beta}$ would be 50 per cent for both the intercept and the slope.

While it is straightforward to quantify how much traits explain of species niches, it is also of interest to ask how much traits explain of the variation in species' occurrences or abundances. We next quantify this by computing the part of the variation in species occurrence that can be predicted by the traits, denoted R^2_{TY}. We first recall the species niches' (the parameters β_{kj}) influence on species' occurrences through the linear predictor $L_{ij} = \sum_{k=1}^{n_c} x_{ik}\beta_{kj}$, where the covariates x_{ik} describe the environmental conditions in the sampling unit i. Following the Supporting Material published in Ovaskainen et al. (2017b), we define R^2_{TYi} as the part of among-species variation that can be predicted by the traits for a particular sampling unit i. We denote by $a_{ij} = \sum_{k=1}^{q} x_{ik}\mu_{kj}$ the expected value of the linear predictor based on trait information, and compute R^2_{TYi} as the squared correlation between the vectors $\boldsymbol{L}_{i\cdot}$ and $\boldsymbol{a}_{i\cdot}$.

$$R^2_{TYi} = \frac{\mathrm{Cov}(\boldsymbol{a}_{i\cdot}, \boldsymbol{L}_{i\cdot})^2}{\mathrm{Var}(\boldsymbol{a}_{i\cdot})\mathrm{Var}(\boldsymbol{L}_{i\cdot})} \tag{6.4}$$

Here the variances and covariance are computed over the species, while the sampling unit i is considered fixed. To obtain an overall measure R^2_{TY} of the amount of among-species variation in species occurrence that can be attributed to traits, we average the sampling unit-specific variances as:

$$R_{TY}^2 = \frac{\sum_{i=1}^{n} \mathrm{Cov}(\boldsymbol{a}_{i\cdot}, \boldsymbol{L}_{i\cdot})^2}{\sum_{i=1}^{n} \mathrm{Var}(\boldsymbol{a}_{i\cdot}) \mathrm{Var}(\boldsymbol{L}_{i\cdot})} \tag{6.5}$$

We note that a weakness of this definition is that the linear predictor L_{ij} does not equal the species occurrence or abundance, as there is a (potentially non-linear) link function and error distribution between these two. Hence, we have made this choice simply for mathematical convenience, and thus R_{TY}^2 more precisely measures the proportion of the among-species variation in the linear predictor that can be attributed to traits. We note that the value of R_{TY}^2 depends not only on how species niches depend on species traits, but also how much variation there is in each environmental covariate.

6.4 Explaining Variation in Species Niches by Phylogenetic Relatedness

Another possible predictor for species niches is given by the phylogenetic relatedness among the species. For example, assume that the 100 species illustrated in Figure 6.1B belong to three families, so that each of the three clusters corresponds to one family. In this case, one could simply include which family the species belongs to as a categorical trait variable, thus sharpening the predictions about species niches. However, this is in general not the best way to include phylogenetic information in a model. For example, in addition to the three families, the 100 species may belong to twenty-six different genera, and adding a categorical variable with twenty-six levels would probably not be wise. Furthermore, sometimes a quantitative phylogenetic tree (illustrated in Figure 6.3) is available, in which case the species are not described by discrete taxonomical levels, but by more complex relationships. Taxonomical classifications can naturally also be viewed in the tree format.

In the context of the regression model explaining realised species niches by their measured traits (Equation 6.3), the species traits can be considered as fixed effects. For the reasons explained above, it can be difficult to incorporate phylogenetic or taxonomic information as a fixed effect. However, it is possible to utilise information contained in a phylogenetic (or taxonomic) tree by incorporating it as a random effect. To do so, one may first convert the phylogenetic tree into a phylogenetic correlation matrix **C**. Assuming that there are n_s species, the matrix **C** is a $n_s \times n_s$ matrix, where the elements $c_{j_1 j_2}$ are the phylogenetic correlations

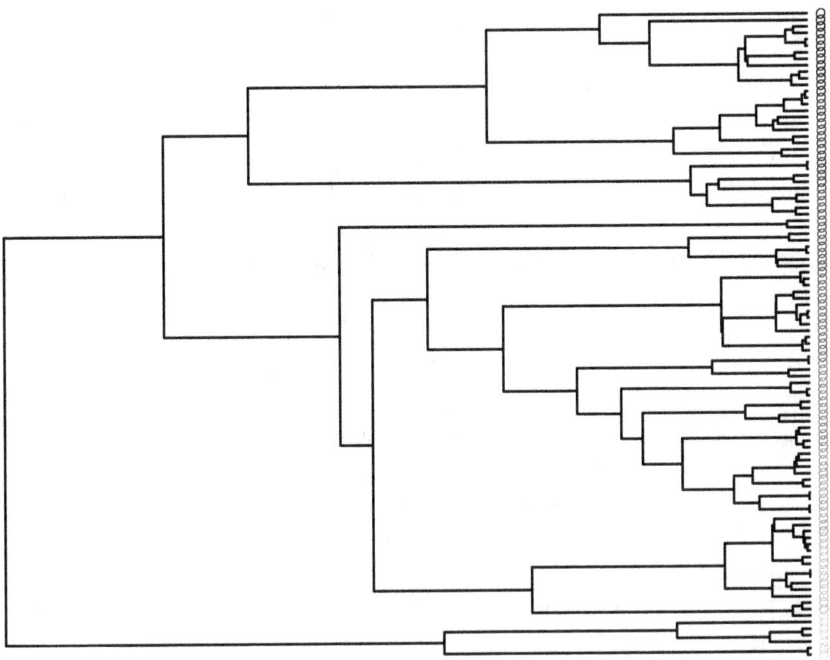

Figure 6.3 A simulated phylogeny of 100 species. The species are at the tips of the tree, and the branch lengths correspond to the duration of time that the species have belonged to the same ancestral state during their evolutionary history. The species have been coded with different shades of grey for later reference. For a colour version of the figure, see the Colour Plate.

among pairs of species j_1 and j_2. The phylogenetic correlation is defined as the fraction of shared evolutionary time between two species, and thus it always holds that the value ranges from 0 to 1 ($0 \leq c_{j_1 j_2} \leq 1$). If the two species are unrelated in the sense that they have diverged already in the root of the phylogenetic tree, then $c_{j_1 j_2} = 0$. In the other extreme, the diagonal elements of the matrix satisfy $c_{jj} = 1$ as the species j has spent the whole evolutionary history with itself.

To simplify our argument, let us first consider that there is only one niche parameter per species, so that the niche of species j is not a vector but a scalar β_j, and the collection of species niches for all species is the vector $\boldsymbol{\beta}$. Assuming further that the species niches are fully phylogenetically structured, one can model the distribution of the species niches with the multivariate normal distribution as $\boldsymbol{\beta} \sim N(\boldsymbol{\mu}, \mathbf{C})$, where the scalar $\boldsymbol{\mu}$ is a common mean (in this section we do not include trait information). This model has the same expectation $\boldsymbol{\mu}$ for all species, but it predicts that phylogenetically close species will on average have more similar traits than phylogenetically distant species. But perhaps the species niches are not fully structured by the phylogeny. At the other extreme, the species niches would be randomly distributed with respect to the phylogeny, which assumption can be written as $\boldsymbol{\beta} \sim N(\boldsymbol{\mu}, \mathbf{I})$, where $\mathbf{I}$ is the $n_s \times n_s$ identity matrix. As a more general case that contains both of these two extremes as special cases, HMSC implements the phylogenetic correlation model of Ives and Helmus (2011). In this model, $\boldsymbol{\beta} \sim N(\boldsymbol{\mu}, \mathbf{W})$, where

$$\mathbf{W} = \rho \mathbf{C} + (1 - \rho)\mathbf{I} \tag{6.6}$$

and the parameter $0 \leq \rho \leq 1$ measures the strength of the phylogenetic signal.

Let us then return to the case of multivariate niche given by the vector $\boldsymbol{\beta}_{\cdot j}$ for species j, and the collection of all species niches given by the $n_c \times n_s$ matrix $\mathbf{B}$. To model this matrix, it is convenient to first collapse it into a vector using the vectorising operation denoted by vec(·). Vectorising means simply arranging the columns of the matrix in succession, so that they form one long vector. Thus, vec($\mathbf{B}$) is a vector of length $n_c n_s$. HMSC models the distribution of this vector with a multivariate normal distribution, the parameters of which are a mean vector of length $n_c n_s$, and a variance-covariance matrix of dimension $n_c n_s \times n_c n_s$. To define the mean vector, we first define the $n_c \times n_s$ matrix $\mathbf{M}$ with elements $\mu_{kj} = \mu_k$, where μ_k is the common expected value for the response of each species j on the covariate k. With this definition, vec($\mathbf{M}$) has the right structure to work as the expected value for vec($\mathbf{B}$). To construct the variance-covariance matrix for vec($\mathbf{B}$), we combine the above model of phylogenetic correlations with Equation 6.1 that uses the variance-covariance matrix $\mathbf{V}$ to model multiple niches simultaneously. Thus, the full model reads as:

$$\text{vec}(\mathbf{B}) \sim N(\text{vec}(\mathbf{M}), \mathbf{W} \otimes \mathbf{V}) \tag{6.7}$$

where the symbol $\otimes$ stands for the Kronecker product and $\mathbf{W}$ is defined by Equation 6.6. The Kronecker product combines the $n_c \times n_c$ matrix $\mathbf{V}$ with the $n_s \times n_s$ matrix $\mathbf{W}$ to result in a $n_s n_c \times n_s n_c$ matrix in the way that Equation 6.7 implies the covariance structure

$$\text{Cov}[\beta_{k_1 j_1}, \beta_{k_2 j_2}] = V_{k_1 k_2} W_{j_1 j_2} \tag{6.8}$$

Thus, modelling species niches by Equation 6.7 includes both the covariance among different environmental covariates (as modelled by $\mathbf{V}$) and the covariance among different species (as modelled by $\mathbf{W}$).

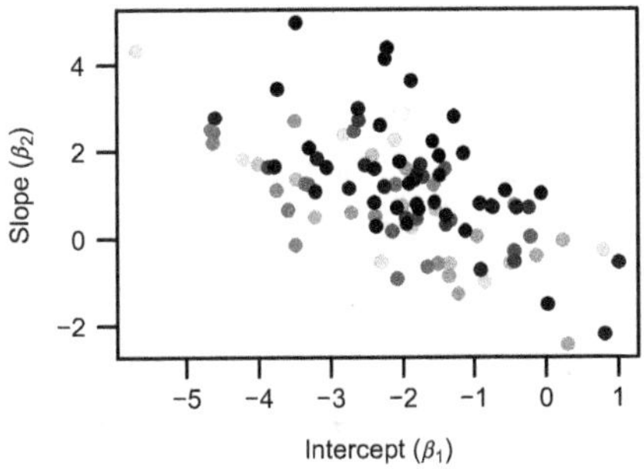

Figure 6.4 Illustration of variation in species niches structured by species phylogeny. The shades of grey of the dots refer to the species location in the phylogeny shown in Figure 6.3. For a colour version of the figure, see the Colour Plate.

We illustrate this in Figure 6.4, where we have generated species traits assuming Equation 6.7, with parameter values $\boldsymbol{\mu} = \boldsymbol{\mu}_1$, $\mathbf{V} = V_1$, and $\rho = 2/3$. The colours of the dots in Figure 6.4 refer to the colour codes of the species used in Figure 6.3, thus they indicate where the species are located within the phylogenetic tree. Species that are close to each other in the phylogeny also appear close to each other in the niche space, illustrating Equation 6.7.

6.5 Explaining Variation in Species Niches by Both Traits and Phylogeny

Before continuing with the modelling approaches, this is a good point to return to the terminology we are using, in particular the relationship between species niche and species trait, as the niche of the species is of course a species trait as well. What we mean by a 'trait' in the context of HMSC is any measured trait for which data are available independently of the species occurrence data, which data are stored in the matrix **T** (Figure 2.1). By realised 'niche' we refer to the estimated B parameters, which map the environmental conditions (matrix **X** in Figure 2.1) to species occurrences or abundances (matrix **Y** in Figure 2.1). By 'phylogeny' we refer to a phylogenetic tree such as that shown in Figure 6.3, the information of which can be converted into a phylogenetic correlation matrix **C** of Figure 2.1. In the context of HMSC, both traits **T** and phylogeny **C** are input parameters that are used to aid the estimation of species niches **B**, as described by Equations 6.7 and 6.3.

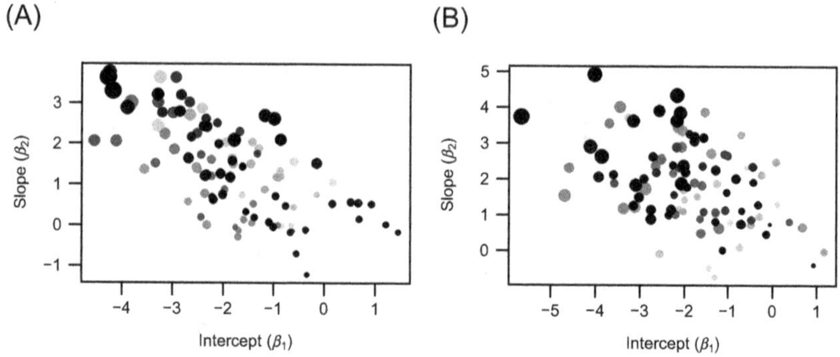

Figure 6.5 Illustration of variation in species niches structured by species phylogeny (panel A) and traits (panel B). In both panels, the colours of the dots refer to the species location in the phylogeny shown in Figure 6.3, and the size of the dot refers to a measured trait of the species. For a colour version of the figure, see the Colour Plate.

Another point that we wish to emphasise is that phylogenies and species traits are not independent of each other, as traits are the result of evolution (Harvey & Pagel 1991). Indeed, traits are often modelled as a function of phylogenies using a similar model to the one we applied above for species niches (Drummond et al. 2012). In Section 6.3, we sampled the species traits as $t_{j2} \sim N(0,1)$ without any reference to the evolutionary history of the species. If we would have assumed that this trait has evolved from a common ancestral species, we could have sampled the traits for all species (denoted by the vector $\boldsymbol{t}_{\cdot 2}$) e.g. from the multivariate normal distribution $\boldsymbol{t}_{\cdot 2} \sim N(0, \mathbf{C})$. By doing so, we would have assumed the so-called Brownian motion (also called random walk or diffusion) model of trait evolution, which is the simplest model among the many alternative models of how traits might evolve (Beaulieu et al. 2012).

Let us now consider the full HMSC model that integrates the influences of both traits and phylogeny on species niches. To do so, we simply combine Equation 6.7 with Equation 6.3, so that now in Equation 6.7 the matrix $\mathbf{M}$ refers to species-specific expected niches, as predicted by the traits in Equation 6.3. In Figure 6.5 we illustrate distributions of species niches generated with this combined model. In both panels of Figure 6.5, we have assumed the same relationship between species traits and species niches as we did in Section 6.3, for which reason the relationship between symbol size (the species trait) and the location of the dot (the species niche) resembles that of Figure 6.2. In the community shown in Figure 6.5A, we have assumed that the species traits are independent of the phylogeny, and thus followed the model $t_{j2} \sim N(0, 1)$. But then we have assumed that the residual variation is fully structured by the phylogeny, and have thus followed Equation 6.7 with $\mathbf{V} = V_2$ and $\rho = 1$. Thus, in this panel there is no relationship between dot size (species trait) and dot colour (position of the species within the phylogeny). Yet, phylogenetically related species are still found near each other in the niche space because the residual variation in species niches (beyond that which is explained by the traits) is phylogenetically structured.

In contrast, we have assumed the opposite case in the community shown in Figure 6.5B. Now the traits are phylogenetically determined as $t_{\cdot 2} \sim N(0, \mathbf{C})$, and thus symbol size and colour are related. But in this case the residual variation is not structured by the phylogeny, as we have assumed $\mathbf{V} = V_2$ and $\rho = 0$. Thus, the reason why species niches (locations of the dots) are related to phylogenetic position (colours of the dots) in Figure 6.5B is that the species trait is phylogenetically structured, and that the species trait influences the species niche. Beyond that, the residual variation is independent of the phylogeny.

The two panels of Figure 6.5 are generated by different underlying assumptions on how traits and phylogenies influence the distribution of species niches in a community. Of these, we view the scenario of Figure 6.5B as a better causal match to why species niches are distributed as they are in real ecological communities. This is because species niches are causally determined by species traits, if accounting for all traits that cover all aspects of the individual's phenotypes. In other words, two species that are phylogenetically close to each other do not have similar realised niches simply because they are phylogenetically related, but because they have similar traits that make their realised niches similar.

The reasoning above suggests that if we could include all the relevant species traits in the model of Equation 6.3, we would not expect to see any phylogenetic signal in Equation 6.7, i.e. the parameter ρ would be estimated to be zero. A positive estimate of a phylogenetic signal parameter $\rho > 0$ thus implies that the relevant traits influencing species niches are missing from the data, and those missing traits are phylogenetically structured. This is a relevant piece of information, as it may hint at the nature of those missing traits. In contrast, if we estimate that $\rho = 0$, then the missing traits can be expected to be randomly distributed within the phylogeny. We note that if all the relevant traits are known, we could expect the residual variation in Equation 6.7 to be very small, and consequently that most of the among-species variation in species niches could be explained by the measured traits, i.e. that $R^2_{T\beta}$ would be close to 100 per cent.

We next illustrate the ideas from this chapter with HMSC-R, using both simulated data (Section 6.6) and real data (Section 6.7).

6.6 Simulated Case Studies with HMSC

In this section, we will generate the simulated data of the two communities depicted in Figure 6.5, and then analyse these data with HMSC. Through this example, we will learn how to include trait and phylogeny into the Hmsc model, and how to interpret the parameter estimates that link traits and phylogenies to species niches. We will also demonstrate how including data on traits and phylogenies can provide more accurate estimates of species niches and thus improve the predictive performance of a model.

6.6.1 Simulating Species Niches

We start by generating the niche variation for the species communities depicted in Figure 6.5. The script below first uses the rcoal function of the ape package to construct a random phylogeny for 100 species, and then the vcv function of the ape package to turn the phylogenetic tree into a phylogenetic correlation matrix **C**.

```
ns = 100
phy = rcoal(n = ns, tip.label =
  sprintf('sp_%.3d',1:ns), br = "coalescent")
C = vcv(phy, model = "Brownian", corr = TRUE)
```

The phylogenetic tree from Figure 6.3 can be visualised by simply calling plot(phy).

For the Community A (Figure 6.5A), the trait values are sampled independently for each species from the standard normal distribution. On the other hand, for the Community B (Figure 6.5B), the trait values are sampled from the multivariate normal distribution, for which the variance-covariance matrix equals the phylogenetic correlation matrix **C**. This can be done with the following script:

```
Tr.A = cbind(rep(1,ns), rnorm(ns))
Tr.B = cbind(rep(1,ns), mvrnorm(n = 1,
  mu = rep(0, ns), Sigma = C))
```

In both cases, we have added the intercept to the trait matrix to enable the application of Equation 6.3 in matrix form. We then define the $\mathbf{\Gamma}$ matrix that describes the link between species traits and niches, and use the matrix product %*% to compute the expected values of species niches according to Equation 6.3.

```
gamma = cbind(c(-2,2), c(-1,1))
mu.A = gamma %*% t(Tr.A)
mu.B = gamma %*% t(Tr.B)
```

We next apply Equation 6.7 to generate the species niches, assuming that in Community A the residual variation is phylogenetically fully structured and that in Community B the residual variation is fully independent among the species.

```
V2 = diag(2)
beta.A = matrix(mvrnorm(n = 1, mu = as.vector(mu.A),
        Sigma = kronecker(C, V2)), ncol = ns)
beta.B = matrix(mvrnorm(n = 1, mu = as.vector(mu.B),
        Sigma = kronecker(diag(ns), V2)), ncol = ns)
```

6.6.2 Simulating Species Data

We assume that our species communities are embedded within the same environmental context that we used to illustrate the univariate models in Section 5.6. Thus, we consider a single environmental covariate x, and use the standard normal distribution to simulate variation in x over $n = 50$ sampling units.

```
n = 50
X = cbind(rep(1, n), rnorm(n))
```

Note that above we have also included the intercept to the $\mathbf{X}$ matrix so that we can compute the linear predictors conveniently in a matrix notation:

```
L.A = X %*% beta.A
L.B = X %*% beta.B
```

What remains is to convert the linear predictors into community data (matrix $\mathbf{Y}$). This example is about occurrence data, thus we convert the linear predictors to presences and absences with the help of the probit model.

```
Y.A = 1*((L.A + matrix(rnorm(n*ns), ncol = ns)) > 0)
Y.B = 1*((L.B + matrix(rnorm(n*ns), ncol = ns)) > 0)
```

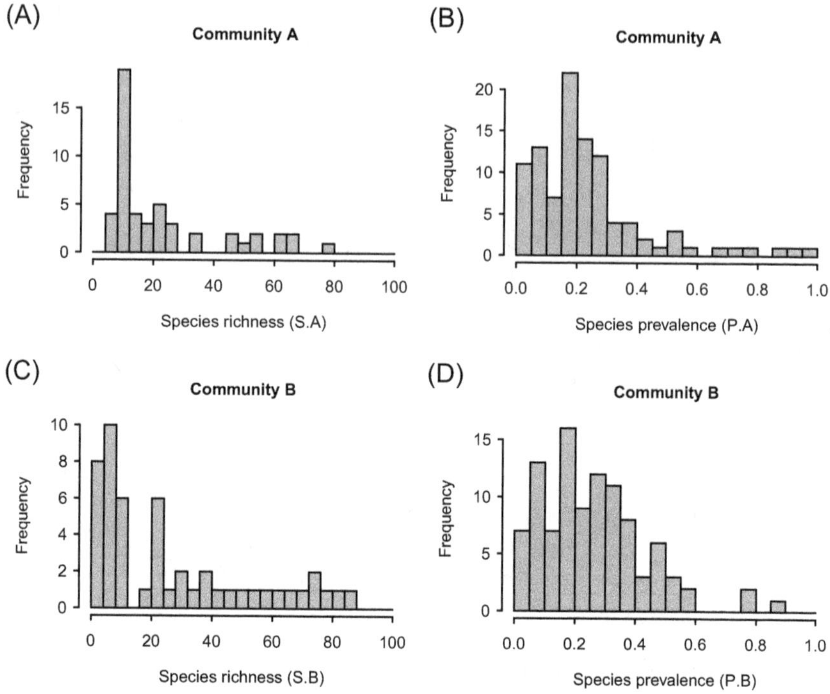

Figure 6.6 Histograms of species richness and species prevalence for the simulated Communities A and B. In the histograms of species richness (panels A and C), the y-axis (Frequency) corresponds to the number of sampling units, and the variable in the x-axis shows the number of species found from each sampling unit. In the histograms of species prevalence (panels B and D), the y-axis (Frequency) corresponds to the number of species, and the variable in the x-axis shows the fraction of sampling units in which each species is present.

6.6.3 Exploring the Raw Data

Before moving to fitting the HMSC model, it is always wise to explore the raw data. Since our data are about species occurrences, we may wish to look at variation in species richness and in species prevalences. The species richness for each sampling unit can be computed as the row sums of the community data matrix **Y**, whereas species prevalences are given by the column means.

```
S.A = rowSums(Y.A)
P.A = colMeans(Y.A)
S.B = rowSums(Y.B)
P.B = colMeans(Y.B)
```

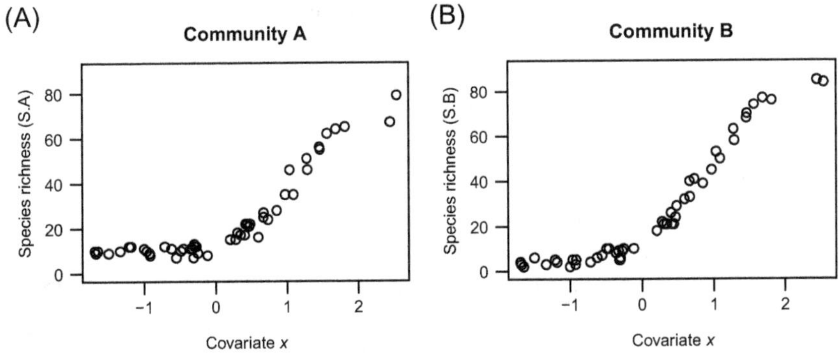

Figure 6.7 Species richness as a function of the environmental covariate x in the simulated Communities A (A) and B (B).

In both communities, there is much variation in species richness among the sampling units, which ranges from two to eighty-five species (Figure 6.6). There is also much variation in species prevalence; many species are present only in a small set of the sampling units, and a few species are present in almost all of the sampling units.

In both communities, species richness increases with increasing value of the environmental covariate x (Figure 6.7). This is to be expected, since we assumed that the species niche parameter β_{2j} is positive for most species. This means that the occurrence probabilities of most species increase with the value of this covariate, and hence so too does species richness.

6.6.4 Fitting an HMSC Model for the Community A with Phylogenetically Structured Species Niches

We start the HMSC analyses by formatting the data so that they are a suitable input for Hmsc.

```
community = "A"
Y = switch(community, "A" = Y.A, "B" = Y.B)
colnames(Y) = phy$tip.label
Tr = switch(community, "A" = Tr.A, "B" = Tr.B)
TrData = data.frame(trait = Tr[,2])
XData = data.frame(x = X[,2])
```

In the script above, we introduced the variable community, which can be set as 'A' or 'B', depending on which community dataset we wish to analyse. By doing so, we can easily replicate the analyses for Community B after first running them for Community A. From now on, we will show the results for Community A, and return to Community B in the end of the section.

We have named the species according to the tip labels of the phylogeny, so that the species in the phylogenetic tree correspond to those in the **Y** matrix. We have placed the trait data into the dataframe TrData and the environmental data into the dataframe XData. Note that in both of these two dataframes we have excluded the intercept, as it is internally added by Hmsc.

We are now ready to define the HMSC model. We model species occurrences as a linear function of the environmental variable x (called x in the dataframe XData), and species niches as a linear function of the trait covariate t_{j2} (called trait in the dataframe TrData). Since our data are on species occurrences, we fit a probit model.

```
m = Hmsc(Y = Y, XData = XData, XFormula = ~x, TrData = TrData,
  TrFormula = ~trait, phyloTree = phy, distr = "probit")
```

We next perform the model fitting.

```
m = sampleMcmc(m, thin = thin, samples = samples,
  transient = transient, nChains = nChains, verbose = verbose)
```

When performing the model fitting, we selected the sampling parameters (choices not shown, in order to keep our treatment compact) so that they lead to satisfactory MCMC convergence, as shown below for the ρ parameter and in Figure 6.8 for the β and γ parameters.

```
effectiveSize(mpost$Rho)
##     var1
## 125.6961

gelman.diag(mpost$Rho,multivariate=FALSE,
  autoburnin=FALSE)$psrf

##     Point est. Upper C.I.
## [1,] 1.013964    1.039106
```

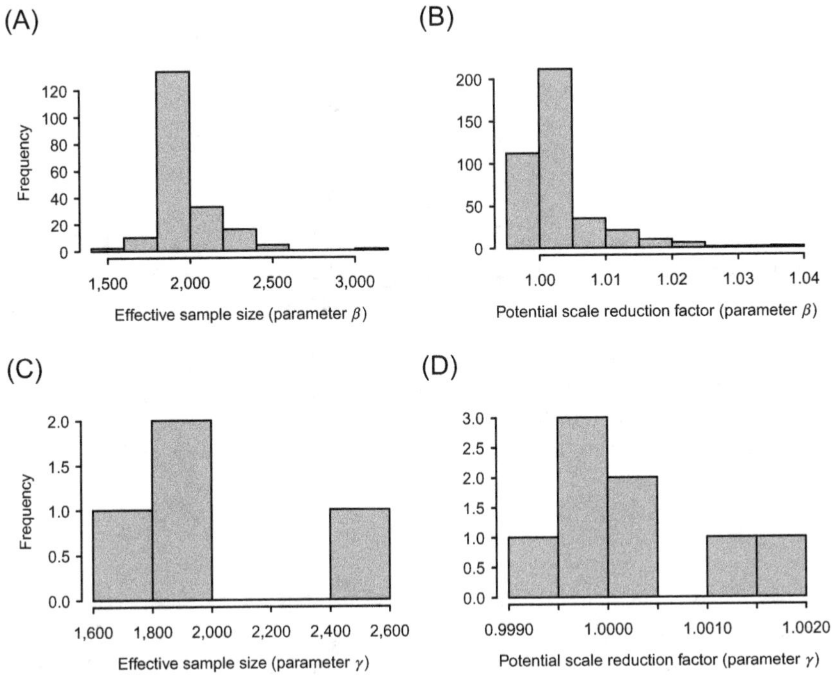

Figure 6.8 MCMC convergence diagnostics for a model with traits and phylogeny. The panels A and B correspond to the β parameters and the panels C and D to the γ parameters. The panels A and C measure MCMC convergence in terms of the effective sample size and the panels B and D in terms of the potential scale reduction factor.

6.6.5 Explanatory and Predictive Powers of the HMSC Model

We next evaluate the explanatory and predictive powers of the model as we have done in the examples of Section 5.6.

```
preds = computePredictedValues(m)
MF = evaluateModelFit(hM = m, predY = preds)

partition = createPartition(m, nfolds = 2)
preds = computePredictedValues(m, partition = partition)
MFCV = evaluateModelFit(hM = m, predY = preds)
```

Unlike in the examples of Section 5.6, there are now multiple species, and thus model fit is assessed separately for each of them. In Figure 6.9, we show model fit in terms of both AUC and Tjur R^2 to make the point that these two measures are correlated, but their absolute values differ

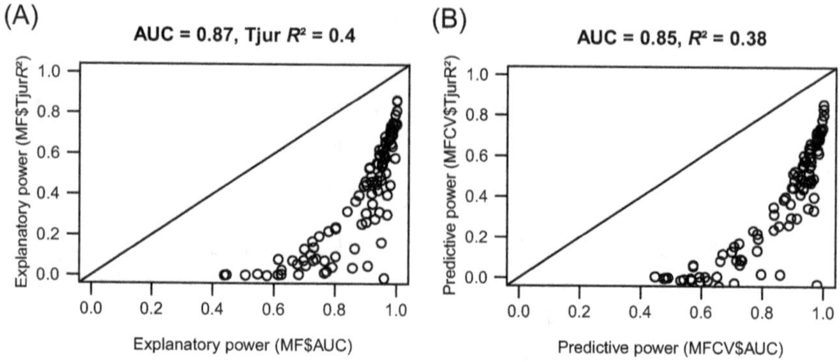

Figure 6.9 Explanatory (A) and predictive (B) power of the model fitted to the simulated data. In both panels, each data point represents one species. The x-axis measures model fit in terms of AUC and the y-axis in terms of Tjur R^2. The AUC and Tjur R^2 values shown in the panel titles indicate the mean values over all species.

greatly. This is because for AUC the baseline that the model prediction is equally good as expected by random is 0.5, whereas for Tjur R^2 the same baseline is 0. For this reason, model fit evaluated in terms of Tjur R^2 may look very bad to readers familiar with the AUC statistic, whereas for readers used to Tjur R^2, a model fit evaluated with AUC may look very good. A more extensive discussion of measures of model fit is given in Section 9.2.

Figure 6.9 shows that the model's predictive power is almost as good as its explanatory power. This is not surprising, since the model includes a single environmental covariate, thus the risk of overfitting is low. Furthermore, because we simulated the community data, we know that the included environmental covariate is highly relevant for explaining the variation in the focal community.

6.6.6 Examining Parameter Estimates

In the present chapter, our main focus is in modelling species niches (the β parameters) as a function of species traits (the matrix **T** and regression parameters γ) and phylogeny (the matrix **C** and phylogenetic signal parameter ρ). We now explore these links by plotting the parameter estimates we obtained for our simulated species community. We first apply the function plotBeta to visualise the estimated species niches.

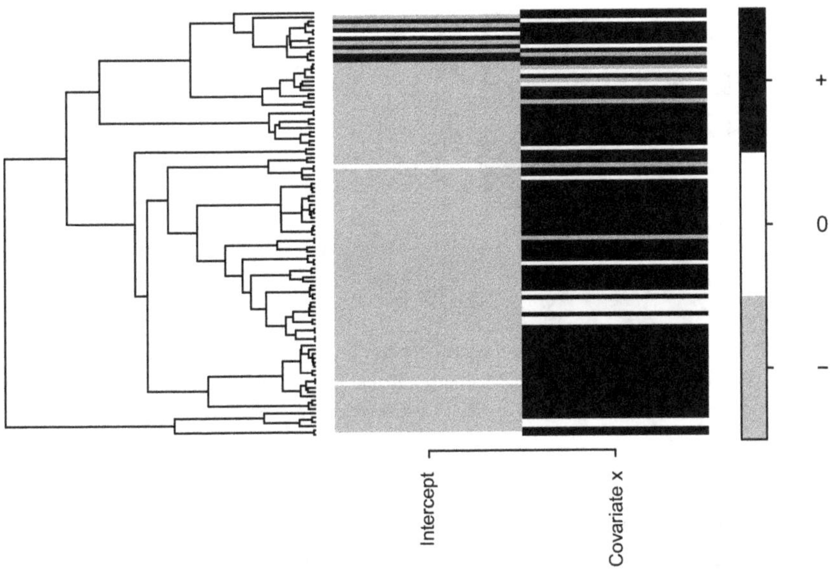

Figure 6.10 Heatmap of estimated species niches. Black and grey colour indicates parameters that are estimated to be positive and negative, respectively, with at least 0.95 posterior probability. In the version of the figure shown in the Colour Plate, the intensity of the colour represents the posterior mean estimate of the parameter, obtained by setting param = "Mean".

```
postBeta = getPostEstimate(m, parName = "Beta")
plotBeta(m, post = postBeta, param = "Sign", plotTree = TRUE,
  supportLevel = 0.95, split = 0.4, spNamesNumbers = c(F,F))
```

We observe that the responses of the species to the covariate x are estimated to be positive for most species (Figure 6.10), which is in line with our assumptions (Figure 6.5A). In the Colour Plate version of the figure, we further observe a clear phylogenetic signal in the β parameter estimates, as the intensity variation in the colours (and thus posterior mean estimates of the parameters) is not randomly distributed with respect to the phylogenetic relationships: entire clades (blocks of related species) show darker or lighter colours than on average.

We next apply the function plotGamma to visualise how species niches are estimated to depend on species traits.

```
postGamma = getPostEstimate(m, parName = "Gamma")
plotGamma(m, post = postGamma, param = "Sign",
  supportLevel = 0.95)
```

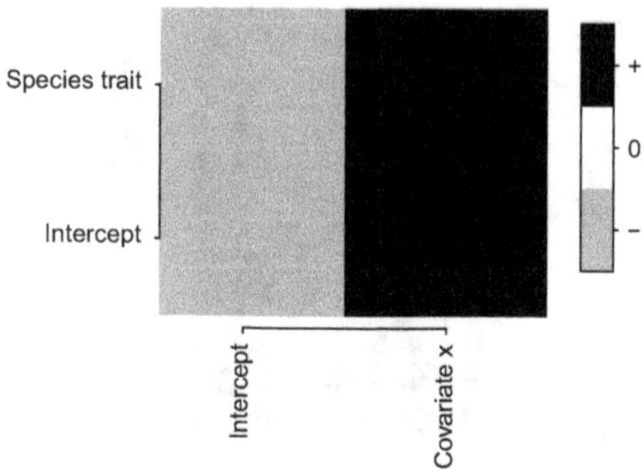

Figure 6.11 Heatmap of estimated gamma parameters linking species traits to species niches. Black and grey colour indicates parameters that are estimated to be positive and negative, respectively, with at least 0.95 posterior probability.

Figure 6.11 shows that the parameter estimate for γ_{22} (influence of the trait on the response to the covariate x) is positive, indicating that species with a high trait value respond particularly positively to the environmental covariate. Furthermore, the parameter estimate for γ_{12} (influence of the trait to the intercept) is negative, indicating that species with a high trait value have a low baseline occurrence probability. These results are in line with our assumptions when generating the simulated data.

We next look at the parameter estimate for the phylogenetic signal parameter ρ.

```
summary(mpost$Rho)$quantiles
## 2.5%   25%  50%   75% 97.5%
## 0.98 1.00 1.00  1.00  1.00
```

The posterior distribution of ρ reveals strong evidence for a phylogenetic signal, as the entire 95 per cent credible interval of ρ is positive and very close to one. This is in line with the fact that we assumed as high of a phylogenetic signal as possible ($\rho = 1$) when generating the data.

6.6.7 Does Including Traits and Phylogenies Help Make Better Predictions?

As illustrated above, including traits and phylogenies helps synthesise the information included in a community dataset, as the community-level parameters γ and ρ integrate the collection of species-level stories into a single community-level story. We next ask whether the inclusion of traits and phylogenies can also help in making better predictions. To address this question, we fit exactly the same model to the same data, but without including information about traits and phylogeny.

```
m = Hmsc(Y = Y, XData = XData, XFormula = ~x,
  distr = "probit")
m = sampleMcmc(m, thin = thin, samples = samples,
  transient = transient,
  nChains = nChains, verbose = verbose)

preds = computePredictedValues(m)
MF.NTP = evaluateModelFit(hM = m, predY = preds)

preds = computePredictedValues(m, partition = partition)
MFCV.NTP = evaluateModelFit(hM = m, predY = preds)

Delta.TjurR2 = MFCV$TjurR2-MFCV.NTP$TjurR2
```

We have added the extension.NTP to the variables measuring model fit (MF) and cross-validation based model fit (MFCV) to indicate that these versions correspond to the model that does not include traits or phylogeny. We next compare the model fits between the original model that includes both phylogeny and traits, and the model that includes neither of these.

A comparison between these models (Figure 6.12) shows that, on average, the model including traits and phylogeny performs somewhat better, especially for the rare species for which it is generally challenging to obtain good predictions. This is because the inclusion of traits and phylogeny allows the model to borrow information from other species, especially from those that have similar traits and are phylogenetically closely related. This can make a major difference for species with only few occurrences in the data. In contrast, species with sufficient data do not need to borrow information from the other species, which explains why the difference between the two models is generally smallest when the prevalence of the species is close to 0.5. The reader may then ask why the model without traits and phylogeny performs better for some species. This is expected due to the inherent stochasticity of the Tjur R^2 measure, especially when measured for species with few occurrences.

Since in this case we know the true parameter estimates, we can also evaluate how well the estimated parameters match the true ones. Since we have defined the HMSC model that follows the same

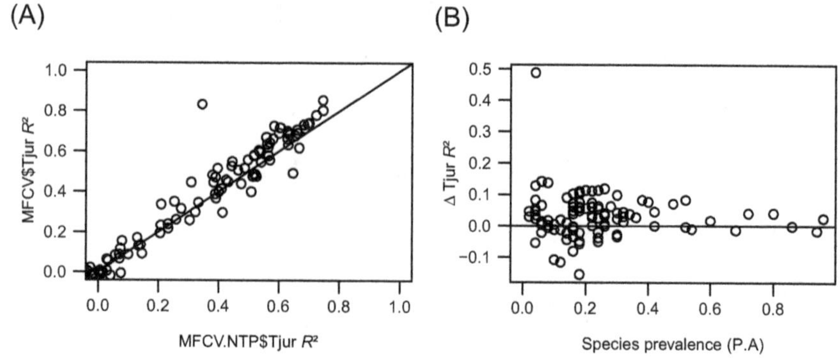

Figure 6.12 Difference in predictive power between models that include traits and phylogeny versus those that do not. In both panels, each dot corresponds to one species. Panel A shows the Tjur R^2 statistic for the models that do (y-axis) and do not (x-axis) include traits and phylogeny. Panel B shows the difference in Tjur R^2 between the two models as a function of species prevalence. Positive values of Δ.TjurR2 indicate that the model with traits and phylogeny performs better in cross-validation.

structural assumptions that we used to simulate the community data, the parameter estimates would ideally be identical to those used to simulate the data.

```
beta.slope.true = beta.A[2,]
beta.slope.est = postBeta$mean[2,]
postBeta.NTP = getPostEstimate(m, parName = "Beta")
beta.slope.est.NTP = postBeta.NTP$mean[2,]
```

The estimated parameter values are close to the true values in both models, particularly in the model including traits and phylogeny (Figure 6.13). For the model without, the estimates deviate from the true values especially when the true value of the slope is large. This is because we assumed that species with a large slope have a small intercept (Figure 6.5A), which correspond to those that are rare and thus benefit most from borrowing information from the other species.

6.6.8 Repeating the Analyses for the Community B Where Species Niches Are Structured by Their Traits

Let us finally repeat some of the analyses above for the Community B. All we need to do is to choose

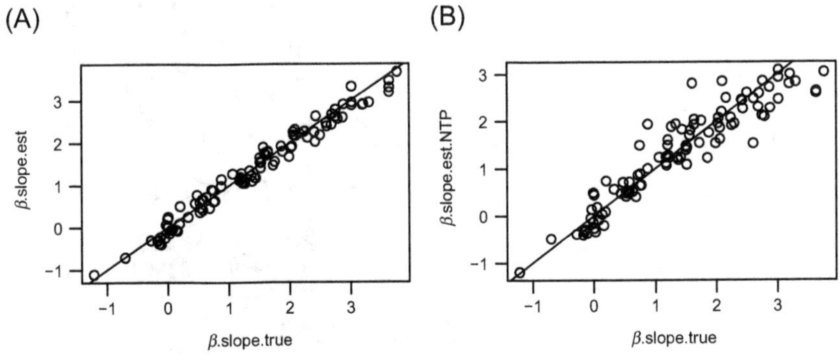

Figure 6.13 Comparison between true and estimated (posterior mean) parameter values. In both panels, the dots show the parameters β_{2j} indicating the response to the environmental covariate x. Panel A corresponds to the model with traits and phylogeny, and panel B corresponds to the model without.

```
community = "B"
```

and then rerun the analyses with this choice:

```
m = sampleMcmc(m, thin = thin, samples = samples,
  transient = transient,
  nChains = nChains, verbose = verbose)

mpost = convertToCodaObject(m)

postBeta = getPostEstimate(m, parName = "Beta")
plotBeta(m, post = postBeta, param = "Sign",
  plotTree = TRUE,  supportLevel = 0.95,
  split = 0.4, spNamesNumbers = c(F,F))
postGamma = getPostEstimate(m, parName = "Gamma")
plotGamma(m, post = postGamma, param = "Sign",
supportLevel = 0.95)
```

As with Community A (Figure 6.10), the β plot for Community B (Figure 6.14) shows that most species respond positively to the environmental covariate x. The Colour Plate version of the Figure 6.14 also reveals that there is a phylogenetic signal in species niches, as the smooth variation of colour in many places indicates that related species have similar parameter estimates.

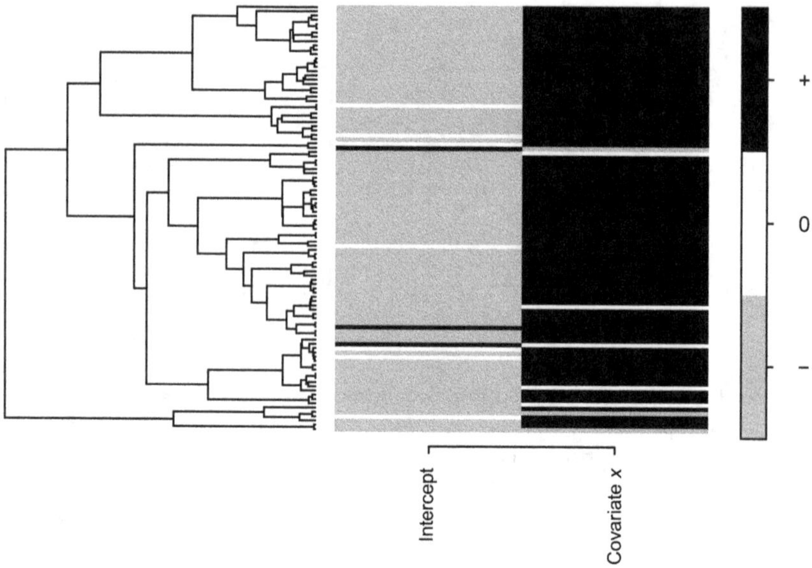

Figure 6.14 Heatmap of estimated species niches. Black and grey show parameters that are estimated to be positive and negative, respectively, with at least 0.95 posterior probability. In the version of the figure shown in the Colour Plate, the intensity of the colour represents the posterior mean estimate of the parameter, obtained by setting param = "Mean". The results shown here are for the Community B, while Figure 6.10 shows the same results for the Community A.

The estimates of the γ parameters of Community B (Figure 6.15) are similar to those for Community A (Figure 6.11). This was expected because we assumed identical γ parameters in both cases.

```
summary(mpost$Rho)$quantiles

##     2.5%      25%      50%      75%    97.5%
## 0.00000 0.00000 0.00000 0.00000 0.08025
```

While the two communities gave very similar results in terms of the β and the γ parameters, their inference about the phylogenetic signal parameter ρ is very contrasting. While for Community A the parameter ρ was estimated to be close to 1, for Community B it is estimated to be close to zero, reflecting the true values that were used when simulating the communities. We recall that these parameter values mean that in Community A, species niches show phylogenetic correlations beyond what can be explained by species traits, whereas in Community B, such correlations can be fully explained by species traits.

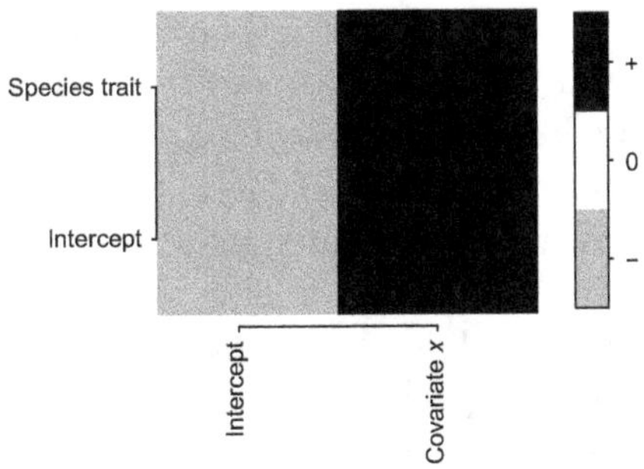

Figure 6.15 Heatmap of estimated gamma parameters linking species traits to species niches. Black and grey shows parameters that are estimated to be positive and negative, respectively, with at least 0.95 posterior probability. The results shown here are for the Community B, while Figure 6.11 shows the same results for the Community A.

6.7 Real Case Study with HMSC: How Do Plant Traits Influence Their Distribution?

As a real data case study, we will re-analyse plant data that Miller et al. (2018; 2019) used to compare several statistical methods for studying trait-environment relationships. These data were collected by Damschen et al. (2010), who revisited Whittaker's historical plant community study sites in Siskiyou Mountains of Southwest Oregon, following the original methods (Whittaker 1960). Whittaker chose the sites to represent the range of topographic variation in the area. In each site, a single 0.1 ha study plot was established, and twenty-five quadrates of size 1 m × 1 m were surveyed along a 50 m transect. The species abundances were recorded as the number of 100 quadrat corners in which each species was found.

Let us start by reading in the data.

```
data = read.csv(file = file.path(data.directory,
  "plant data\\whittaker revisit data.csv"))
head(data)

##    site        species    trait        env  value
## 1 109 Abies concolor 1.404362 -0.3508278      0
## 2 113 Abies concolor 1.404362 -0.7740247      0
## 3  12 Abies concolor 1.404362  1.6674961      0
## 4 156 Abies concolor 1.404362  1.0164239     25
## 5 157 Abies concolor 1.404362  1.3419600      7
## 6 160 Abies concolor 1.404362 -1.3925433      4
```

The data are in the long format, where each row corresponds to one plant species on one site. The column value is the count of plant abundance. The column env is Whittaker's index describing the site's position along the topographic moisture gradient (TMG). Sites on mesic, north-facing slopes receive lower TMG values than sites on warmer, south-facing slopes (Damschen et al. 2010; Miller et al. 2019). The column trait is the functional trait that Miller et al. (2019) selected for their analyses: leaf tissue carbon-to-nitrogen ratio (C:N). This ratio can be considered as a surrogate of competitive ability: plants with low C:N grow faster but have lower stress tolerance than plants with high C:N (Cornelissen et al. 2003; Miller et al. 2019; Poorter & Bongers 2006). It can thus be expected that species occurring on dry and warm sites have on average higher C:N ratios, resulting in a positive relationship between TMG and C:N. Miller et al. (2019) applied several statistical methods to examine the association between the C:N ratio and the environmental gradient, which we will now readdress by reanalysing the data with HMSC.

We first reformat the data so that it will work as input for Hmsc. To do so, in the script below we construct the matrix **Y** of species abundances, the dataframe XData of the environmental variable TMG, and the dataframe TrData of the trait C:N ratio.

```
data$site = factor(data$site)
sites = levels(data$site)
species = levels(data$species)
n = length(sites)
ns = length(species)
Y = matrix(NA, nrow = n, ncol = ns)
env = rep(NA, n)
trait = rep(NA, ns)
for (i in 1:n){
  for (j in 1:ns){
      row = data$site==sites[i] & data$species==
        species[j]
      Y[i,j]  = data[row,] $value
      env[i]  = data[row,] $env
      trait[j]  = data[row,] $trait
  }
}
colnames(Y) = species
XData = data.frame(TMG = env)
TrData = data.frame(CN = trait)
```

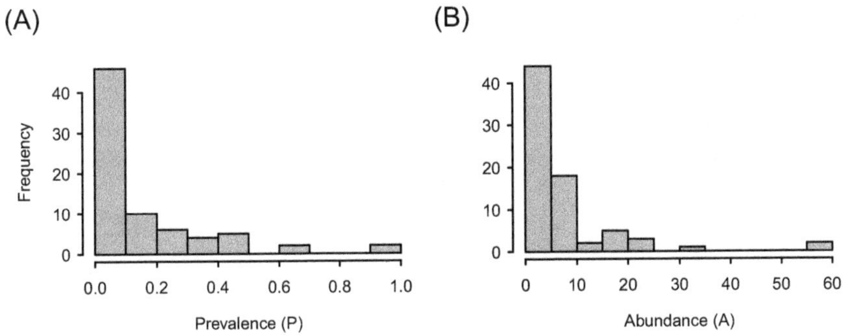

Figure 6.16 Species prevalences (A) and abundances (B) in the plant data. Prevalence is measured as the fraction of occupied sampling units, and abundance as the mean number of individuals over sites where the species is present. The y-axis (frequency) refers to the number of species with a given prevalence (A) or abundance (B).

Let us explore the raw data by plotting histograms of the species prevalences and abundances.

```
P = colMeans(Y > 0)
A = colSums(Y)/colSums(Y > 0)
```

Typical of community ecology data, most species are rare in the sense that they are present only in a minority (on average 15 per cent) of the study sites (Figure 6.16A). When present, the average count ranges from 1 to 57.7 individuals per site (Figure 6.16B).

To account for relatedness among species, we will use taxonomy as a proxy for phylogeny. To do so, in the script below we read in a classification of the species into families and genera, and then use the function as.phylo from the ape package to construct a taxonomical tree. We assume equal branch lengths among families, among genera within a family, and among species within a genus.

```
taxonomy = read.csv(file = file.path(data.directory,
                    "plant data\\taxonomy.csv"))
plant.tree = as.phylo(~family/genus/species,
  data = taxonomy, collapse = FALSE)
plant.tree$edge.length = rep(1, length(plant.
  tree$edge))
```

6.7.1 Setting up and Fitting HMSC Models

We are now ready to set up the HMSC models. To examine the robustness of the results, we will fit two models: a probit model to the presence–absence data, and a lognormal Poisson model to the full count data. In the script below, we include both models into a single list named models.

```
XFormula = ~TMG
TrFormula = ~CN

models = list()
models[[1]] = Hmsc(Y=1*(Y > 0), XData = XData,
  XFormula = XFormula,TrData = TrData,
  TrFormula = TrFormula, phyloTree = plant.tree,
  distr = "probit")

models[[2]] = Hmsc(Y = Y, XData = XData,
  XFormula = XFormula,TrData = TrData,
  TrFormula = TrFormula, phyloTree = plant.tree,
  distr = "lognormal poisson")
```

We next move on to fit the models as usual.

```
for (i in 1:2){
     models[[i]] = sampleMcmc(models[[i]], thin=thin,
  samples = samples, transient = transient,
  nChains = nChains, verbose = verbose)
}
```

Figure 6.17 evaluates MCMC convergence for the β parameters in terms of effective sample sizes and the potential scale reduction factors. MCMC convergence can be considered satisfactory, so we move on to examine the results.

6.7.2 Do Species that Occur on Dry, Warm Sites Have a High Carbon-to-Nitrogen Ratio?

The parameter estimates of species niches (Figure 6.18) show that many species respond negatively to the TMG. This means that species are more likely to be present (Figure 6.18A) and be more abundant (Figure 6.18B) in sites with low TMG values, i.e. in sites located on mesic, north-facing slopes.

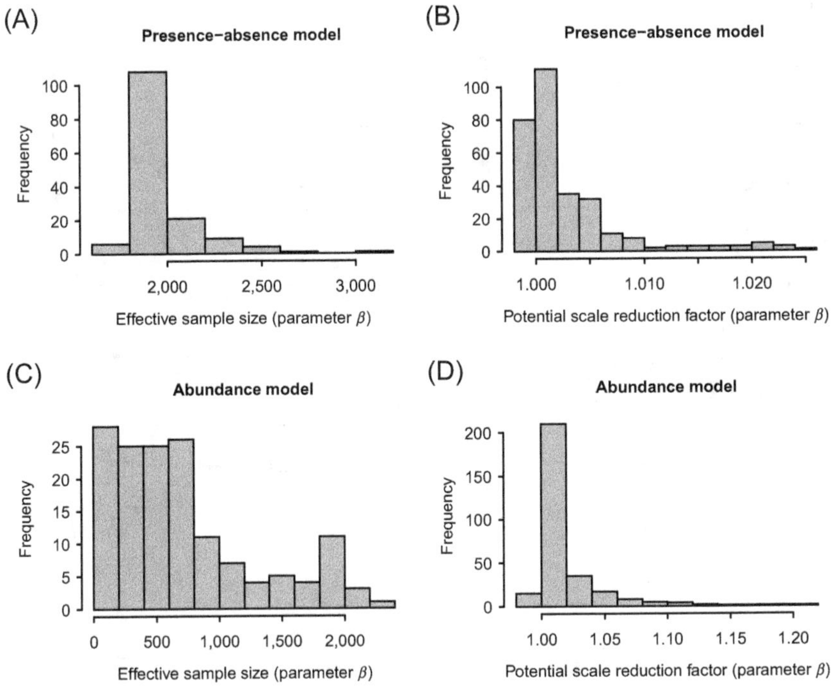

Figure 6.17 MCMC convergence diagnostics for the β parameters measured for the presence–absence (panels A and B) and abundance (panels C and D) models in terms of effective sample sizes (panels A and C) and potential scale reduction factors (panels B and D).

Figure 6.19 addresses the main study question: is there an association between environmental conditions and species traits? For both models, we find a negative relationship between the 'species trait intercept' and TMG, and a positive relationship between the species trait C:N and TMG. As the species traits are scaled to a mean of zero, the result related to the intercept can be interpreted as that an average species (i.e. that with zero C:N) responds negatively to TMG. This is reflected in Figure 6.18, where most species respond negatively to TMG. The positive relationship between C:N and TMG indicates that those species with low C:N respond especially negatively to TMG, whereas species with high C:N may respond even positively to it.

Figure 6.19 further shows a positive relationship between C:N and the 'environmental covariate intercept'. As the environmental covariates

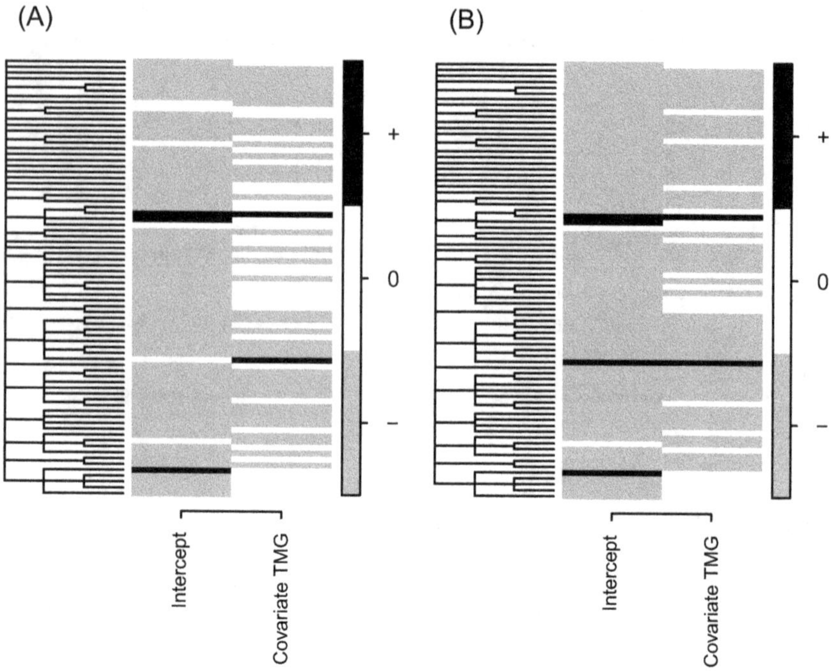

Figure 6.18 Heatmap of estimated species niches. Black and grey colour indicates parameters that are estimated to be positive and negative, respectively, with at least 0.95 posterior probability. Panel A corresponds to the presence–absence model and panel B to the abundance model.

have been scaled to have a zero mean, this result indicates that species with high C:N are on average more common (in terms of both occurrence and abundance) than species with low C:N.

We next construct gradient plots to examine how species richness and the community-weighted mean of C:N vary over the environmental gradient of TMG.

```
for (i in 1:2){
  m = models[[i]]
  Gradient = constructGradient(m, focalVariable = "TMG")
 predY = predict(m, Gradient = Gradient, expected = TRUE)
  q = c(0.25,0.5,0.75)
  plotGradient(m, Gradient, pred = predY, measure = "S",
```

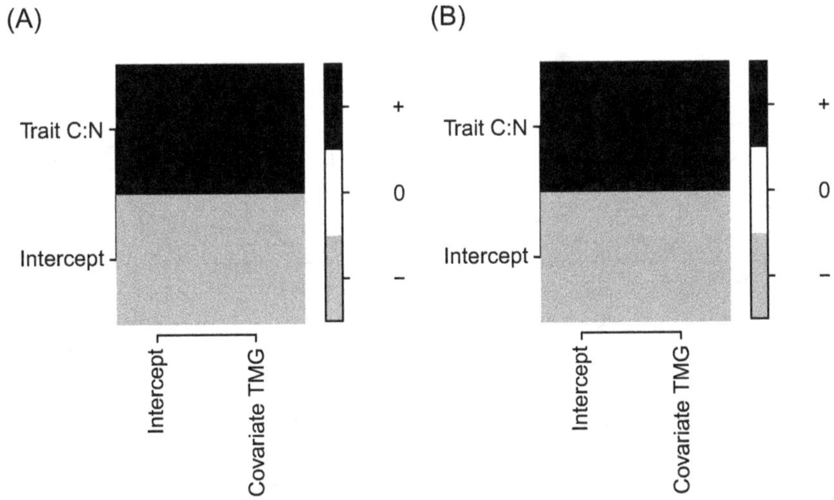

Figure 6.19 Heatmap of estimated γ parameters linking species traits to species niches. Black and grey colour indicates parameters that are estimated to be positive and negative, respectively with at least 0.95 posterior probability. Panel A corresponds to the presence–absence model and panel B to the abundance model.

```
    showData = TRUE, q = q)
plotGradient(m, Gradient, pred = predY, measure
    = "T", index = 2,showData = TRUE, q = q)
}
```

The results of Figure 6.20 are in agreement with those of Figure 6.19, showing that species richness and abundance decreases with TMG, and that community-weighted mean C:N is on average positive and increases with TMG.

To ask how much of the variation in species niches and occurrences is explained by C:N ratio, we utilise the function compute VariancePartitioning.

```
VP = computeVariancePartitioning(models[[1]],
  group = c(1,1),  groupnames = "TMG")
VP$R2T
## $Beta
## (Intercept)        TMG
##   0.2386144 0.1297235
##
## $Y
## [1] 0.2479599
```

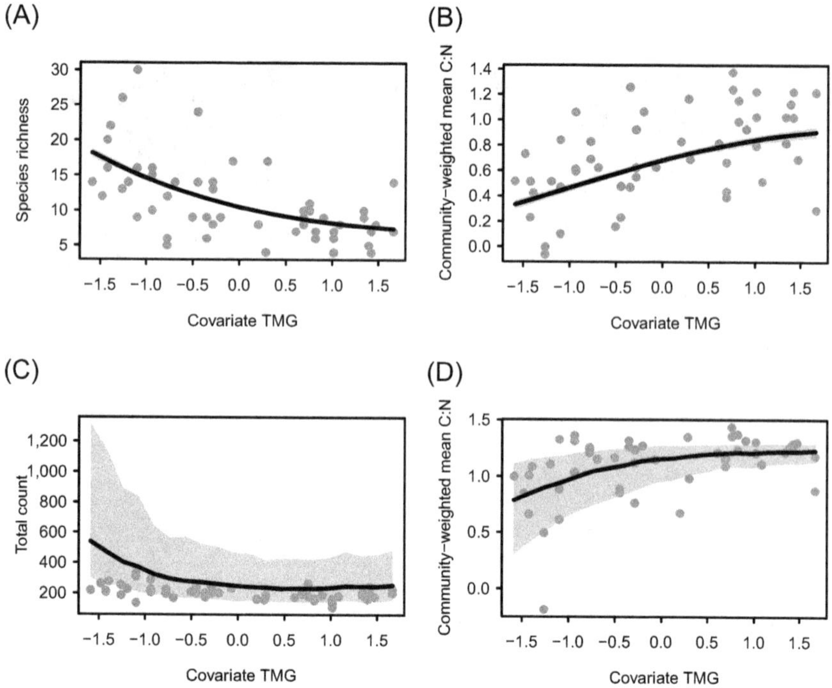

Figure 6.20 Model predictions of species richness (panel A), number of individuals (panel C), and community-weighted mean of C:N ratio (panels B and D) over the environmental TMG gradient. The upper panels (A and B) are based on the probit model of presence–absence data, and the lower panels (C and D) on lognormal Poisson model of abundance data.

```
VP = computeVariancePartitioning(models[[2]],
  group = c(1,1),groupnames = "TMG")
VP$R2T
## $Beta
##  (Intercept)        TMG
##    0.2178736 0.1882688
##
## $Y
## [1] 0.2352205
```

We observe that C:N explains not only how species respond to TMG, but also a substantial part of the variation among the species in their intercepts, in line with the discussion above.

Let us finally ask if there is evidence of phylogenetic signal in the residual variation in species niches, on top of what can be explained by the trait C:N.

```
mpost = convertToCodaObject(models[[1]])
summary(mpost$Rho)$quant

## 2.5%  25%  50%  75% 97.5%
## 0.00 0.00 0.07 0.38  0.83

mpost = convertToCodaObject(models[[2]])
summary(mpost$Rho)$quant

## 2.5%  25%  50%  75% 97.5%
## 0.00 0.00 0.00 0.33  0.79
```

We do not find such evidence, and hence conclude that related species do not respond more similarly to TMG than unrelated species, beyond what can be expected based on their C:N.

In line with our results, Miller et al. (2019) found a strong negative main effect of TMG on abundance, and a strong positive main effect of C:N on abundance. However, they found support for the relationship between C:N and TMG with only some of the applied methods. Based on our analyses, there is good evidence for a positive relationship between C:N and TMG, both for species occurrences (Figure 6.19A) and abundances (Figure 6.19B). Thus, our HMSC analyses give support for the original hypothesis that species occurring on drier and warmer sites have on average higher C:N than those occurring in moister and cooler sites.

7 · *Joint Species Distribution Modelling*

Biotic Interactions

In the previous chapter we discussed how HMSC models species niches as a function of measured traits and phylogenetic relationships. This hierarchical structure is one reason why HMSC is a JSDM rather than a stacked distribution model, as traits and phylogenies connect the species-specific niche models with each other. But there is another reason why HMSC is a JSDM rather than a stacked distribution model: HMSC can be used to model residual co-occurrences (or more generally, residual associations) among the species, which is the topic of this chapter. In terms of the community assembly processes, residual associations are aimed at capturing biotic interactions.

We begin this chapter with a brief overview of the different modelling strategies that can be used for estimating biotic interactions in species distribution models (Section 7.1). We then build the statistical approach, first discussing the relationship between occurrence and co-occurrence probabilities (Section 7.2), and then describing how latent variables can be used to compactly model co-occurrences in species-rich communities (Section 7.3). After introducing the baseline model, we extend it to hierarchical, spatial and temporal study designs (Section 7.4), as well as to cases where the biotic interactions depend on the environmental conditions (Section 7.5). After introducing the statistical approaches, we focus on interpretation, recalling that residual associations can be caused by many processes other than biotic interactions, and thus great caution must be taken when interpreting associations as biotic interactions (Section 7.6). As a final theoretical section, we illustrate when and how the estimated species associations can be used to make improved predictions (Section 7.7). Like the previous chapter, we conclude this chapter with case studies, of which the first is based on simulated data (Section 7.8) and the second on sequencing data on dead-wood-inhabiting fungi (Section 7.9).

7.1 Strategies for Estimating Biotic Interactions in Species Distribution Models

JSDMs model biotic interactions by including residual covariance structure that captures species-to-species associations in their occurrence or abundance (Clark et al. 2014; Ovaskainen et al. 2017b; Pollock et al. 2014; Thorson et al. 2015). As discussed in Chapter 1, biotic interactions can be expected to result in non-random co-occurrence patterns in the data even after accounting for environmental variation. Interactive relationships such as mutualism, parasitism and facilitation can be expected to cause aggregated distributions between the interactive species, and thus lead to positive species-to-species associations. In contrast, competitive interactive relationships can be expected to lead to segregated distributions between the interactive species, and thus to negative species-to-species associations. JSDMs generally model these associations by measuring the covariance structure of the model's residuals, i.e. after accounting for the effect of the environmental covariates. For small species communities, it may be possible to estimate the entire covariance matrix without any further structural assumptions (Ovaskainen et al. 2010; Pollock et al. 2014). However, for large species communities, this would require an enormous amount of data. For example, in 100 species there are 4,950 pairs of species, and thus 4,950 parameters to be estimated. For this reason, JSDMs aimed at large species communities utilise latent variable approaches that make it possible to estimate large association matrices (Clark et al. 2017; Ovaskainen et al. 2016a; Warton et al. 2015). The use of a latent variable approach for estimating species associations can be seen as a further ecological assumption, as it models the residual associations for all pairs of species jointly in attempt to find leading axes of residual variation, similar to how ordinations extract the main signals from multivariate data. For this reason, JSDMs with latent variable structures can be viewed as model-based ordinations (Hui et al. 2015; Warton et al. 2015).

In the context of single-species distribution modelling, it is possible to account for interspecific associations by using the occurrences of some species as predictors (Araújo & Luoto 2007; Kissling et al. 2012; Meier et al. 2011; Mod et al. 2015; Pellissier et al. 2010). However, doing so for large communities requires prior information or hypotheses about which are the influential species, as not all of the other species can be included as predictors. In Arctic–Alpine tundra for example, some dwarf shrub species are known to have positive effects on many other species by offering beneficial microclimate. Therefore, their inclusion as a predictor can be a more straightforward strategy to assess their influence as biotic filters than to estimate residual associations between all species pairs (le Roux et al. 2014; Mod et al. 2015).

Species associations are estimated in the random effect part of the HMSC model. There are many reasons why random effects might be included in a model. First, sometimes the study design is hierarchical, so that sampling units are nested within higher levels such as plots. With such a sampling design, one should include a plot-level random effect already in a univariate analysis, to control for the dependency structure in the data and thus avoid biased inference (see Section 5.4). For exactly the same reasons, one should also include a plot-level random effect in a multivariate analysis. In this case, the plot-level random effect may give additional information, as it can be used to identify which species co-occur more or less often than expected by random in the same plots. Similarly, in spatial or temporal study designs a spatial or temporal random effect should already be included with univariate analyses, as otherwise one assumes that the sampling units are independent. Adding a spatial or temporal random effect into HMSC also controls for such dependencies, but it may additionally reveal which species pairs co-occur spatially or temporally, and at which spatial or temporal scales such co-occurrences take place. In addition to the above cases, a random effect may be included in the multivariate HMSC model even if the sampling units can be viewed as independent of each other – something that would not be meaningful to do in a univariate model. In the multivariate HMSC model, the random effect defined at the sampling-unit level will model species co-occurrences.

7.2 Occurrence and Co-occurrence Probabilities

To explain what we mean by co-occurrence in the context of species distribution modelling, let us consider a toy example that includes presence–absence data on only two species, modelled by a probit model. Assume that we know the species niches of species 1 (the $n_c \times 1$ column vector $\boldsymbol{\beta}_{\cdot 1}$) and species 2 (the $n_c \times 1$ column vector $\boldsymbol{\beta}_{\cdot 2}$), and that we wish to predict species occurrences under some particular environmental conditions (the $1 \times n_c$ row vector $\boldsymbol{x}^T$). To do so, we compute the linear predictors for these two species as $L_1 = \boldsymbol{x}^T\boldsymbol{\beta}_{\cdot 1}$ and $L_2 = \boldsymbol{x}^T\boldsymbol{\beta}_{\cdot 2}$, and convert them by the probit model as probabilities of species occurrence as $p_1 = Pr(y_1 = 1) = \Phi(L_1)$ and $p_2 = Pr(y_2 = 1) = \Phi(L_2)$. Assume that the probit model predicts that both species occur with probability 0.5, so that $p_1 = p_2 = 0.5$, which is the case if $L_1 = L_2 = 0$. Knowing that species 1 occurs at probability 0.5 under environmental conditions $\boldsymbol{x}^T$ means that if we survey 100 sampling units having the same environmental conditions $\boldsymbol{x}^T$, we would expect to see this species fifty times on average (the same consideration also holds for species 2).

While knowing the species-specific occurrence probabilities p_1 and p_2 is sufficient to tell the complete story from the species-specific point of view, this is only part of the story from the species co-occurrences point of view. This is because when considering the two species simultaneously, there are four co-occurrence outcome possibilities: either both species are present (the probability of which we denote by q_{11}), both species are absent (with probability q_{00}), species 1 is present and species 2 absent (with probability q_{10}), or species 1 is absent and species 2 present (with probability q_{01}). If the two species occur independently of each other, the probability both being present is given by the product $q_{11} = p_1 p_2$, which would be 0.25 in the case of our toy example. Similarly, the probability of species 1 being present and species 2 absent would be $q_{10} = p_1(1 - p_2)$, which would also be 0.25 with our toy example.

As discussed in Chapter 1, there are many reasons why species may not occur independently of each other. For example, they might compete for common resources. Let us assume that competition is so severe that the two species show competitive exclusion. In this case, each sampling unit can be occupied by only one of the two species. Let us assume that species 1 is present and species 2 absent in half of the sampling units whereas, in the remaining sampling units, species 2 is present and species 1 is absent. In other words, let us assume that $q_{11} = q_{00} = 0.5$, and $q_{10} = q_{01} = 0$. The species-specific occurrence probabilities can be computed from these co-occurrence probabilities as $p_1 = q_{11} + q_{10} = 0.5$ and $p_2 = q_{11} + q_{01} = 0.5$. Thus, the species-specific occurrence probabilities are the same as in our toy example above, but now there is a strong negative co-occurrence pattern. The difference between independent and dependent occurrences can be detected only if both species are considered simultaneously (jointly). For example, with data on species 1 only, there would be no signal that would indicate competitive exclusion from species 2. Rather, it would just appear to be missing from half of the sampling units 'by chance'.

7.2.1 Raw versus Residual Co-occurrence

Let us then return to the meaning of co-occurrence. If considering a community of two species, negative co-occurrence means that the species co-occur less often than expected by chance, whereas positive co-occurrence means that the two species co-occur more often than expected by chance. With this general definition in mind, it is important to be specific by what is meant by 'more or less often than expected by chance', i.e. what is the underlying null model.

One possible null model is obtained by recording the expected number of co-occurrences that would result when reshuffling the species occurrences randomly across the sampling units. This approach corresponds to the classical tests of null-model analysis of co-occurrence by Gotelli (2000) which we briefly discussed in Section 3.2 and that we will illustrate in Section 11.5.3. We will refer to co-occurrences derived from this approach as raw co-occurrences. It is important to note that raw co-occurrences do not necessarily have anything to do with biotic interactions, such as competitive exclusion in the toy example above. To illustrate this, let us assume that the researcher has surveyed the occurrences of capercaillies (*Tetrao urogallus*) and white-backed woodpeckers (*Dendrocopos leucotos*) in a set of forest patches. Let us further assume that some of the forest patches consist of coniferous forest, and others of deciduous forest. As capercaillies inhabit primarily coniferous forest and white-backed woodpeckers inhabit deciduous forests, most forest patches would be inhabited just by one of the species. Using a null-model approach, the raw co-occurrences would identify the negative association between the species, but this would reflect their differential habitat preferences rather than their interactions.

It is also possible to extend the above explained null-model analysis to account for the co-occurrences that result from differential habitat preferences. This can be done by reshuffling the species occurrences among the sampling units in a more restricted way that maintains the relationship between species occurrence and environmental conditions. This approach is called the environmentally constrained null-model analysis (Peres-Neto et al. 2001) that we briefly introduced in Section 3.2. With the example above of capercaillies and white-backed woodpeckers, the environmentally constrained null model would reshuffle the capercaillie occurrences within coniferous forests, and the white-backed woodpecker occurrences within deciduous forests. As a result, the amount of co-occurrence between the two species would be equally low in both the original and reshuffled data, suggesting that after controlling for the environmental conditions the co-occurrence between the two species is zero. We will refer to co-occurrence derived from this approach as residual co-occurrence. As residual co-occurrences control for variation caused by the match between species niche and the environmental conditions, they are more likely to indicate biotic interactions than raw co-occurrences do.

7.2.2 Species Association: Co-occurrence or Co-variation in Abundance

In our toy example of the two species, we have used the terms raw and residual co-occurrence, although we will more generally use the terms raw and residual association. The reason for these terminological choices is that our toy example was specifically about the probit model of species occurrence, whereas the general model can be supplemented with any link function and error distribution. Specifically, the probit model generates non-random patterns of co-occurrence, whereas an abundance model generates non-random patterns of co-variation in abundance. For example, one species can be predicted to be especially abundant in sampling units where another species is especially abundant as well (e.g. when species facilitate each other). Hence, we use species association as a more general term that involves either co-occurrence or co-variation in abundance.

7.3 Using Latent Variables to Model Co-occurrence

In this section, we will describe how residual co-occurrences are modelled in HMSC with a latent variable approach.

7.3.1 The Simplest Case: Co-occurrence of Two Species

Let us return to the case of the two species that occur under some environmental conditions $\boldsymbol{x}^T$ with probabilities p_1 and p_2. As explained above, knowing these two probabilities is not sufficient to infer the co-occurrence patterns. The four co-occurrence outcome probabilities between the two species are constrained by the condition $q_{11} + q_{10} + q_{01} + q_{00} = 1$, thus there are three degrees of freedom. This means that to fully model their co-occurrence probabilities, we need three parameters. One intuitive way of parameterising these degrees of freedom is by first fixing two of them with the species-specific occurrence probabilities p_1 and p_2, and then defining a third parameter that describes the extent to which the two species co-occur more or less often than expected from independent occurrences. This is not exactly how residual co-occurrences are modelled in HMSC, but it represents the basic idea. Before describing exactly how they are modelled in HMSC for the case of many species, we will discuss in detail the simplest case of two species. Although these explanations are quite technical, understanding the statistical machinery underlying co-occurrence modelling – at least to some extent – is important for correct interpretation of the HMSC results.

We use L_1 and L_2 to denote the linear predictors for the two species for some particular sampling unit. In HMSC, co-occurrences are modelled through a latent variable approach. What this means in practice is that a latent factor η from the standard normal distribution $\eta \sim N(0, 1)$ is simulated for each of the sampling units, and then factor loadings λ are estimated for each of the species. Thus, in our example, λ_1 is the factor loading for species 1, and λ_2 is the factor loading for species 2. While the terminology of latent factors and factor loadings is the one used in the original statistical literature (Bhattacharya & Dunson 2011), we will follow a more ecologically intuitive terminology by calling henceforth the latent factors η as site loadings (called site scores in ordination literature), and the factor loadings λ as species loadings (called species scores in ordination literature) (Section 3.1).

The site and species loadings are multiplied together, and then added to the linear predictors. Thus, under the probit model, the occurrence probability of species 1 becomes $p_1 = \Phi(L_1 + \eta\lambda_1)$, and for species 2 it becomes $p_2 = \Phi(L_2 + \eta\lambda_2)$. If we have estimated the parameters from training data and we wish to make predictions for a new sampling unit, then the species loadings would be known, but the site loading unknown for the new sampling unit. To obtain species-specific occurrence probabilities for the new sampling unit, we need to marginalise the site loading out by integrating over its distribution. Thus, under the latent variable model, species-specific occurrence probabilities are given by the following equations:

$$\begin{aligned} p_1 &= \int_{-\infty}^{\infty} \Phi(L_1 + \eta\lambda_1)\phi(\eta)d\eta \\ p_2 &= \int_{-\infty}^{\infty} \Phi(L_2 + \eta\lambda_2)\phi(\eta)d\eta \end{aligned} \tag{7.1}$$

In Equation 7.1, we have denoted by $\phi(\eta)$ the probability density of the standard normal distribution. The latent variable η is shared between the two species, and thus it influences their co-occurrence. For example, the probability of both species being present (thus co-occurring) can be computed as:

$$q_{11} = \int_{-\infty}^{\infty} \Phi(L_1 + \eta\lambda_1)\Phi(L_2 + \eta\lambda_2)\phi(\eta)d\eta \tag{7.2}$$

To understand what Equation 7.1 and Equation 7.2 mean in practice, let us consider a few numerical examples. First, to set the baseline, if $L_1 = L_2 = 0$, and if $\lambda_1 = \lambda_2 = 0$, it holds that $p_1 = p_2 = 0.5$ and $q_{11} = 0.25$, corresponding to independent occurrences. If we keep $L_1 = L_2 = 0$ but change $\lambda_1 = \lambda_2 = 1$, then it still holds that $p_1 = p_2 = 0.5$ but now $q_{11} = 0.33$. Thus, the species-specific occurrence probabilities remained unchanged, but the co-occurrence probability increased. If instead of $\lambda_1 = \lambda_2 = 1$ we set $\lambda_1 = \lambda_2 = -1$, the result is exactly the same: $p_1 = p_2 = 0.5$ and $q_{11} = 0.33$. While reverting the sign of the species loadings for both species at the same time did not change the result, reverting the sign for one species only does change the result. If we still keep $L_1 = L_2 = 0$ but now assume $\lambda_1 = 1$ and $\lambda_2 = -1$, it holds that $p_1 = p_2 = 0.5$ while $q_{11} = 0.17$, and thus the two species co-occur less often than expected by chance. That is, $\lambda_1\lambda_2 = 0$ implies independent occurrence ($q_{11} = p_1p_2$), whereas $\lambda_1\lambda_2 > 0$ implies positive co-occurrence ($q_{11} > p_1p_2$), and $\lambda_1\lambda_2 < 0$ implies negative co-occurrence ($q_{11} < p_1p_2$).

What somewhat complicates the interpretation of the modelling results is that the species loadings can also influence the marginal species-specific occurrence probabilities. This did not happen with our numerical examples above, since $L_1 = 0$ always resulted in $p_1 = 0.5$. But unfortunately, the case where the linear predictor is zero is a very special case. To illustrate this, let us next assume that $L_1 = L_2 = -1$; in this case, independent occurrences generated by $\lambda_1 = \lambda_2 = 0$ leads to $p_1 = p_2 = 0.16$. If we wish to generate a positive co-occurrence by setting $\lambda_1 = \lambda_2 = 1$, the species-specific occurrence probabilities change to $p_1 = p_2 = 0.24$ due to the non-linear probit link function Φ in Equation 7.1. The fact that the species loadings influence species-specific occurrence probabilities may not be desirable in terms of interpreting the results. However, the main advantage is that they influence the co-occurrences. In our example with $L_1 = L_2 = -1$ and $\lambda_1 = \lambda_2 = 1$, the same sign of the species loadings generates a positive co-occurrence: $q_{11} = 0.11$, which is greater than what would be expected by chance, as $p_1p_2 = 0.06$.

Importantly, the model given by Equations 7.1 and 7.2 is able to predict any feasible combination of species-specific occurrence probabilities and co-occurrence probabilities. For example, if we wish to keep $p_1 = p_2 = 0.16$ but have the amount of positive co-occurrence implied by

$\lambda_1 = \lambda_2 = 1$, we can do so by setting $L_1 = L_2 = -1.4$. With these parameters, $p_1 = p_2 = 0.16$ and $q_{11} = 0.06 > p_1 p_2 = 0.03$. Therefore, no matter what the values of the species loadings are, the model can predict any marginal occurrence probabilities by adjusting the value of the linear predictor. This means that when we fit the latent variable model to the data, the species-specific occurrence probabilities can be estimated from the data, even if the presence of latent variables modifies the link between the linear predictors and occurrence probabilities.

7.3.2 The Full Story: Co-occurrence of Many Species

Our explanation above considered a single sampling unit and two species. Let us next return to the general case of n sampling units indexed by i, and n_s species indexed by j. We further assume that there can be several latent factors, the number of which we denote by n_f, and which we index by h. We denote by η_{ih} the site loading number h for the sampling unit i, and by λ_{hj} the species loading number h for species j. This means that the site loading of each sampling unit is a vector, as is the species loading of each species, similar to the ordination approaches in which the site scores and species scores are vectors.

The full HMSC model with both fixed and random effects can now be written as:

$$L_{ij} = L_{ij}^F + L_{ij}^R \tag{7.3}$$

where L_{ij}^F models the fixed effects as before:

$$L_{ij}^F = \sum\nolimits_{k=1}^{n_c} x_{ik}\beta_{kj} \tag{7.4}$$

and L_{ij}^R models the random effects through the latent variable approach as:

$$L_{ij}^R = \sum\nolimits_{h=1}^{n_f} \eta_{ih}\lambda_{hj} \tag{7.5}$$

As we may interpret the latent variable model as a random effect model, Equation 7.3 can be viewed as a multivariate mixed model. Denoting the $n \times n_f$ matrix of site loadings η_{ih} by $\mathbf{H}$, and the $n_f \times n_s$ matrix of species loadings by $\mathbf{\Lambda}$, we may write Equation 7.3 compactly in the matrix notation as $\mathbf{L} = \mathbf{L}^{\mathrm{F}} + \mathbf{L}^{\mathrm{R}}$, where $\mathbf{L}^{\mathrm{F}} = \mathbf{XB}$ and $\mathbf{L}^{\mathrm{R}} = \mathbf{H\Lambda}$.

The equations for the fixed (Equation 7.4) and random (Equation 7.5) effects are very similar to each other. By comparing these two equations, we can see that the number of latent factors n_f is analogous to the number of covariates n_c, that the species loadings η_{ih} are analogous to the environmental covariates x_{ik} and that the species loadings λ_{hj} are analogous to the responses of the species β_{kj} to the environmental covariates. Indeed, the site loadings η_{ih} can be interpreted as hidden environmental covariates and the species loadings as the species response to those hidden environmental covariates. The difference between the fixed and random effect parts of the model is that while the environmental covariates **X** are known and included in the model as input data, the hidden environmental covariates **H** are unknown. Thus, while only the species responses **B** are estimated for the fixed effects, both the site loadings **H** and the species loadings **Λ** are estimated for the random effect part.

While the number of the environmental covariates included in the model n_c is known, the relevant number of latent factors n_f is not known *a priori*. Thus, a good question to ask is how many latent factors one should include in the model. In the Bayesian context of HMSC, this question is actually part of the more general question about what prior should be assumed for the species loadings **Λ**. We will postpone addressing this question until Chapter 8, which is devoted to explaining how Bayesian inference is conducted in the HMSC model. In Chapter 8 we will explain what the prior distribution is for each of the model parameters, what the default prior parameters are in Hmsc and how the choice of the prior is likely to influence the results.

In Section 7.3.1, we discussed in length how the latent variable approach can be used to infer a non-random pattern of co-occurrence in the case of two species. The same principles also hold for the case of any number of species under the model of Equation 7.3. This is because of the assumption that the site loadings η_{ih} follow the standard distribution $\eta_{ih} \sim N(0, 1)$ independently for each factor h and each sampling unit i, which implies that the covariance between the two species is:

$$\mathrm{Cov}\left[L^R_{i_1 j_1}, L^R_{i_2 j_2}\right] = \sum\nolimits_{h=1}^{n_f} \lambda_{hj_1} \lambda_{hj_2} \delta_{i_1 i_2} \tag{7.6}$$

The term $\delta_{i_1 i2}$ in Equation 7.6 is the Kronecker delta, with a value of 1 if $i_1 = i_2$ and 0 if $i_1 \neq i_2$, thus indicating that there is no covariance between different sampling units. If the species loadings of two species have the same sign for a given factor h, then the product $\lambda_{hj_1}\lambda_{hj_2}$ makes a positive contribution to the covariance between the two species, whereas if the species loadings are of opposite sign, $\lambda_{hj_1}\lambda_{hj_2}$ makes a negative contribution.

We can write the equation above more compactly in matrix notation as:

$$\boldsymbol{L}_{i\cdot}^{R} \sim N(0, \boldsymbol{\Omega}) \tag{7.7}$$

where the species-by-species variance-covariance matrix $\boldsymbol{\Omega}$ has the elements defined by Equation 7.6, which we may write in matrix form as $\boldsymbol{\Omega} = \boldsymbol{\Lambda}^{\mathrm{T}}\boldsymbol{\Lambda}$.

In the context of Equation 7.3, the random effect part of the model generates covariances on top of the fixed effects, and thus the random effect part models residual associations (i.e. those associations that cannot be explained by the environmental covariates x_{ik} included in the model). The matrix $\boldsymbol{\Omega}$ defines the species-to-species residual associations as a variance-covariance matrix, and we may scale it as a correlation matrix $\mathbf{R}$ as $R_{j_1j_2} = \Omega_{j_1j_2} / \sqrt{\Omega_{j_1j_1}\Omega_{j_2j_2}}$. If $R_{j_1j_2} = 0$, then the two species j_1 and j_2 occur independently of each other after accounting for their fixed effects. If $0 < R_{j_1j_2} \leq 1$, then they show a positive residual correlation, and if $-1 \leq R_{j_1j_2} < 0$, then they show a negative residual correlation.

7.4 Accounting for the Spatio-temporal Context through Latent Variables

In Equation 7.5 we applied the latent variable approach at the sampling-unit level, which means that each sampling unit i has its specific site loading η_{ih} for each of the factors h. Before moving on, let us note that for a univariate model it would not be meaningful to include latent variables at the sampling-unit level. This is because latent variables can be interpreted as random effects, and a random effect at the sampling-unit level would be fully confounded with residual variation. In contrast, it is meaningful to include a latent variable at the sampling-unit level in a multivariate model. This is because in the multivariate case, the random effect models the residual associations among the species that are not confounded with independent residual variation among the species. Yet, the latent variables do not necessarily need to be set for the sampling-unit level, as they can also be set for higher hierarchical levels such as plots (Ovaskainen et al. 2016a), or they can be spatially or temporally structured (Ovaskainen et al. 2016b). Thus, we now use the latent variable approach to extend the univariate case with hierarchical, spatial, and temporal study designs (Section 5.4) to the multivariate case.

7.4.1 Hierarchical Latent Variables

We first discuss how to set up latent variables to a hierarchical level, for example to the plot level if the sampling units are nested within plots. To do so, we follow the notation of Section 5.4.1 and thus denote by $p(i)$ the plot involving the sampling unit i. While the index of the sampling unit ranges as $i = 1, \ldots, n$, the index of the plot ranges as $p = 1, \ldots, n_p$, where the number of plots is typically much smaller than the number of sampling units, $n_p \ll n$. To define a plot-level random effect, we follow Ovaskainen et al. (2016a) and write:

$$L^R_{p(i)j} = L^R_{pj} = \sum\nolimits_{h=1}^{n_f} \eta_{ph}\lambda_{hj} \tag{7.8}$$

Here, η_{ph} is the site loading number h for plot p, and L^R_{pj} is the random effect part of the linear predictor that is now identical for all sampling units that belong to the plot p. The covariance $\mathrm{Cov}\left[L^R_{p_1j_1}, L^R_{p_2j_2}\right]$ between the two species j_1 and j_2 and the two plots p_1 and p_2 can be computed with Equation 7.6, as was done for the sampling-unit level latent variables. Now the covariance describes whether the two species occur more or less often than expected by chance (or are more or less abundant than expected by chance) in the same plots instead of in the same sampling units.

Let us finally simplify the notation by writing Equation 7.8 in matrix form. This can be done as $\mathbf{L}^{\mathrm{R}} = \mathbf{\Pi H \Lambda}$, where $\mathbf{H}$ and $\mathbf{\Lambda}$ denote the site loadings and species loadings, respectively. The new component here is the matrix $\mathbf{\Pi}$, which was already introduced in Figure 2.1 as the matrix describing the study design. The matrix $\mathbf{\Pi}$ is a $n \times n_p$ matrix, with elements $\pi_{ip} = 1$ if sampling unit i belongs to the plot p, and $\pi_{ip} = 0$ if this is not the case. This matrix is needed to copy the plot-level random effects to the sampling units that belong to each focal plot.

7.4.2 Spatial and Temporal Latent Variables

In spatially explicit study designs, it is natural to define that the latent variables have a spatially explicit structure (Ovaskainen et al. 2016b). This can be done with Equation 7.5 by assuming that the site loadings η_{ih} are not independent among the sampling units. In Section 5.4.2, we defined spatially structured residuals with the help of the exponentially decaying covariance function $f(d) = \sigma^2_S \exp(-d/\alpha)$, where d is the distance between sampling units, σ^2_S is the spatial variance and α is the spatial scale of autocorrelation. In the multivariate case we will assume the same covariance structure for the site loadings η_{ih}. In the case of the latent variables, however, we wish for the marginal distribution of each site loading to be $N(0,1)$, and thus we fix the spatial variance to $\sigma_S^2 = 1$. Thus, for factor number h, we assume that the $\boldsymbol{\eta}_{\cdot h}$ (the vector of latent factors for all sampling units i) follows the multivariate normal distribution $\boldsymbol{\eta}_{\cdot h} \sim N(0, \boldsymbol{\Sigma}_h)$, where $\Sigma_{h,i_1 i_2} = \exp(-d_{i_1 i_2}/\alpha_h)$, and α_h is the spatial scale of site loadings corresponding to factor h.

Spatial latent variables provide a very powerful tool for many kinds of analyses, because they model both spatial variation as well as species-to-species associations at the same time. Namely, the spatial latent variables generate the covariance structure

$$\mathrm{Cov}\left[L_{i_1 j_1}^R, L_{i_2 j_2}^R\right] = \sum\nolimits_{h=1}^{n_f} \lambda_{hj_1}\lambda_{hj_2} \exp\left(-d_{i_1 i_2}/\alpha_h\right) \tag{7.9}$$

between the sampling unit i_1 and the species j_1, and the sampling unit i_2 and the species j_2. With the help of this covariance structure, spatial random effects can improve model predictions by borrowing information across both species and space, as we will see in the simulated and real-data examples.

We note that Equation 7.9 can equally well be used to model temporal data (i.e. repeated visits), as the distance d_{ij} can be the one-dimensional temporal distance, i.e. the duration of time between the sampling events for sampling units i_1 and i_2.

7.4.3 Multiple Random Effects in the Same Model

As discussed in Chapter 1, different assembly processes act simultaneously at different spatio-temporal scales. Therefore, the spatio-temporal context of the data used in an SDM framework is of great importance for adequately understanding how the underlying processes shape communities.

For example, in fungal communities inhabiting dead wood, macroclimatic variables such as mean annual temperature affect species distributions at large scales (e.g. continental), whereas conditions such as wood moisture affect their distributions already at local scales (e.g. within a forest) (Abrego et al. 2017b). Therefore, for understanding the full set of drivers behind the realised distributions of wood-inhabiting fungi, one needs data simultaneously from large and small scales, as well as the inclusion of the relevant environmental predictors from both of these spatial scales. Of course, having data from a single scale is still informative, but it only tells a part of the story. When data from multiple scales are available, the hierarchical and spatial nature of the data can be accounted for in SDMs through multiple hierarchical, spatial or temporal random effects (Ovaskainen et al. 2016a; Ovaskainen et al. 2016b; Thorson et al. 2015).

The HMSC model allows for any number of random effects. Denoting with n_r the number of random effect types, different random effects are simply added together as:

$$L_{ij}^{R} = \sum\nolimits_{r=1}^{n_r} L_{ij}^{r,R} \tag{7.10}$$

where $L_{ij}^{r,R}$ is the linear predictor related to the random effect r. For example, if the data are collected from multiple plots at multiple years, we may set $r=1$ as the random effect of the plot, $r=2$ as the random the effect of the year, and $r=3$ to model species associations at the sampling-unit level.

When defining the hierarchical random effect in Equation 7.8, the higher-level units were called plots for the sake of illustration. To use more general notation here, we will call henceforth both the sampling units as well as any higher-level units simply 'units', and denote the unit behind sampling unit i by $u^r(i)$. Note that the units are specific to the random effect r because the units (e.g. plots or years) can vary among them. For example, if each sampling unit i (i.e. row of the data matrix **Y**) has been collected from some particular plot in some particular year, then we may wish to set up random effects at the levels of plots, years and sampling units. In this case, the units relating to each of these are defined by setting $u^1(i)$ as the plot from where the data point i was collected, by setting $u^2(i)$ as the year of when the data point i was collected, and by setting $u^3(i) = i$ to indicate that the third random effect is at the sampling-unit level. The random effect number r is then defined as:

$$L_{ij}^{r,R} = \sum\nolimits_{h=1}^{n_f^r} \eta_{u^r(i)h}^{r} \lambda_{hj}^{r} \tag{7.11}$$

Here the summation goes over n_f^r, the number of factors included for random effect r. As before, the $\eta_{u^r(i)h}^{r}$ are the site loadings and λ_{jh}^{r} are the species loadings, but now many subscripts and superscripts are needed to identify their random effect r, the number of the factor h, the unit $u^r(i)$ and the species j. We note that while we could call the $\eta_{u^r(i)h}^{r}$ more generally 'unit loadings', we will keep the more intuitive terminology of site loadings even if the 'sites' can now actually refer to any spatial or temporal descriptor.

The site loadings $\eta^r_{u^r h}$ are assumed to be independent between random effects r and factors h. They can also be independent among the units u^r, in which case we denote them as unstructured. Alternatively, they may have the exponentially decaying correlation structure $\boldsymbol{\eta}^r_{\cdot h} \sim N\left(0, \boldsymbol{\Sigma}\right)$, where $\sum_{u^r_1 u^r_2} = \exp\left(-d_{u^r_1 u^r_2}/\alpha^r_h\right)$, and where $d_{u^r_1 u^r_2}$ is the distance (in space or time) between the units u^r_1 and u^r_2, and α^r_h is the spatial (or temporal) scale associated with the factor number h of random effect r.

In matrix notation, Equations 7.10 and 7.11 can be written as $\mathbf{L}^{\mathrm{R}} = \sum_{r=1}^{n_r} \boldsymbol{L}^{r,R}$, where $\mathbf{L}^{r,\,\mathrm{R}} = \boldsymbol{\Pi}^r \mathbf{H}^r \boldsymbol{\Lambda}^r$. Here, $\mathbf{H}^r$ and $\boldsymbol{\Lambda}^r$ denote the site loadings and species loadings, respectively, for the random effect r, and $\boldsymbol{\Pi}^r$ is a $n \times n^r_u$ matrix of zeros and ones describing to which units of the random effect r the sampling units belong.

When applying Hmsc to nested study designs, it is necessary to give unique identifiers for all levels. For example, assume that the study design consists of sampling units within plots, and that one wishes to include the effects of both the sampling unit and the plot. Assume also that within each plot there are ten sampling units, indexed from 1 to 10. The nested effects of sampling units and plots could then be modelled e.g. with the lme4 package using the (1| sampling_unit/plot) syntax. However, this syntax is not available in Hmsc, and thus one should use a different index for sampling units belonging to different plots (e.g. 1 to 10 for plot 1, 11 to 20 for plot 2 and so on). We note that in this case, the lme4 syntax (1| sampling_unit/plot) would be equivalent to the syntax (1| sampling_unit) + (1| plot), both corresponding to the appropriate way of implementing a nested design.

7.5 Covariate-Dependent Species Associations

The way described in the previous section of estimating biotic interactions through residual species associations assumes that the associations are constant in space, time and environmental gradients. Yet an issue gaining increasing attention in ecological literature is that the direction and strength of interspecific interactions can co-vary with environmental conditions (Pellissier et al. 2018). For instance, when resources become scarce, competition among species might be intensified (Goldberg 1990; Grime 1973), whereas under abiotically stressful environmental conditions, facilitation might become particularly important (Brooker et al. 2008; He et al. 2013). Changes in the outcomes of interspecific interactions in relation to changing environmental conditions have been found for a wide array of taxonomical groups (Pellissier et al. 2018).

When fitting single-species distribution models to communities in which the influential species are known, it is possible to account for changing species associations along environmental gradients by including an interaction term between the influential species and the environmental variables in the predictors (Mod et al. 2014). In the context of JSDMs, the analogous approach is to estimate environmentally dependent species associations by assuming that the latent variable structure behind the species associations co-varies with the environmental variables. In this section, we will follow Tikhonov et al. (2017) to describe how this possibility has been implemented in HMSC.

We recall from the above sections that the latent variables lead to the covariance structure $\boldsymbol{L}_{i\cdot}^{R} \sim N(0, \boldsymbol{\Omega})$, where the matrix of residual associations among the species is given by $\boldsymbol{\Omega} = \boldsymbol{\Lambda}^T\boldsymbol{\Lambda}$. One 'hidden' assumption behind this formulation is that the matrix $\boldsymbol{\Omega}$ is constant, and thus that it does not depend for example on the environmental conditions at which the community is sampled. It thus assumes that ecological interactions remain constant over environmental conditions, or more generally over space and time. To account for associations that vary with environmental conditions $\boldsymbol{x}$, we will assume that $\boldsymbol{\Omega}$ is a function of those conditions, so that $\boldsymbol{\Omega} = \boldsymbol{\Omega}(\boldsymbol{x})$. Before introducing how this can be done, let us first note that the core HMSC model with constant $\boldsymbol{\Omega}$ can already be used to predict e.g. whether there is a higher fraction of negative associations among the species when resources are limited than when resources are abundant. This can happen when the species that are present when resources are limited have more negative associations than those species that are present when resources are abundant. Thus, even if the associations $\boldsymbol{\Omega}$ among all pairs of species remains constant, what may change is the subset of species being present and thus the subset of associations that are realised. This is important to keep in mind: while $\boldsymbol{\Omega}$ contains an estimate of a residual association among all pairs of species, some of those species pairs may have such contrasting responses to the environmental covariates (as modelled by the fixed effects) that they do not co-occur simply for this reason. In such a case, a residual species association between the two species is not of practical relevance.

In addition to the possibility that the realised associations change because the identity of the species that are present changes, the associations themselves can change. For example, let us assume that the species A and B occur both under conditions where resources are limited and where resources are abundant. When resources are abundant, the species A and B occur independently of each other, but when they are limited, they display competitive exclusion and thus show a negative pattern of co-occurrence. This phenomenon could not be captured by the core HMSC, and this is where the extension by Tikhonov et al. (2017) is needed. In this extension, the species loadings λ_{hj} are modelled as a function of some environmental covariates $\boldsymbol{x}$, so that $\lambda_{hj} = \lambda_{hj}(\boldsymbol{x})$, and thus the whole matrix of species loadings depends on the environmental conditions, $\boldsymbol{\Lambda} = \boldsymbol{\Lambda}(\boldsymbol{x})$. This also implies that the association matrix depends on the environmental conditions, as now $\boldsymbol{\Omega}(\boldsymbol{x}) = \boldsymbol{\Lambda}(\boldsymbol{x})^T \boldsymbol{\Lambda}(\boldsymbol{x})$.

The environmental covariates that are used to model variation in the species loadings can be the same as or different from the environmental covariates that are used to model species niches. To separate the former from the latter, we denote the matrix of environmental covariates that are used to model species associations by $\mathbf{X}^{\Omega}$. We assume that there are n_c^{Ω} such covariates indexed by $k = 1, .., n_c^{\Omega}$, so that x_{ik}^{Ω} is the covariate number k for sampling unit i, and $\boldsymbol{x}_{i\cdot}^{\Omega}$ denotes the vector of all covariates for sampling unit i. As with the matrix $\mathbf{X}$, we include the intercept in the matrix $\mathbf{X}^{\Omega}$, so that $x_{i1}^{\Omega} = 1$ for all sampling units i.

In the notation above, the model for the covariate-dependent loading for species j and factor h under the environmental conditions of sampling unit i can be written:

$$\lambda_{hj}\left(\boldsymbol{x}_{i\cdot}^{\Omega}\right) = \sum\nolimits_{k=1}^{n_c^{\Omega}} \lambda_{hjk} x_{ik}^{\Omega} \tag{7.12}$$

where λ_{hjk} models how the species loading λ_{hj} depends on the covariate k. Thus, the covariate-dependent random effect model for the sampling unit i becomes:

$$L_{ij}^{R} = \sum\nolimits_{h=1}^{n_f} \eta_{ih} \sum\nolimits_{k=1}^{n_c^{\Omega}} \lambda_{hjk} x_{ik}^{\Omega} \tag{7.13}$$

Under the model of Equation 7.13, the researcher is likely to be interested in quantifying the level of statistical evidence about whether the associations indeed vary with the environmental conditions. To do so, one can compute the species-to-species matrix of posterior probabilities $\mathbf{S}(\boldsymbol{x}_1, \boldsymbol{x}_2) = Pr[\boldsymbol{\Omega}(\boldsymbol{x}_1) > \boldsymbol{\Omega}(\boldsymbol{x}_2)]$, where the inequality is assessed element by element. A value of $S_{j_1 j_2}(\boldsymbol{x}_1, \boldsymbol{x}_2)$ close to one indicates that there is a high level of statistical support for the association between species j_1 and j_2 being more positive under environment $\boldsymbol{x}_1$ than under environment $\boldsymbol{x}_2$, whereas a value of $S_{j_1 j_2}(\boldsymbol{x}_1, \boldsymbol{x}_2)$ close to zero indicates the opposite.

7.6 A Cautionary Note about Interpreting Residual Associations as Biotic Interactions

Let us return to the question of why residual co-occurrences do not provide direct evidence for species interactions. As discussed in Chapter 1, biotic interactions are one of the principal assembly processes, and thus it is plausible that they leave their signature on data about species occurrence or abundance. For example, if two species facilitate each other, we might expect to find both abundant in the same locations, whereas if two species show competitive exclusion, we would not expect to find them from the same locations.

However, species distributions are simultaneously influenced by a myriad of factors, and thus what remains in the residual associations depends on what is and is not controlled for by the model. This is because in the SDM framework the user decides on which environmental predictors are accounted for, and thus the estimated residual species-to-species associations may indicate that the user has missed some important environmental predictor. This is one of the many important reasons why ecologists should get to know the ecology of their study system before fitting any model. To conclusively relate residual co-occurrences to species interactions, one should complement the results based on observational data with experimental approaches (Dormann et al. 2018). Another way of understanding the link from species interactions to the resulting co-occurrence patterns is to fit SDMs to data simulated from mechanistic models, allowing one to assess if and how the estimated associations relate to the underlying interactions (e.g. Zurell et al. 2018, Ovaskainen et al. 2019). We will follow the latter approach in Chapter 10.

In our example of the capercaillie and the white-backed woodpecker, one would identify a strong negative association between the two species if not controlling for habitat type differences. Thus, a researcher who is poorly informed about the ecology of these two species could erroneously suggest that they competitively exclude each other. In contrast, a researcher who would control for the influence of habitat type would find that there are no residual associations between these species, and thus conclude that they do not interact.

More generally, observational data on species occurrences or abundances cannot be used to test the alternative hypotheses that species associations are generated by biotic interactions or species responding to some environmental covariates that are not controlled for in the model. This is not a shortcoming of the statistical approach itself, but a result of the fact that different underlying processes can lead to identical patterns in the data (Cale et al. 1989). In the context of HMSC, the fact that residual associations can be due to missing covariates rather than biotic interactions is evident in the latent variable implementation, where the site loadings **H** can represent missing environmental covariates to which the species respond through the loadings **Λ**.

For a researcher interested in identifying biotic interactions with HMSC, it is recommended to first record the raw associations in the data, and then see which of these can be explained away when including different potential environmental or spatial predictors in the model. But most importantly, the researcher should first investigate which are the potential environmental variables influencing the focal community and gain information about the underlying mechanisms. The results from species-association analyses should always be interpreted with caution, and in light of ecological knowledge on the study system.

7.7 Using Residual Species Associations for Making Improved Predictions

SDMs that account for biotic filters can be used to generate improved predictions if the distributions of the interactive species are known. This can be done equally well with single-species distribution models where the interactive species are included directly as predictors (Kissling et al. 2012), or with JSDMs where the interactive species are included as response variables (e.g. Ovaskainen et al. 2016a). Before we explain how the latter is technically done in HMSC, let us first discuss when accounting for species associations is expected to improve predictions and when it is not. To do so, we assume that a HMSC model has been fitted to training data originating from a number of sampling units, and the aim is to use it to make a prediction for a new sampling unit not included in the training data.

Let us first assume that the data originate from a spatially explicit context, and that the prediction task relates to spatial interpolation, so that the new sampling unit for which the predictions are to be made is located close to some of the sampling units from the training data. In this case, one would expect that a model that includes a spatial random effect will make better predictions than a model that does not include a spatial random effect. There are two reasons for this. The first is the same as why a univariate model with a spatial random effect would be expected to make better predictions than a non-spatial one. This is because the spatial model bases its predictions not only on the environmental covariates of the new sampling unit, but also on whether or not the species occurs in the nearby sampling units included in the training data. The second is that in the case of the multivariate HMSC model, the predictions are additionally based on the observed occurrences of the other species in the nearby sampling units. For example, if the model estimates from the training data that there is a positive association between species A and B that decays at the spatial scale of one kilometre, then knowing that species B is present in some sampling unit that is 500 metres away will increase the predicted probability that species A is present there.

In a spatial model, residual species associations can be expected to generally improve model predictions; however, this is not the case with non-spatial models. When predicting the occurrence probability or the abundance of species A in a new sampling unit, it does not help to know that this species is positively associated with species B if we do not know whether species B is present in the new sampling unit. However, if the presence or abundance of species B in the new sampling unit is known, then a known association between species A and B may help improve the prediction. Predicting the occurrences (or abundances) of some species conditional on known occurrences (or abundances) of other species is called a conditional prediction.

7.7.1 Conditional Prediction

To describe how conditional prediction works, we recall from Equation 7.7 that the vector of linear predictors capturing a random effect is distributed according to a multivariate normal distribution $\boldsymbol{L}^R \sim N(\mathbf{0}, \boldsymbol{\Omega})$. Here we consider just one sampling unit to which predictions are to be made, so $\boldsymbol{L}^R$ is not a matrix but a vector with one value per species. Before making predictions, we assume that we have fitted the HMSC model to training data, and hence that we have estimated the matrix $\boldsymbol{\Omega}$. When developing our argument below, for simplicity we assume that the matrix $\boldsymbol{\Omega}$ would be known without any estimation error. Yet, as we will exemplify in Section 11.2, propagating parameter uncertainty into predictions is straightforward in the Bayesian context. When making predictions for a new sampling unit without knowing the occurrences or abundances of any of the species, the best one can do is simply to sample the random effect as $\boldsymbol{L}^R \sim N(\mathbf{0}, \boldsymbol{\Omega})$. This will result in the correct association structure among the species in the predictions, but not better predictions for some individual species, in the sense that the expected value of the random effect will be zero. However, let us assume that we know the linear predictor $\boldsymbol{L}^R$ for some of the species. Let us denote by A the collection of species for which the value of the linear predictor is not known, and by B the collection of species for which its value is known. We next partition the linear predictor $\boldsymbol{L}^R$ and the matrix $\boldsymbol{\Omega}$ as:

$$\boldsymbol{L}^{\mathrm{R}} = \begin{pmatrix} \boldsymbol{L}_A^{\mathrm{R}} \\ \boldsymbol{L}_B^{\mathrm{R}} \end{pmatrix}, \boldsymbol{\Omega} = \begin{pmatrix} \boldsymbol{\Omega}_{AA} & \boldsymbol{\Omega}_{AB} \\ \boldsymbol{\Omega}_{BA} & \boldsymbol{\Omega}_{BB} \end{pmatrix} \tag{7.14}$$

The basic mathematical theory of multivariate normal distributions now tells that, conditional on the known values of $\boldsymbol{L}_B^R$, the linear predictor $\boldsymbol{L}_A^R$ for the subset A of species is distributed as:

$$\boldsymbol{L}_A^{\mathrm{R}} \sim N\left(\boldsymbol{\Omega}_{AB}\boldsymbol{\Omega}_{BB}^{-1}\boldsymbol{L}_B^{\mathrm{R}}, \boldsymbol{\Omega}_{AA} - \boldsymbol{\Omega}_{AB}\boldsymbol{\Omega}_{BB}^{-1}\boldsymbol{\Omega}_{BA}\right) \tag{7.15}$$

Given that the species in groups A and B are associated, this makes the matrix $\boldsymbol{\Omega}_{AB}$ non-zero, thus changing our inference about $\boldsymbol{L}_A^R$. Conditional on $\boldsymbol{L}_B^R$, the expectation of $\boldsymbol{L}_A^R$ is $\boldsymbol{\Omega}_{AB}\boldsymbol{\Omega}_{BB}^{-1}\boldsymbol{L}_B^R$, and thus the mean prediction of $\boldsymbol{L}_A^R$ can now be non-zero, unlike in the unconditional prediction. Further, while for an unconditional prediction the variance-covariance matrix of $\boldsymbol{L}_A^R$ is $\boldsymbol{\Omega}_{AA}$, for the conditional prediction it is $\boldsymbol{\Omega}_{AA} - \boldsymbol{\Omega}_{AB}\boldsymbol{\Omega}_{BB}^{-1}\boldsymbol{\Omega}_{BA}$, and thus it is generally smaller than $\boldsymbol{\Omega}_{AA}$. This means that conditional on $\boldsymbol{L}_B^R$, the uncertainty associated with $\boldsymbol{L}_A^R$ is smaller, and hence the predictions will be sharper. In the context of HMSC, knowing the occurrences or abundances of the subset B of species does not mean that the linear predictor $\boldsymbol{L}_B^R$ would be directly known; however, knowing the occurrences or abundances of the subset B of species will yield an estimate of $\boldsymbol{L}_B^R$, and thus Equation 7.15 can be applied.

While Equation 7.15 is helpful for showing that the knowledge of some species can change the mean prediction and decrease residual variance for some other species, the actual implementation in Hmsc is not based on Equation 7.15, but on the underlying latent variable approach. Thus, the known occurrences or abundances of the subset B of species are used to estimate the site loadings (the hidden environmental variables) in the new sampling unit. From the pragmatic point of view, the fact that conditional prediction involves estimation through MCMC means that making conditional predictions can take a lot of time. After the site loadings have been estimated, the species loadings of the species in group A (their responses to the hidden environmental variables) can be utilised when predicting their occurrence or abundance in the new sampling unit.

Predictions in HMSC can be made conditional on any knowledge about the occurrence of the same or other species in the same or other sampling units. We note that conditional prediction is conceptually similar to using the non-focal species as predictors. While including the other species as predictors is a viable option for communities with small numbers of species, it is not straightforward to do if there are tens or hundreds of species. In contrast, conditional predictions can be made for large species communities.

7.7.2 Conditional versus Unconditional Cross-validation

As one important application of conditional prediction, we will introduce conditional cross-validation. Figure 7.1 illustrates the difference between unconditional and conditional cross-validation. For both cases, we assume two-fold cross-validation across the sampling units. Both unconditional and conditional cross-validations loop over the folds of sampling units to make predictions one by one. Assume that the current task is to make predictions for sampling units belonging to fold 2, indicated by white in Figure 7.1. To do so, the model is first fitted to the data from fold 1, indicated by grey in Figure 7.1. In standard cross-validation, predictions for the sampling units in fold 2 are then made in the usual, unconditional way, and thus community data $\mathbf{Y}$ is utilised only from those sampling units that belong to fold 1 (Figure 7.1B). In conditional cross-validation, there is an additional step that loops over the folds of species – each fold is represented by a different symbol in Figure 7.1A. When making predictions for species belonging to e.g. fold 3 (circles), the occurrence or abundance of the other species belonging to folds 1 (squares) and 2 (triangles) are assumed to be known also to the sampling units in fold 2, for which the predictions are to be made. Hence, the predictions are made conditional upon this information, so that in the context of Equation 7.15, the species in fold 3 form the subset A, whereas the species in folds 1 and 2 form the subset B.

One would expect that conditional cross-validation generally yields a higher predictive power than unconditional cross-validation, as conditional cross-validation can take the advantage of species associations. We recall that the usual unconditional cross-validation can be used to assess the extent of overfitting of the fixed effects, which happens when the predictive power of the model is much lower than the explanatory power. In an analogous way, conditional cross-validation can be used to assess the extent of overfitting of the random effects. Namely, if the estimate of an association matrix $\mathbf{\Omega}$ is accurate, and if the matrix $\mathbf{\Omega}$ shows

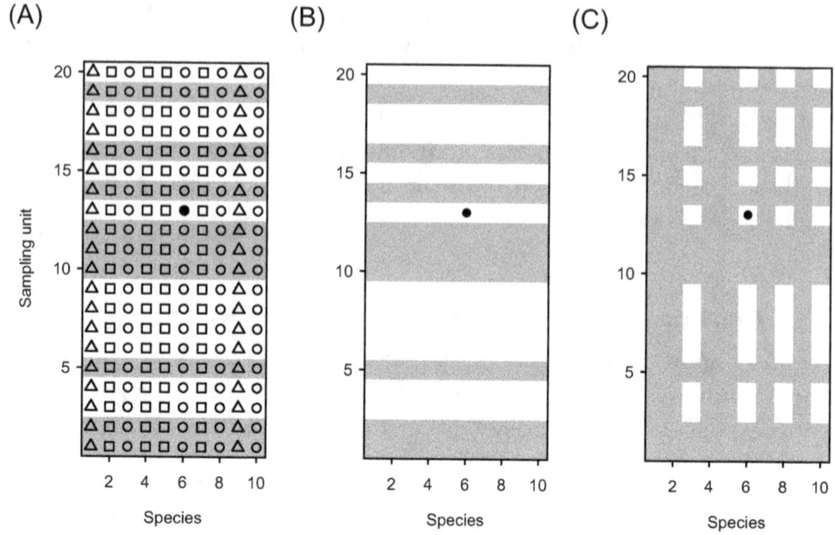

Figure 7.1 Information used in cross-validation and conditional cross-validation. Panel A shows how the twenty sampling units and ten species are partitioned into two folds of sampling units (shown by grey and white) and three folds of species (shown by the square, triangle and circle shapes). One of the symbols is filled, indicating that the prediction is to be made for that particular species in that particular sampling unit. The panels B and C show which parts of the community matrix **Y** are included as training data for model fitting, when making predictions related to either cross-validation (B) or conditional cross-validation (C). In panels B and C, the combinations of species and sampling units shown in grey are included as training data, whereas the ones shown in white are excluded. In this example, the species and sampling units have been randomly assigned to their folds.

a strong pattern of species-to-species associations, conditional cross-validation should yield a higher predictive power than unconditional cross-validation. In contrast, if the associations in the matrix $\mathbf{\Omega}$ are based on noise rather than signal, conditional cross-validation is not expected to yield any higher predictive power than unconditional cross-validation. In fact, it can then even lead to lower predictive power.

Ideally, one may conduct a leave-one-out cross-validation both in terms of the sampling units and the species. However, even with moderately sized data this is computationally very demanding, as it means that the conditional predictions should be done $n_s n$ times, where n_s is the number of species and n is the number of sampling units. For this reason, cross-validations are usually conducted only over a small number of folds, concerning both the sampling units and the species.

7.8 Simulated Case Studies with HMSC

As a simulated case study, we consider a small community consisting of five species. We first generate data for these species, and then use HMSC to estimate the parameters back. When generating the data, we will assume that the species respond differently to variation in two environmental covariates. When fitting models to these data, we will consider three different models: one that contains no environmental covariates, one that contains one of the two environmental covariates and one that contains both of the environmental covariates. Our focus is thus on examining how the covariates included in the model influence the results.

7.8.1 Generating Simulated Data

We will generate presence–absence data for $n_s = 5$ species in $n = 200$ sampling units. The environmental predictors x_1 and x_2 are assumed to be continuous covariates that follow the standard normal distribution. In the script below, we define beta1, beta2 and beta3 as the true parameters for the intercept and the slopes associated with the two covariates. We then combine the three true parameters into the matrix beta. The data are generated by assuming the probit model.

```
n = 200
ns = 5
X = cbind(rep(1, n), rnorm(n), rnorm(n))
beta1 = rep(0, ns)
beta2 = c(2,2,-2,-2,0)
beta3 = c(1,-1,1,-1,0)
beta = cbind(beta1, beta2, beta3)
L = X %*% t(beta)
Y = 1*((L + matrix(rnorm(n*ns), ncol = ns)) > 0)
```

In this script we have assumed that species 1 and 2 respond positively to the covariate x_1 whereas species 3 and 4 respond negatively. We have further assumed that species 1 and 3 respond positively to the covariate x_2 whereas species 2 and 4 respond negatively. Species 5 does not respond to either of the covariates.

Note that when generating the data, we have assumed no residual species associations, meaning that the species occur independently of each other after accounting for their responses to the covariates x_1 and x_2.

Let us explore the data we have generated by computing the species prevalences as the column means of the community matrix **Y**.

```
colMeans(Y)

## [1]  0.470 0.515 0.450 0.515 0.445
```

The prevalences of all five species are close to 0.5, as to be expected from the fact that the intercepts have been set to zero and the covariates have been assumed to have zero mean. The data are as informative as possible for presence–absence data, in the sense that with a prevalence of 0.5 they contain roughly as many zeros as ones, and hence as much variation as possible.

7.8.2 Defining and Fitting Three Alternative HMSC Models

We next formulate three alternative HMSC models for these data. To be able to run the analyses for all models in a compact way, we collect the model objects in a list.

```
XData = data.frame(x1 = X[ ,2] , x2 = X[ ,3] )
studyDesign = data.frame(sample = as.factor(1:n))
rL = HmscRandomLevel(units = studyDesign$sample)

models = list()
for (i in 1:3){
  XFormula = switch(i, ~1, ~x1, ~x1+x2)
  m = Hmsc(Y = Y, XData = XData, XFormula = XFormula,
           studyDesign = studyDesign, ranLevels = list
             (sample = rL),distr = "probit")
  models[[i]] = m
}
```

Model 1 (models[[1]]) has no environmental predictors, hence it is an intercept-only model. Model 2 (models[[2]]) has the covariate x_1 as the sole predictor, whereas Model 3 (models[[3]]) has both covariates x_1 and x_2 as the environmental predictors.

Note that in the above script, we have defined XData as a dataframe that contains both of the environmental covariates. While XData is common to all models, we have selected which variables to include in each model by varying XFormula.

All models include a random effect at the sampling unit level. To do so, we have included a single column the dataframe studyDesign that provides a unique identity number for each sampling unit. We then use the function HmscRandomLevel to define a random effect where the units are the samples.

We loop over the three models to fit them to the data.

```
for (i in 1:3){
  models[[i]] = sampleMcmc(models[[i]], thin = thin,
                samples = samples,transient = transient,
                nChains = nChains,verbose = verbose)
}
```

Let us next check MCMC convergence diagnostics to assess whether the sampling parameters were sufficient. To keep our treatment compact, we show it here for the full Model 3 only.

```
mpost = convertToCodaObject(models[[3]])
ess.beta = effectiveSize(mpost$Beta)
psrf.beta = gelman.diag(mpost$Beta, multivariate = FALSE)$psrf
ess.omega = effectiveSize(mpost$Omega[[1]])
psrf.omega = gelman.diag(mpost$Omega[[1]], multivariate =
  FALSE)$psrf
```

In Figure 7.2, we have evaluated MCMC mixing both for the β parameters (species niches) and $\mathbf{\Omega}$ parameters (species associations). Even if we have run 10^6 iterations, the convergence diagnostics are not ideal for the $\mathbf{\Omega}$ parameters, demonstrating the challenges with fitting non-normal models in general, particularly models including latent variables.

7.8.3 Parameter Estimates in the HMSC Models

We first use the function plotBeta to illustrate the estimated β parameters for Models 2 and 3 (Figure 7.3). The parameter estimates are consistent with those assumed when simulating the data. To see this, we note that both Models 2 and 3 correctly identified those species that respond positively (species 1 and 2), negatively (species 3 and 4) or not at all (species 5) to the covariate x_1. Additionally, Model 3 correctly identified the species that responded positively (species 1 and 3), negatively (species 2 and 4) or not at all (species 5) to the covariate x_2.

We next move to the main focus of our analyses, which is to estimate the residual species-association matrix. To do so, we extract the association matrix $\mathbf{\Omega}$ from the model object with the computeAssociations function, which also converts it to the scale of a correlation matrix. In the

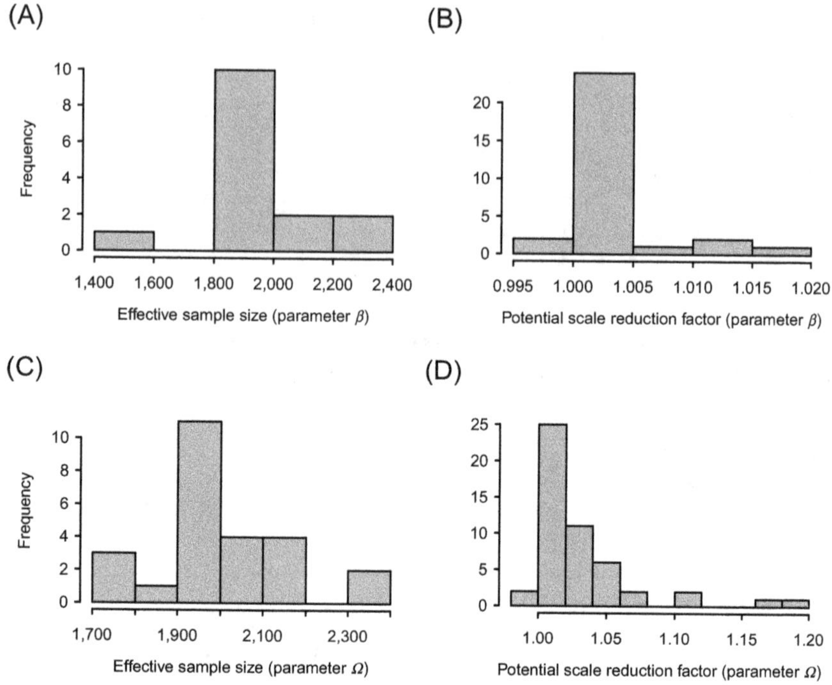

Figure 7.2 MCMC convergence diagnostics for the β (panels A and B) and $\mathbf{\Omega}$ parameters (panels C and D) evaluated in terms of effective sample size (panels A and C) and the potential scale reduction factor (panels B and D). The results are shown for Model 3 that includes both covariates x_1 and x_2.

script below, we choose to plot only those associations for which the posterior probability of being negative or positive is at least 0.95. There is no specific function for plotting species associations in HMSC, but such plots can be generated straightforwardly with the corrplot function of the corrplot package (Wei et al. 2017).

```
for (i in 1:3){
   OmegaCor = computeAssociations(models[[i]])
   supportLevel = 0.95
   toPlot = ((OmegaCor[[1]]$support > supportLevel)
          + (OmegaCor[[1]]$support < (1-supportLevel)) > 0)
          * OmegaCor[[1]]$mean
   corrplot(toPlot, method = "color", col = c("grey","white",
    "black"))
}
```

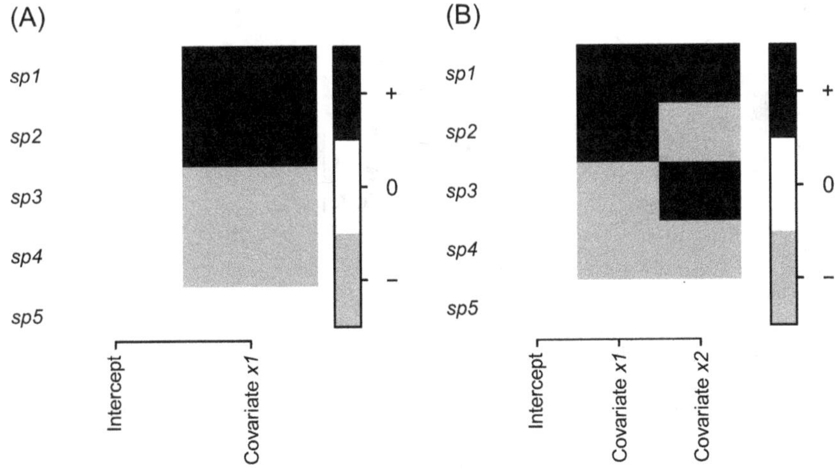

Figure 7.3 Heatmaps of the β parameters. Panel A corresponds to Model 2 containing x_1 as the sole environmental covariate, whereas panel B corresponds to Model 3 containing both x_1 and x_2. In both panels, the colours indicate the parameters that are estimated to be positive (black) or negative (grey) with at least 0.95 posterior probability.

Let us start the interpretation of the associations shown in Figure 7.4 from Model 3. The only associations estimated in Model 3 (Figure 7.4C) are the diagonal within-species associations, for which the correlations are always one, by definition. The fact that there are no associations among the species is fully in line with the fact that we assumed independent occurrences when simulating the data: all species-to-species associations due to the species occurring on similar or dissimilar environmental conditions are accounted for, as Model 3 includes both covariates x_1 and x_2.

In contrast to Model 3, Model 2 reveals residual associations among the species (Figure 7.4B). Based on this model, species 1 and 3 are positively correlated with each other, and so are species 2 and 4, but these two groups of species are negatively correlated with each other. Further, species 5 is not correlated with any other species. The reason for these residual correlations is the species responses to the covariate x_2, which was not included in Model 2. The influence of the covariate x_2 is reflected in the residual associations. For example, species 1 and 3 were assumed to respond similarly (both positively) to the missing covariate x_2, and thus failing to account for this covariate creates a positive residual association between them. As another example, species 1 and 2 were assumed to respond dissimilarly (one positively and one negatively) to the

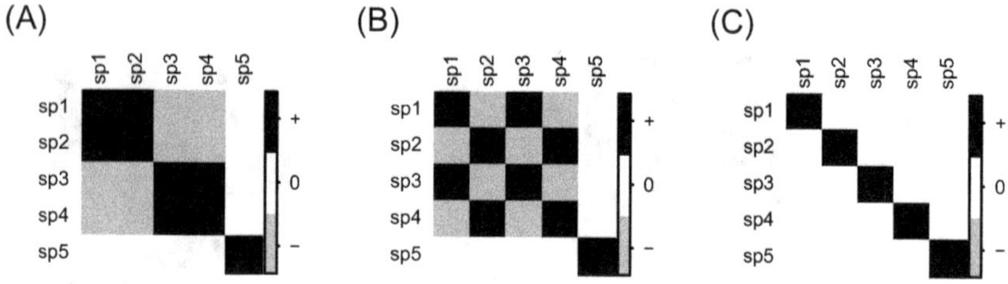

Figure 7.4 Residual species associations. The three panels correspond to Model 1 (A), Model 2 (B) and Model 3 (C). In each panel, associations that are estimated to be positive with at least probability 0.95 are shown in black, and associations that are estimated to be negative with at least probability 0.95 are shown in grey.

missing covariate x_2, and thus failing to account for this covariate creates a negative residual association between them.

Model 1 also reveals residual associations among the species (Figure 7.4A), but these are different from those identified by Model 2. This is because these residual associations reflect the responses of the species to both the covariates x_1 and x_2. As we assumed the species would respond more strongly to x_1 than to x_2, their summed influence is dominated by their responses to the covariate x_1, which explains why species 1 and 2 now appear to be positively associated with each other.

This example demonstrates the importance of the choice of environmental covariates for the interpretation of residual associations. In our example, the inference of species associations is different depending on the covariates that are accounted for. This makes the point that while residual associations among species can be an exciting starting point for formulating hypotheses on species interactions, the results of such analyses must always be carefully interpreted in light of the environmental covariates that are and are not controlled for in the model.

7.8.4 Explanatory Power, Predictive Power and Conditional Predictive Power

We next evaluate the explanatory power, predictive power and conditional predictive power of the three alternative models.

```
partition = createPartition(m, nfolds = 2, column = "sample")
partition.sp = c(1,2,3,4,5)
result = matrix(NA, nrow = 3, ncol = 3)
for (i in 1:3){
  m = models[[i]]
```

```
  #Explanatory power
  preds = computePredictedValues(m)
  MF = evaluateModelFit(hM = m, predY = preds)
  result[1,i] = mean(MF$TjurR2)

  #Predictive power based on cross-validation
  preds = computePredictedValues(m, partition = partition)
  MF = evaluateModelFit(hM = m, predY = preds)
  result[ 2,i] = mean(MF$TjurR2)

  #Predictive power based on conditional cross-validation
  preds = computePredictedValues(m, partition = partition,
                                 partition.sp = partition.sp,
                                 mcmcStep = 100)
  MF = evaluateModelFit(hM = m, predY = preds)
  result[3,i] = mean(MF$TjurR2)
}
```

In this script we compute the explanatory and predictive powers of the models as we have done before, the latter being based on two-fold cross-validation across the sampling units. Additionally, we evaluate the predictive power based on conditional cross-validation. To do so, we have performed leave-one out cross-validation over the species, as we have assigned each species to its own fold in the species partition partition.sp. Conditional cross-validation is based on conditional prediction, which in turn requires estimating the site loadings by MCMC, as explained in Section 7.7.1. In the script above, we have selected to do so with 100 MCMC iterations.

```
##                                  Model 1    Model 2    Model 3
## Explanatory power                  0.444      0.490      0.517
## Predictive power
   (cross-validation)               -0.014      0.339      0.479
## Predictive power
   (conditional cross-
   validation)                       0.132      0.382      0.479
```

As expected, both the explanatory and the predictive powers are highest for the model that includes both covariates (Model 3), followed by the model that includes only one of the covariates (Model 2), and lowest for the model that contains no covariates (Model 1). Initially, it may be surprising that the explanatory power of Model 1 is greater than zero, since this model has no covariates. It does, however, have the random effect part, which also contributes to explanatory power. Nevertheless, the random effect part does not help when making predictions for new sampling units, as shown by the zero predictive power of Model 1 based on cross-validation. But as discussed in Section 7.7.1, the random effect part of the model can help to make conditional predictions. This is illustrated in our results as Model 1 indeed has some predictive power for conditional cross-validation. This is because in this approach, we assume that the observed occurrences of the other species are known in the sampling units for which predictions on the focal species are to be made (Figure 7.1), and hence the estimated associations (Figure 7.4) can be utilised when making those predictions.

In all three models, the explanatory power is the highest, the predictive power based on usual cross-validation is the lowest and the predictive power based on conditional cross-validation is somewhere in the middle. This is not a coincidence; it is to be expected in the general case, since it reflects the amount of training data that are used. Specifically, all data are used to train the model when evaluating explanatory power, and the least amount of training data are used for the usual cross-validation. An intermediate amount of data is used for the conditional cross-validation (Figure 7.1).

7.9 Real Case Study with HMSC: Sequencing Data on Dead Wood-Inhabiting Fungi

As a real-data case study, we consider sequencing data on dead wood-inhabiting fungi.

7.9.1 The Data and the Ecological Context

The data that we will use as an example originate from Ovaskainen et al. (2013). In the original paper, the fungal data is recorded both by visual fruit-body surveys and by sequencing sawdust samples of Norwegian spruce logs (Ovaskainen et al. 2013). Here we only consider the sequencing data. Furthermore, while the original study included two sawdust samples per log (representing the basal and middle parts of the log), and performed molecular species identification based on two barcoding regions (ITS1 and ITS2, Schoch et al. 2012), here we restrict the analyses to one sample per log (representing the basal part), and we base molecular species identification on the ITS2 region only.

Let us read in the data and examine the first rows and columns.

```
data = read.csv(file = file.path(data.directory,
    "fungal data\\data.csv"))
n = dim(data)[1]
head(data[, 1:6])
##     LogID  DC  readcount   unk    Absidia     Alutaceodontia
                                      _glauca          _alutacea
## 1   R001   1        2206  1701          0                  0
## 2   R002   1        2053  1637          0                  0
## 3   R003   2        2624  2428          0                  0
## 4   R004   2        4208  4177          0                  0
## 5   R005   2        1756  1528          0                  0
## 6   R006   2        4829  2930          0                  0
```

For each of $n = 99$ logs, the decay class (the column DC) of the log is classified as 1–4, with 1 representing freshly fallen logs and 4 representing heavily decayed logs. The column readcount is the total number of sequences obtained for each sample (commonly called sequencing depth), which can be viewed as representing observation effort. The sequences were identified using the probabilistic taxonomic placement algorithm PROTAX-Fungi (Abarenkov et al. 2018). The sequences that could not be assigned with at least 50 per cent probability to any species are pooled as unknown (category unk). This category typically contains the majority of the sequences. This is because species-level molecular identification of fungi is challenging e.g. due to incompleteness of the reference databases, which the PROTAX-Fungi method explicitly accounts for (Abarenkov et al. 2018). The remaining columns consist of those 413 fungal species that were detected in the data at least once with at least 50 per cent of identification probability. Using 50 per cent as a threshold means that many of the species-level names may be wrong, in which case the correct species names are likely to belong to some closely related species.

Let us first organise the data. We construct the dataframe XData, which includes the variables to be used as fixed effects in Hmsc, namely decay class and the total sequence count. We then construct the dataframe YData of the community data, which also includes the unknown class of species.

```
XData = data.frame(DC = as.factor(data$DC),
    readcount = data$readcount)
YData = data[, 4:dim(data)[2]]
sel.sp = colSums(YData > 0) >= 10
YData = YData[, sel.sp]
```

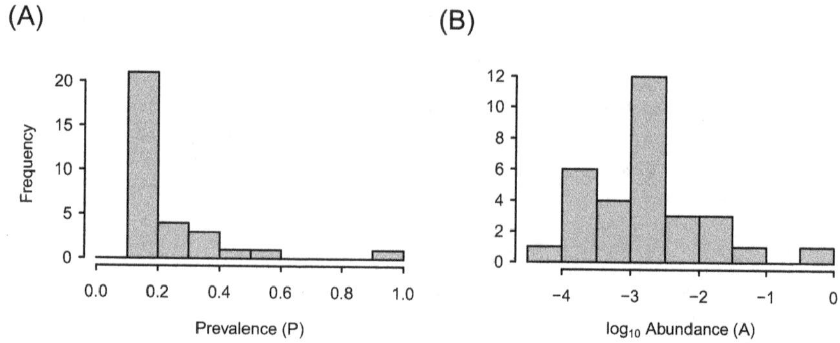

Figure 7.5 Species prevalence and abundance in the fungal data. Prevalence is measured as the fraction of occupied sampling units (i.e. logs). Abundance is measured as the log_{10} transformed proportion that the species represents out of the total sequence count. The y-axis (frequency) refers to the number of species with a given prevalence (panel A) or abundance (panel B).

Since in this example we are primarily interested in the co-occurrences among the species, we select those species that were found in at least ten logs. This is because detecting co-occurrences requires a substantial amount of information. This leaves us with thirty-one species.

Let us explore the raw data by plotting histograms of species prevalences and abundances.

```
P = colMeans(YData > 0)
A = colSums(YData)/sum(YData)
```

As usual, there is great variation among the species, where some are common and others are rare (Figure 7.5). Excluding the unknown class that is present in all sampling units, the most prevalent species are *Fomitopsis pinicola* (prevalence = 0.54), *Capronia kleinmondensis* (0.42), *Phellopilus nigrolimitatus* (0.37), *Antrodia serialis* (0.31) and *Heterobasidion parviporum* (0.30). Out of these, *C. kleinmondensis* has been described as a microfungi occurring on the Proteaceae in the fynbos (shrublands and heathlands located in South Africa) (Marincowitz et al. 2008). Hence, either this species has simply not been recorded in Finland yet, or it has been misidentified and the true species is some relative of *C. kleinmondensis*. The four other species mentioned above are Basidiomycota, which were recorded abundantly as fruit-bodies and are thus likely to be correctly identified from the molecular data (Ovaskainen et al. 2013). In terms of abundance, the unknown class represents the vast majority

(86 per cent) of all the 324,000 sequences included in our analyses. The most abundant species are *F. pinicola* (5 per cent), *P. nigrolimitatus* (1.8 per cent) and *A. serialis* (1.5 per cent). While the species *C. kleinmondensis* and *H. parviporum* have a high prevalence, they are not abundant, as both are represented by only 0.1 per cent of all the sequences.

7.9.2 Fitting Six Alternative HMSC Models to the Data

We will fit six different HMSC models to the fungal data, resulting from the combination of three model types and two choices of explanatory variables. In the scripts below, the object models will include all six models, organised as a list of lists, so that models[[i]][[j]] is the model type $i = 1, 2, 3$ for the choice of explanatory variables $j = 1, 2$.

The model type $i = 1$ is a lognormal Poisson model that is fitted to the sequence count data as they are. This model will simultaneously account for whether the species is present or not, as well as how abundant it is when it is present. The model types $i = 2$ and $i = 3$ form together a hurdle model that separates variation in presence–absence from variation in abundance. Thus, model $i = 2$ is a probit model that is fitted to sequence counts truncated to presence–absence, whereas model $i = 3$ is a normal model that is fitted to log-transformed sequence counts conditional on presence. This means that for $i = 3$, the sampling units where the species are not present are shown in the **Y** matrix as missing data (NA), not as zeros.

We note that the most natural model type for the full sequencing data (now modelled with the lognormal Poison distribution model) would be a multinomial model, with sample size equalling sequencing depth, and the probability vector being modelled as a function of fixed and random effects, through Dirichlet distributions. Such multinomial models have been developed in the context of microbial metagenomics (Holmes et al. 2012), but their integration with HMSC is yet to be done. Compared to the lognormal Poisson model, the multinomial model accounts for the dependency among the species generated by the total sequence count being fixed. Ignoring such dependency can create negative associations among the species as an artefact.

Concerning the choices of the explanatory variables, we choose them with the aim of estimating raw associations ($j = 1$) or residual associations ($j = 2$) among the species. We always include log-transformed sequencing depth, as this variable measures the total observation effort rather than environmental filtering. Thus, we will estimate raw associations from the models including sequencing depth $j = 1$ as the only variable. With $j = 2$, we additionally include the categorical variable of the decay class, and thus from these models we will estimate residual associations. We note that there are also many properties of the log other than the decay class that are likely to influence fungal occurrences, such as the size of the log. However, we ignore these other variables, as the sampling was specifically designed to include much variation in decay class, but minimise variation in other properties such as log size.

In all models, we also add a random effect at the sampling-unit level. The random effect models associations among the species, which is what we are primarily interested in.

```
studyDesign = data.frame(sample = data$LogID)
rL = HmscRandomLevel(units = studyDesign$sample)
models = list()
for (i in 1:3){
   Y = as.matrix(YData)
   if (i==2) {Y = 1*(Y > 0)}
   if (i==3) {
      Y[Y==0] = NA
      Y = log(Y)
   }
   tmp = list()
   for (j in 1:2){
      XFormula = switch(j, ~1 + log(readcount),~DC +
       log(readcount))
      m = Hmsc(Y = Y, XData = XData, XFormula = XFormula,
          studyDesign = studyDesign, ranLevels =
          list(sample = rL), distr = switch(i,
          "lognormal poisson", "probit", "normal"),YScale=TRUE)
      tmp[[j]] = m
   }
   models[[i]] = tmp
}
```

In the above script, we have used the option Yscale = TRUE to scale the response data to zero mean and unit variance. As we discuss in more detail in Section 8.3.1, this scaling influences only the normal model, and it is done to make the default priors of Hmsc compatible with the data.

We loop over both model types as well as the selections of explanatory variables to fit all six models.

```
for (i in 1:3){
  for (j in 1:2){
    models[[i]][[j]] = sampleMcmc(models[[i]][[j]], thin = thin,
                             samples = samples, transient = transient,
                             nChains = nChains, verbose = verbose,
                             initPar = "fixed effects")
  }
}
```

Let us evaluate MCMC convergence for the β and $\mathbf{\Omega}$ parameters in terms of the potential scale reduction factor in the model versions that contains environmental covariates ($j = 2$).

```
for (i in 1:3){
  mpost = convertToCodaObject(models[[i]][[2]])
  psrf.beta = gelman.diag(mpost$Beta,
     multivariate = FALSE)$psrf
  psrf.omega = gelman.diag(mpost$Omega[[1]],
     multivariate = FALSE)$psrf
}
```

With these data, good MCMC convergence is the most difficult to obtain in the case of the lognormal Poisson model (Figure 7.6). While usually the normal model shows the best MCMC convergence, here it behaves essentially the same as the probit model. This is because now the normal model has a large fraction of missing data, which makes MCMC convergence more challenging.

7.9.3 Inference on Abiotic and Biotic Species Niches

Figure 7.7 shows the responses of the species to the environmental covariates in the models that include both sequencing depth and decay class ($j = 2$). For all model types, some species are estimated to respond positively to sequencing depth, whereas no species respond negatively. This is to be expected, as having more sequences increases the probability of recording a focal species.

As decay class is a categorical variable, the responses of the species are somewhat hard to interpret from Figure 7.7. To improve the visualisation of the species responses to a given covariate, an option is to construct gradient plots, as we next do for the probit model:

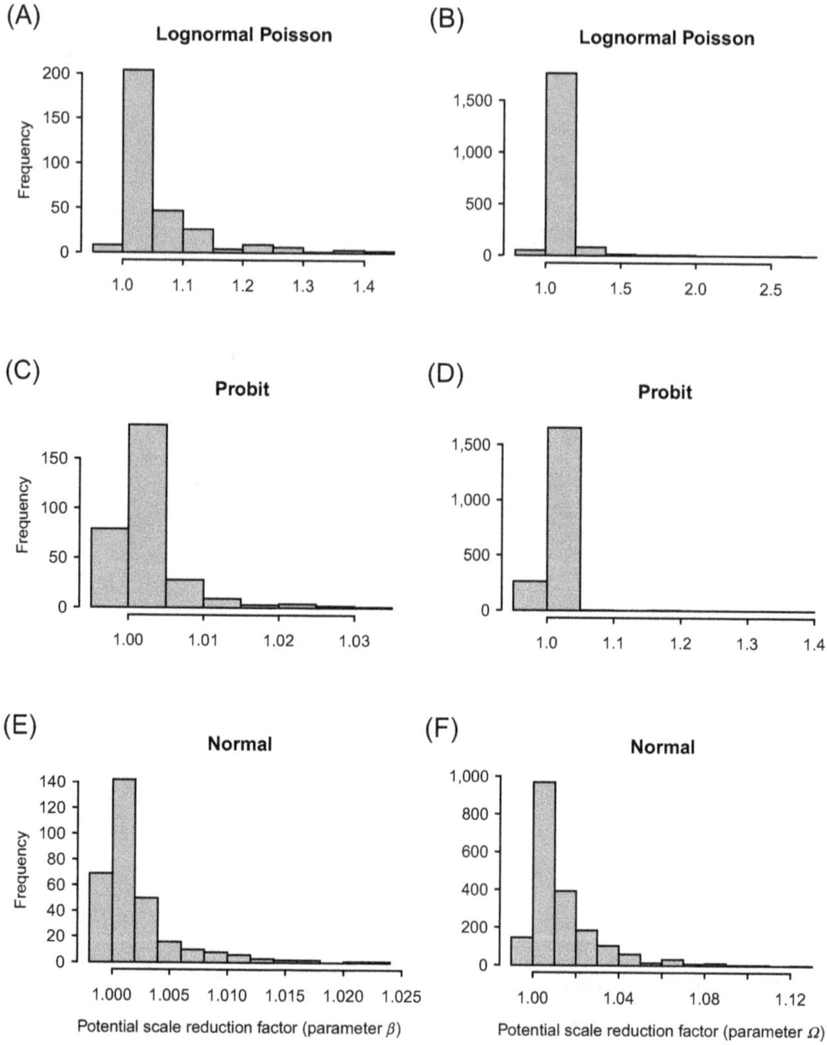

Figure 7.6 MCMC convergence diagnostics for the β and $\boldsymbol{\Omega}$ parameters measured in terms of the potential scale reduction factor. Panels A, C and E and panels B, D and F show the β and $\boldsymbol{\Omega}$ parameters, respectively. The rows of panels correspond to model type: lognormal Poisson (panels A and B), probit (panels C and D) and normal (panels E and F). All results are evaluated for models that include both sequencing depth and decay class as covariates.

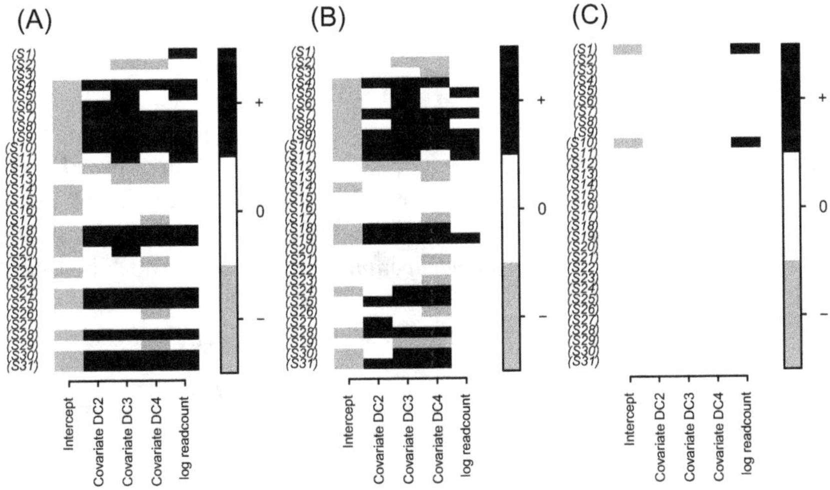

Figure 7.7 Heatmaps of estimated species niches. Black and grey show parameters that are estimated to be positive and negative, respectively, with at least 0.95 posterior probability. The panels show the results of the lognormal Poisson model on sequence counts (A), the probit model on presence–absence data (B) and the normal model on log-transformed sequence count conditional on presence (C). All results are evaluated for models that include both sequencing depth and decay class as covariates.

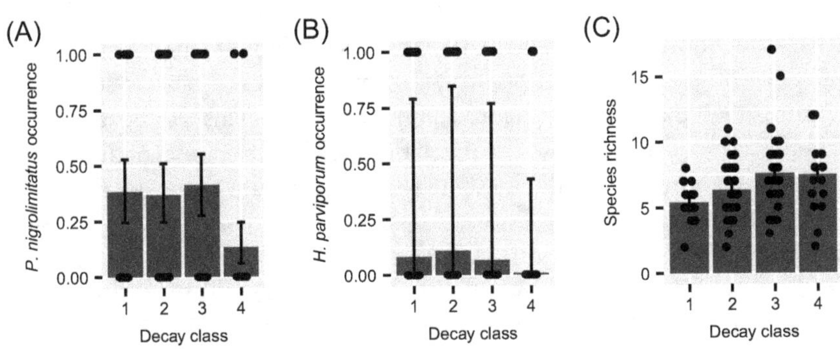

Figure 7.8 Predicted responses to variation in decay class. The panels show the occurrence probabilities of the species *Phellopilus nigrolimitatus* (A) and *Heterobasidion parviporum* (B), as well as the expected species richness (C). The bars show posterior median and the whiskers shows the posterior interquartile range.

```
m = models[[2]][[2]]
Gradient = constructGradient(m, focalVariable = "DC",
           non.focalVariables = list("readcount" = list(1)))
predY = predict(m, Gradient = Gradient, expected = TRUE)
```

Figure 7.8, which is plotted with the plotGradient function, shows that the overall species richness increases with the decay class of the logs. Yet the occurrence probabilities of some species, for example *Phellopilus nigrolimitatus* and *Heterobasidion parviporum*, are lowest in the last decay class. We will return to these two species later.

We next use move to the analysis of species-to-species associations.

```
for (i in 1:3){
  for (j in 1:2){
    OmegaCor = computeAssociations(models[[i]][[j]])
    supportLevel = 0.95
    toPlot = ((OmegaCor[[1]]$support > supportLevel)
             + (OmegaCor[[1]]$support < (1-supportLevel))
                > 0)
             * OmegaCor[[1]]$mean
    corrplot(toPlot, method = "color",
    col = c("grey","white","black")
    }
}
```

Most associations are revealed by the presence–absence model, while the two other model types show almost no associations (Figure 7.9). In the presence–absence model, there are many more raw associations (Figure 7.9C) than residual (Figure 7.9D) associations. This is to be expected, as the raw associations are also generated by differential responses of the species to the environmental conditions, here to their responses to the decay class.

Sometimes the residual associations can reveal patterns that are not present in the raw associations. As one example of this, let us consider the species pair *H. parviporum* and *P. nigrolimitatus*. The posterior probability of a negative association between these two species is 0.8065 in the raw associations and 0.999 in the residual associations. Thus, the residual associations suggest that the species co-occur less than expected by chance. In the raw associations, the negative association was masked by the fact that both species show similar responses to the decay class (Figure 7.8). We note that *H. parviporum* and other *Heterobasidion* species represent a major threat to timber production in areas of intensive forest management. As a biological control, cut stumps can be treated with the

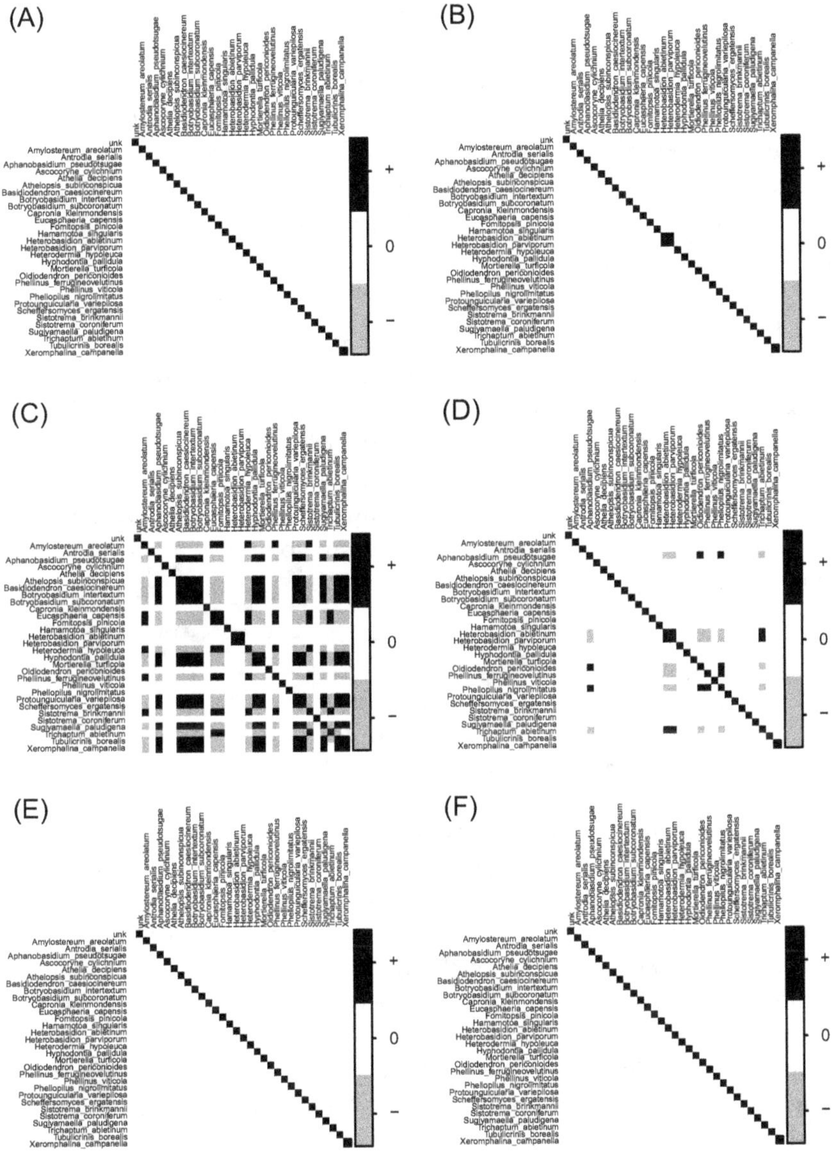

Figure 7.9 Residual species associations. The three rows of panels show the results of the lognormal Poisson model on sequence counts (A and B), the probit model on presence–absence data (C and D) and the normal model on log-transformed sequence count conditional on presence (E and F). Panels A, C and E show the raw associations and panels B, D and F show residual associations. In each panel, associations that are estimated to be positive and negative with at least 0.95 probability are shown in black and grey, respectively.

species *Phlebiopsis gigantea*, which competes for the substrate and applies hyphal interference with *Heterobasidion species* (Garbelotto & Gonthier 2013). The negative associations that we recorded with *P. nigrolimitatus* suggest that these two species may also show strong competitive interactions. However, as discussed throughout the book, results from association analyses should be taken as data-driven hypotheses of ecological interactions rather than as direct evidence of those.

7.9.4 Latent Variables as Model-Based Ordination

Latent variables in joint species distribution models can be interpreted as a model-based ordination (Hui et al. 2015; Warton et al. 2015). With Hmsc, such ordinations can be illustrated with the biPlot function. As the input, biPlot takes the site scores (the Eta variables) and the species scores (the Lambda variables). With the script below, we generate biplots of the raw and residual associations for the presence–absence model.

```
for (j in 1:2){
  m = models[[2]][[j]]
  biPlot(m, etaPost = getPostEstimate(m, "Eta"),
  lambdaPost = getPostEstimate(m, "Lambda"), colVar = 2)
}
```

In the ordination plot corresponding to the raw associations (Figure 7.10A), much of the variation related to the dominating latent variable 1 is due to the decay class of the log, as reflected by the non-random ordering of the shades of grey. In the residual associations where the dominating effect of the decay class is removed, logs representing different decay classes become randomly distributed (Figure 7.10B).

Whether to illustrate species associations as pairwise species matrices (Figure 7.9) or as ordinations (Figure 7.10) is mainly a choice of style, as both of these contain the same qualitative information. Species pairs that are positively associated with each other will appear close to each other in the ordination space, whereas species pairs that are negatively associated with each other will appear far from each other. For example, in both the raw and residual ordinations, the species *H. parviporum* and *P. nigrolimitatus* are located on opposite sides of the ordination space, reflecting the fact that they are negatively associated. An advantage of illustrating

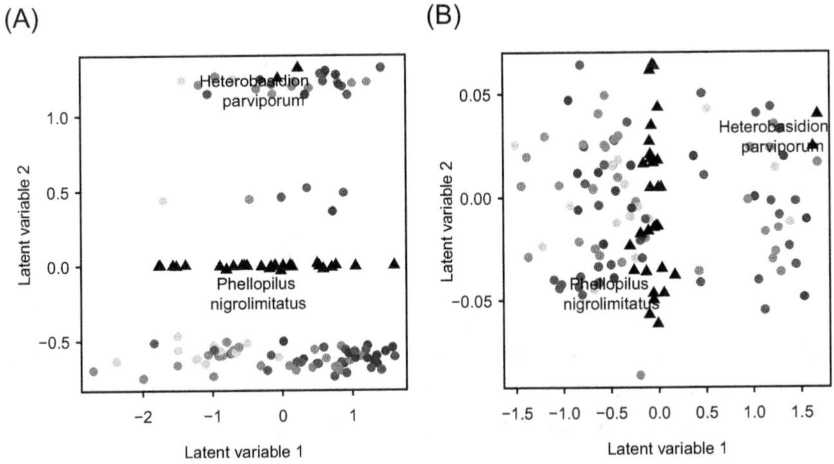

Figure 7.10 Biplots illustrating the latent variables modelling the random effect part of the HMSC model. In the biplots, the site scores (each corresponding to one sampling unit) are shown by the circles, the shade of grey showing the decay class of the log. The triangles show to the species scores (each corresponding to one species), of which only *H. parviporum* and *P. nigrolimitatus* are labelled in the figure, to avoid excessive overlap of text. Panel A corresponds to the model that includes only sequencing depth as a covariate and thus shows raw associations. Panel B corresponds to the model that includes both sequencing depth and decay class as covariates and thus shows residual associations. For a colour version of the figure, see the Colour Plate.

species associations by pairwise species matrices (Figure 7.9) is that this plot can include information about the level of statistical support. An advantage of the ordination plot (Figure 7.10) is that it shows both the information about species scores and the site scores. Thus, these two ways of illustrating the random effect part of the model are complementary.

8 · *Bayesian Inference in HMSC*

In Chapters 5–7, we have gradually built the core part of the JSDM, HMSC. In those chapters, we have already illustrated with many examples how HMSC is fitted to data using Bayesian inference. However, we have so far skipped many details related to how Bayesian inference is actually applied, which is what this chapter will focus on. Most importantly, we will describe in greater detail the prior distribution of HMSC, which has hardly been mentioned thus far.

We start by summarising the structure of the core HMSC model in Section 8.1. In Section 8.2 we briefly recall some of the fundamentals of Bayesian inference, aimed primarily at those readers who are not very familiar with it. As our treatment of these fundamentals is very superficial, we recommend those readers not familiar with Bayesian inference to read about this framework more broadly, e.g. using the excellent textbook on Bayesian inference by Gelman et al. (2013). Sections 8.3–8.5 form the core part of this chapter, in which we describe the structure of the prior distribution of HMSC, and explain in particular how the default prior has been chosen. In Section 8.6, we briefly discuss how posterior sampling is conducted in HMSC through MCMC. We discuss the prior distributions in greater detail than for how posterior sampling is done, because the prior distribution involves choices that are important in terms of interpretation of the results, whereas different ways of performing MCMC sampling all lead (theoretically) to the same end result and thus do not involve choices that an ecologist should worry about, as long as MCMC convergence is achieved. In Section 8.7, we demonstrate how the prior distribution can be sampled with Hmsc, and recall that samples from the prior distribution are identical to posterior samples if the model is fitted without any data. Finally, in Section 8.8 we discuss how the computational time needed to fit an HMSC model depends on the size and type of the data.

8.1 The Core HMSC Model

We recall that in Section 4.2 we graphically illustrated the core HMSC model as a DAG (Figure 4.1), and that the notation of the indices, data matrices and parameters is summarised in Tables 4.2–4.4. We further recall that in the DAG, the continuous arrows denote stochastic links defined with the help of equations that involve statistical distributions, while the dashed arrows denote deterministic links. These relationships were introduced in Chapters 5–7, and are next briefly summarised to present the core HMSC in a compact notation.

8.1.1 Fixed Effects

In Chapter 6 we defined the distribution of species niches as a function of species traits and phylogenetic relationships as:

$$\text{vec}(\mathbf{B}) \sim N(\text{vec}(\mathbf{\Gamma T}^{\mathrm{T}}), [\rho\mathbf{C} + (1-\rho)\mathbf{I}] \otimes \mathbf{V}) \tag{8.1}$$

This equation is represented by the continuous arrows in the upper left part of the DAG of Figure 4.1. The fixed effects model the link from species niche to the linear predictor as $\mathbf{L}^{\mathrm{F}} = \mathbf{XB}$. This is a deterministic relationship and thus it is represented by dashed arrows in the DAG.

8.1.2 Random Effects

To keep the notation simple, we will consider only a single random effect. We thus drop the index r, even if the model is capable of involving a sum of an arbitrary number n_r of random effects. In Section 7.4.2, we defined spatial site loadings as vec($\mathbf{H}$), where the vector of site loadings for each factor is distributed as $\boldsymbol{\eta}_{\cdot h} \sim N(0, \mathbf{\Sigma})$, where $\Sigma_{i_1 i_2} = \exp(-d_{i_1 i_2}/\alpha_h)$. This equation is represented by continuous arrows in the upper right part of the DAG of Figure 4.1. The random-effect part also contains a stochastic link from the parameters $\boldsymbol{\Phi}$ and $\boldsymbol{\delta}$ to the species loadings $\mathbf{\Lambda}$. In fact, we have not mentioned these two parameters before. They relate to the prior distribution of species loadings, which we will explain in detail in Section 8.4. The random effects are defined as $\mathbf{L}^{\mathrm{R}} = \mathbf{\Pi H \Lambda}$, which is a deterministic link and thus represented as dashed arrows in the random-effect part of the DAG. We note that the species association matrix $\mathbf{\Omega}$ differs from the other parameters in the DAG in the sense that no arrows flow out from it.

This is because $\mathbf{\Omega}$ is not a primary parameter of the model, but a derived parameter. The deterministic relationship between species loadings and the association matrix $\mathbf{\Omega} = \mathbf{\Lambda}^T\mathbf{\Lambda}$ is illustrated in the DAG as a dashed arrow. We note that it is generally not necessary to include derived parameters such as $\mathbf{\Omega}$ in a DAG; instead, we could include other derived parameters, such as those related to variance partitioning. We include $\mathbf{\Omega}$ in the DAG because of its central role in the interpretation of the results, and because we often use it to visualise the primary parameter behind it, namely $\mathbf{\Lambda}$.

8.1.3 Data Models

To enable the use of normal, probit, Poisson and lognormal Poisson models (defined in Section 5.3) in a unified notation, we write the data model as $\mathbf{Y} \sim \boldsymbol{D}(\mathbf{L}, \mathbf{\Sigma})$, where $\boldsymbol{D}$ refers to a vector of statistical distributions for the data model. In the general model, the data model can vary among the species, so that it is possible to define the probit model for one species and the lognormal Poisson model for another species.

8.1.4 The Vector of All HMSC Parameters

Following the standard notation used in Bayesian inference, we denote by $\boldsymbol{y}$ in this chapter the data and by $\boldsymbol{\theta}$ the primary model parameters. In the context of HMSC, $\boldsymbol{y}$ is the collection of all data matrices and $\boldsymbol{\theta}$ is the collection of all primary model parameters. Thus, for the core HMSC model depicted in Figure 4.1 we may write:

$$\boldsymbol{y} = (\mathbf{Y}, \mathbf{X}, \mathbf{T}, \mathbf{C}, \mathbf{\Pi}, \mathbf{S})$$

$$\boldsymbol{\theta} = (\mathbf{B}, \boldsymbol{\rho}, \mathbf{\Gamma}, \mathbf{V}, \mathbf{H}, \boldsymbol{\alpha}, \mathbf{\Lambda}, \mathbf{\Phi}, \boldsymbol{\delta}, \mathbf{\Sigma}) \tag{8.2}$$

We recall that the model may involve multiple random effects, in which case $\mathbf{H}$ for example is a list of matrices, each of which may have a different dimension.

Many of the parameters included in Equation 8.2 appear as the outermost tips of the DAG, so that they are not shot by any arrow. These are the parameters for which the priors should be defined. The priors of those parameters that are shot by arrows coming from other parameters (species loadings $\mathbf{\Lambda}$ and species niches $\mathbf{B}$) are defined through the priors of the other parameters. In Sections 8.3–8.5, we will define these prior distributions. But before that, for readers not familiar with

Bayesian inference, we summarise some fundamentals in Section 8.2 – a section that a reader well-grounded in Bayesian inference may feel free to skip.

8.2 Basics of Bayesian Inference: Prior and Posterior Distributions and Likelihood of Data

Bayesian inference is similar to maximum likelihood inference, in the sense that both are based on the likelihood of the data. We denote the likelihood of observing the data $\boldsymbol{y}$ under the parameters $\boldsymbol{\theta}$ by $p(\boldsymbol{y} \mid \boldsymbol{\theta})$. As implied by the term 'maximum likelihood', the parameters in this inference are estimated by finding the parameter combination that maximises the likelihood of observing the data. The maximum likelihood estimate $\boldsymbol{\theta}^{ML}$ can thus be mathematically defined as $\boldsymbol{\theta}^{ML} = \arg\max_{\boldsymbol{\theta}} p(\boldsymbol{y} \mid \boldsymbol{\theta})$.

In Bayesian inference, the interest is in the probability distribution of the parameters conditional on the data, and thus in $p(\boldsymbol{\theta} \mid \boldsymbol{y})$ instead of $p(\boldsymbol{y} \mid \boldsymbol{\theta})$. In the Bayesian context, $p(\boldsymbol{\theta} \mid \boldsymbol{y})$ is called the density of the posterior distribution. Informally speaking, the posterior distribution describes what one knows about the parameter $\boldsymbol{\theta}$ after seeing the data $\boldsymbol{y}$, and thus in Bayesian inference estimating the parameters is the same thing as finding their posterior distribution.

In Bayesian inference, the parameter estimates (i.e. the posterior distribution) are not only influenced by the data, but also by the so-called prior distribution. The density of the prior distribution is denoted by $p(\boldsymbol{\theta})$, and this describes what one knows about the parameters $\boldsymbol{\theta}$ before seeing the data $\boldsymbol{y}$. The main content of this chapter will be about explaining the prior distribution of HMSC (Sections 8.3–8.5), and discussing how the choices related to the prior parameters may influence the results.

The so-called Bayes theorem describes that the posterior density is proportional to the product of the prior density and the likelihood of the data, $p(\boldsymbol{\theta} \mid \boldsymbol{y}) \propto p(\boldsymbol{y} \mid \boldsymbol{\theta})p(\boldsymbol{\theta})$. The Bayes theorem thus shows how the posterior distribution depends on both the prior distribution and the data. The theorem is important because all the different algorithms by which the parameters can be estimated in Bayesian inference are fundamentally derived from the Bayes theorem. We will return to this point in Section 8.6, where we briefly discuss how the posterior distribution is sampled in HMSC.

The major part of this chapter focuses on defining the prior distribution $p(\boldsymbol{\theta})$ for the HMSC parameters, which are the parameters listed in

Equation 8.2. While we discuss the prior choices for each parameter in turn, there is only one prior distribution, which is a joint distribution defined simultaneously for all parameters. For understanding how the parameters are linked to each other in a hierarchical Bayesian model, it is very useful to draw the DAG, that describes the model graphically, as we have done for the core HMSC model in Figure 4.1. With the help of the DAG, we can decompose the prior density $p(\boldsymbol{\theta}) = p(\mathbf{B}, \rho, \boldsymbol{\Gamma}, \mathbf{V}, \mathbf{H}, \boldsymbol{\alpha}, \boldsymbol{\Lambda}, \boldsymbol{\Phi}, \boldsymbol{\delta}, \boldsymbol{\Sigma})$ to a product of three parts as:

$$p(\boldsymbol{\theta}) = p(\mathbf{B}, \rho, \boldsymbol{\Gamma}, \mathbf{V})p(\mathbf{H}, \boldsymbol{\alpha}, \boldsymbol{\Lambda}, \boldsymbol{\Phi}, \boldsymbol{\delta})p(\boldsymbol{\Sigma}) \tag{8.3}$$

In Equation 8.3, the term $p(\mathbf{B}, \rho, \boldsymbol{\Gamma}, \mathbf{V})$ relates to the fixed effect part of the model, i.e. to species niches. The term $p(\mathbf{H}, \boldsymbol{\alpha}, \boldsymbol{\Lambda}, \boldsymbol{\Phi}, \boldsymbol{\delta})$ relates to the random-effect part of the model, i.e. to species associations. The last term $p(\boldsymbol{\Sigma})$ relates to the residual variation, and hence to the data model. We will discuss the prior assumptions for each of these in a separate section. We expect that most users might wish to fit the HMSC model without putting too much thought into the kind of prior that would be best to assume for the model parameters, but rather to 'let the data decide' about the parameters. This line of thinking is compatible with the idea of an uninformative prior distribution. However, it is not possible to simply say that an 'uninformative prior is chosen', as the choice needs to be specified. In HMSC, the default priors have been selected to be uninformative in the ways that we specify below.

8.3 The Prior Distribution of Species Niches

The prior for species niches $p(\mathbf{B}, \rho, \boldsymbol{\Gamma}, \mathbf{V})$ can be further decomposed with the help of the DAG of Figure 4.1 as:

$$p(\mathbf{B}, \rho, \boldsymbol{\Gamma}, \mathbf{V}) = p(\mathbf{B} \mid \rho, \boldsymbol{\Gamma}, \mathbf{V})p(\rho)p(\boldsymbol{\Gamma})p(\mathbf{V}) \tag{8.4}$$

The product structure of Equation 8.4 implies that the prior distributions are defined independently for ρ, $\boldsymbol{\Gamma}$ and $\mathbf{V}$, and that they then imply a prior for $\mathbf{B}$. Thus, the prior for the species niches $\mathbf{B}$ is defined implicitly by defining the priors of how traits influence species niches ($\boldsymbol{\Gamma}$), what kind of residual variation there is on top of the influence of species niches ($\mathbf{V}$), and how strongly the residual variance is influenced by phylogenetic relationships (ρ). The priors of these define the prior for $\mathbf{B}$ through Equation 8.1.

Sometimes the model does not have species traits or phylogenetic relationships. However, these are just a special case of the general case, and thus the parameters $\mathbf{\Gamma}$, ρ and $\mathbf{V}$ also exist even if there are no input data for species traits or phylogenetic relationships. If there are no traits, then the trait matrix $\mathbf{T}$ is defined to contain only the intercept, in which case the HMSC model assumes that all species have the same expected niche. If there is no phylogenetic correlation matrix $\mathbf{C}$, then the ρ parameter is fixed to $\rho = 0$, and thus the residual variation among species is assumed to be independent.

The prior distributions for the link between species traits and species niches are defined through the multivariate normal distribution

$$\text{vec}(\mathbf{\Gamma}) \sim N(\boldsymbol{\mu}_\gamma, \mathbf{U}_\gamma) \tag{8.5}$$

where vec(·) again stands for vectorising a matrix, i.e. placing its columns one after the other into one long vector. The mean $\boldsymbol{\mu}_\gamma$ and the variance-covariance matrix $\mathbf{U}_\gamma$ are prior parameters that are not estimated but chosen by the user. As the $\mathbf{\Gamma}$ matrix involves a parameter γ_{kl} for each combination of environmental covariate and trait, the length of vec($\mathbf{\Gamma}$) is $n_t n_c$, and hence $\boldsymbol{\mu}_\gamma$ is a vector of length $n_t n_c$ and $\mathbf{U}_\gamma$ is a matrix of dimensions $n_t n_c \times n_t n_c$. These two parameters describe what is assumed about how species traits influence species niches before seeing the data. Therefore, these parameters relate to the assumptions about the underlying community assembly processes.

In the case of the prior parameters $\boldsymbol{\mu}_\gamma$ and $\mathbf{U}_\gamma$, the default values are a zero vector for $\boldsymbol{\mu}_\gamma$, and an identity matrix for $\mathbf{U}_\gamma$. This means that the default prior for each element γ_{lk} of the matrix $\mathbf{\Gamma}$ is distributed according to the standard normal distribution, $\gamma_{lk} \sim N(0, 1)$, and that the different elements of the matrix $\mathbf{\Gamma}$ are independent of each other. We will soon explain why this choice has been made as the default in HMSC.

The default prior in HMSC for the phylogenetic signal parameter ρ is the following:

$$\begin{cases} \text{with probability } 0.5, & \rho = 0, \\ \text{with probability } 0.5, & \rho \sim \text{Uniform}(0, 1) \end{cases} \tag{8.6}$$

This choice for the prior is made to avoid forcing a phylogenetic signal if the data do not have it. If we assume $\rho \sim \text{Uniform}(0, 1)$, then the prior probability for the presence of a phylogenetic signal would be $\Pr(\rho > 0) = 1$, and thus the posterior probability would necessarily be $\Pr(\rho > 0 \mid \mathbf{y}) = 1$, even for data without any phylogenetic signal. With the choice of the prior shown in Equation 8.6, the prior probability for the presence of a phylogenetic signal is $\Pr(\rho > 0) = 0.5$. The posterior probability $\Pr(\rho > 0 \mid \mathbf{y})$ can be compared to this value in order to evaluate the extent to which the data support the presence or absence of a phylogenetic signal. A loose interpretation of Equation 8.6 is that before seeing the data, one assumes that it is equally likely for the data to have or not have a phylogenetic signal, and if there is such a signal, it can have any strength.

For technical reasons related to how the MCMC sampling is conducted (see Section 8.6), the prior for ρ is not implemented exactly according to Equation 8.6, but instead as a discrete grid approximation of it. As a default, 101 different values of ρ are assumed, starting from $\rho = 0$ and advancing in increments of 0.01 until $\rho = 1$. For each of these, a prior probability is given. In the default prior, this is 0.5 for $\rho = 0$, and 0.005 for each of the remaining 100 values of ρ.

If the model does not include a phylogenetic correlation matrix **C**, the phylogenetic parameter ρ is fixed to $\rho = 0$. This choice can be viewed as a prior distribution where all the probability mass is assigned to this single value.

The parameter **V** determines (together with ρ) the residual variation in species niches that cannot be explained by the traits. As the prior of this parameter, HMSC assumes the inverse-Wishart not in distribution

$$\mathbf{V} \sim W^{-1}(\mathbf{V}_0, f_0) \tag{8.7}$$

where the prior parameters to be chosen are the scale matrix $\mathbf{V}_0$ and the degrees of freedom f_0. The inverse-Wishart distribution is a distribution of matrices. Further, samples from it are matrices that are symmetric and positive definite, and thus suitable as variance-covariance matrices, such as the matrix **V**.

We recall that the dimensions of **V** are $n_c \times n_c$ where n_c is the number of environmental covariates, and thus $\mathbf{V}_0$ is a $n_c \times n_c$ matrix as well. As a default choice, HMSC sets $\mathbf{V}_0$ as the identity matrix, and the degrees of freedom parameter to $f_0 = n_c + 1$. The degrees of freedom parameter controls the amount of uncertainty in the distribution. To make the distribution mathematically well-defined, this parameter must satisfy $f_0 > n_c - 1$. To make the mean of the distribution finite, it must satisfy $f_0 > n_c + 1$. Thus, the choice we have made can be considered as uninformative, but still informative enough to make the posterior distribution well-behaved in the sense of a finite mean, for example.

After explicitly defining the prior distributions for $\mathbf{\Gamma}$ (Equation 8.5), ρ (Equation 8.6) and **V** (Equation 8.7), we have implicitly defined the prior for the species niches **B** (see Equation 8.1). However, as shown by Equation 8.1, the prior of **B** also depends on the trait data **T** and the phylogenetic data **C**. Let us describe the prior of **B** by considering first the simplest possible case, which does not include traits, phylogeny or environmental covariates. Thus, we assume that for each species the trait matrix **T** only includes the intercept, and thus $n_t = 1$. Similarly, for each sampling unit the environmental covariate matrix **X** only includes the intercept, and thus $n_c = 1$. Due to the lack of phylogenetic component or equivalently the assumption of $\rho = 0$, the species are independent of each other in the prior, and thus we can without loss of generality consider the case of a single species. As we did not include any environmental covariates, the species has a single β parameter that models the intercept.

It can be tedious to determine analytically the prior distribution for **B**, especially in the case of the full HMSC model. But one can always sample the prior distribution simply by simulating it. Thus, to simulate the prior of the single β parameter in our simplistic toy example, based on Equation 8.5 we simulate the single γ parameter from the standard normal distribution $\gamma \sim N(0, 1)$. We then sample the matrix V – which has the dimensions 1×1 and is thus actually a scalar – from the inverse-Wishart distribution of Equation 8.7, with the default choice of the prior parameters being $V_0 = 1$ and $f_0 = 2$. We note that in this toy example with $n_c = 1$, the inverse-Wishart distribution reduces to an inverse gamma distribution. Finally, the β parameter can be sampled as $\beta \sim N(\gamma, V)$.

For the sake of illustration, let us next assume that there are $n_s = 100$ species. In this case, the parameters γ and V are the same for all species and thus they are not resampled from the prior when the parameters β are sampled for each species from $\beta \sim N(\gamma, V)$ one after the other. Let us further consider how the distribution of the β parameters translates into species occurrence, which we model by the probit model. As we considered the case where the intercept is the only 'trait', the linear predictor can be written simply as $L = \beta$, which results in the occurrence probability $\Pr(y = 1) = \Phi(L)$. Using the above recipe, we may sample community data from the prior distribution with the following R-script:

```
f0 = 2
V0 = 1
ns = 100
gamma = rnorm(n = 1)
V = riwish(f0, V0)
beta = rnorm(n = ns, mean = gamma, sd = sqrt(V))
L = beta
p = pnorm(L)
```

Figure 8.1 illustrates two communities sampled from the prior. The species belonging to Community 1 shown in the upper two panels

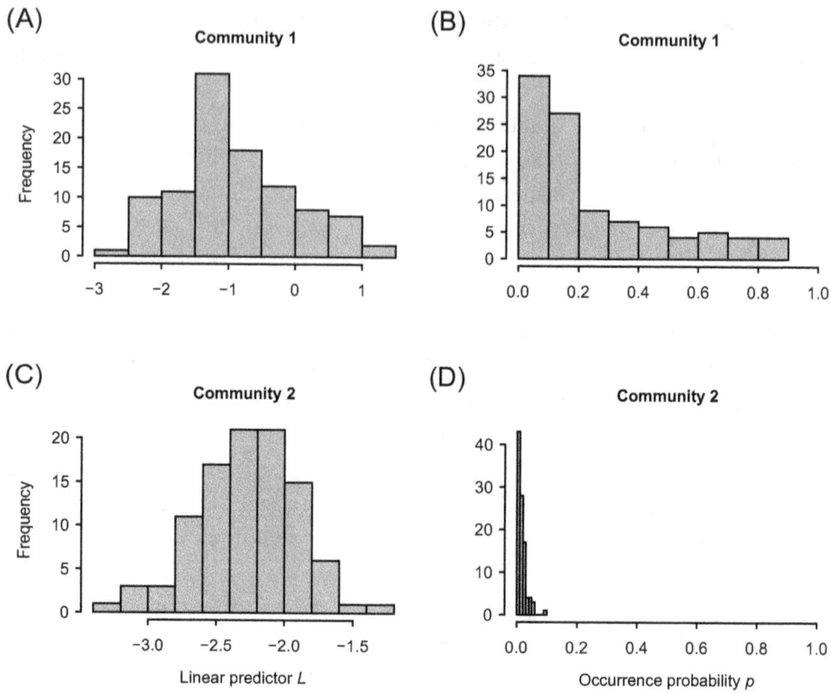

Figure 8.1 Prior samples of linear predictors (L, panels A and C) and occurrence probabilities (p, panels B and D) in a probit model containing no traits, no phylogenetic data, and no environmental covariates. Two rows of panels correspond to two simulated communities, each consisting of 100 species.

(A and B) vary greatly in terms of their occurrence probabilities; some are very rare and others are very common. In contrast, the species belonging to Community 2 shown in the lower two panels (C and D) are all rare. This is because the γ parameter is very low for this community, as reflected by the low mean of the linear predictors. These two samples illustrate the prior distribution, which describes by definition the kind of communities one would expect before seeing any data. The value of the γ parameter creates variation in the mean occurrence probability over all species, whereas the value of the V parameter determines how much the species differ from each other. Variation in both of these parameters creates variation in community structure, as reflected by the two different communities that we sampled from the prior in Figure 8.1.

Let us then slightly complicate the above example by adding one environmental covariate x. We still assume that there are no species traits, so that the **T** matrix contains an intercept only, and that species niches are independent of each other ($\rho = 0$). Now each species has two β parameters, one related to the intercept and one to the slope. Assuming the default prior parameters, we may sample a species community from the prior with the following script:

```
f0 = 3
V0 = diag(2)
ns = 100
gamma = rnorm(n = 2)
V = riwish(f0, V0)
beta = mvrnorm(n = ns, mu = gamma, Sigma = V)
L1 = beta %*% c(1,0)
L2 = beta %*% c(1,1)
L3 = beta %*% c(1,10)
p1 = pnorm(L1)
p2 = pnorm(L2)
p3 = pnorm(L3)
```

In this script, we have sampled a single community, as we have sampled from the prior only once. We have then computed the linear predictors and the corresponding occurrence probabilities of the species belonging to this community under three environmental conditions: when $x = 0$, when $x = 1$ and when $x = 10$. These three predictions are illustrated in Figure 8.2.

When $x = 0$, the species occurrence probabilities are quite uniformly distributed from zero to one (Figure 8.2A), due to the β parameters for the intercept and thus the linear predictors L having values close to zero (Figure 8.2B). When $x = 1$, the occurrence probability distributions of the species becomes more bimodal, with many being close to zero and others close to one (middle row of panels in Figure 8.2). This makes sense, as now the environmental condition $x = 1$ deviates from its average, and thus one would expect the environmental conditions to be either suitable or unsuitable for the species. In other words, this reduces uncertainty in species occurrence, placing occurrence probabilities close to zero or close to one. The uncertainty in species occurrence almost disappears when $x = 10$, as in this case the occurrence probabilities are very close to zero or one (lowest two panels in Figure 8.2). This is because in the slope parameter β has been multiplied with a very high value of the covariate ($x = 10$), resulting in linear predictors that have high absolute values.

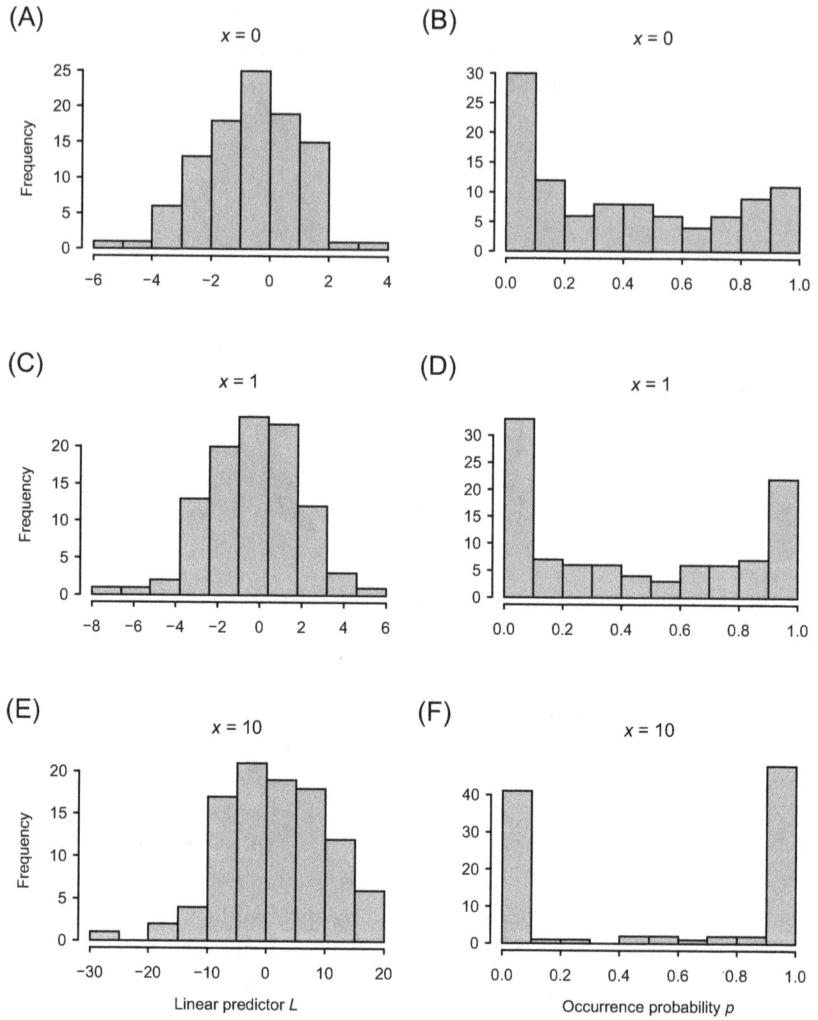

Figure 8.2 Prior samples of linear predictors (L, panels A, C and E) and occurrence probabilities (p, panels B, D and E) of a probit model containing no traits, no phylogenetic data and a single environmental covariate x. All panels correspond to the same simulated community consisting of 100 species. The rows of panels differ in terms of the value of the environmental covariate: in panels A and B, $x = 0$, in panels C and D, $x = 1$ and in panels E and F, $x = 10$.

The above example illustrates that not only the prior parameters but also the values of the environmental covariates influence the prior distributions of model predictions. One might not be very comfortable with the kind of communities the prior distribution implies for the environmental conditions $x = 10$. This is because one might not often be so sure about species presences or absences before seeing the data, as community assembly typically involves a high amount of stochasticity. Thus, the default prior seems to be more suitable for $x = 0$ or $x = 1$ than for $x = 10$, and definitely more so than for $x = 1{,}000$, in which case the predictions would become essentially deterministic.

In the above example, the covariate x could for example represent altitude. If one has measured altitude in the units of kilometres, then $x = 0$ may represent low altitude and $x = 1$ high altitude, and the default prior seems appropriate. But if one has measured altitude in the units of metres, then high altitude corresponds to $x = 1{,}000$, for which the default prior may not be appropriate, for the reasons discussed above. Thus, the priors need to be adjusted to account for the scale of variation in the data. As we explain in more detail below, this is done in Hmsc by scaling the **X** and **T** matrices, and sometimes also the **Y** matrix.

8.3.1 Scaling of Data Matrices to Make Default Priors Generally Applicable

To make the default priors more generally applicable, Hmsc scales both the **X** and **T** matrices by default so that they have zero mean and unit variance over the columns. These scalings are invisible to the user, in the sense that the estimated parameters are back-transformed to the original scale. Thus, if making predictions by the model for some specific values of covariates or traits, these can be given in the original units, not in the scaled units. To illustrate how this works in practice, let us consider again a model with two parameters only, i.e. the intercept β_1 and the slope β_2. The linear predictor of this model reads as $L = \beta_1 + \beta_2 x$, where x is the environmental covariate, for example the altitude measured in metres. Within Hmsc, a variable x is scaled to zero mean and unit variance before the estimation starts. We denote the scaled variable by $\hat{x}$, so that $\hat{x} = (x - m)/s$, where m is the mean and s the standard deviation of x, both computed over the sampling units. Estimation is then conducted with the scaled environmental covariate matrix $\hat{\mathbf{X}}$ instead of the original environmental covariate matrix **X**. During the estimation, when storing a posterior sample, the effect of the scaling is reverted. To see how this can be done, we denote by $\hat{\beta}_1$ and $\hat{\beta}_2$ the intercept and the slope of the model with scaled covariates, and note that

$$\hat{\beta}_1 + \hat{\beta}_2\hat{x} = \hat{\beta}_1 + \hat{\beta}_2(x - m)/s = \left[\hat{\beta}_1 - \hat{\beta}_2 m/s\right] + \left[\hat{\beta}_2/s\right]x \quad (8.8)$$

Thus, $\beta_1 = \hat{\beta}_1 - \hat{\beta}_2 m/s$ and $\beta_2 = \hat{\beta}_2/s$ are the intercept and the slope of the model with original covariates. In the posterior distribution, the values of β_1 and β_2 are stored rather than the values of $\hat{\beta}_1$ and $\hat{\beta}_2$. Thus, the estimated parameter estimates are compatible with the original units, and in this sense a user of Hmsc does not need to make any special consideration because of the scaling. However, it is good to be aware of the scaling, because it makes the default prior more generally applicable: whether altitude is measured in kilometres or in metres, the results of HMSC analyses will be identical.

The scaling of the **X** matrix does not influence only the parameters **B**, but also the parameters **Γ** and **V**, so these are back-transformed as well. Further, the columns of the trait matrix **T** are also scaled to zero mean and unit variance, for exactly the same reasons as why the **X** matrix are scaled. For this reason, the default priors are equally appropriate, whether e.g. in the **T** matrix the body masses of the species are measured in grams or in kilograms. The scaling of **X** and **T** matrices is done only for continuous covariates, as categorical variables are coded into dummy variables, and thus their scale (zero or one) is compatible with the default priors as such. We note that scaling cannot be done if the model does not include an intercept, but we expect this to rarely be the case.

While scaling the **X** and **T** matrices is generally recommended, and is done by default in Hmsc, the case of the community data matrix **Y** is a different story. When the probit model is applied to presence–absence data, the values of **Y** are zeros and ones, and the default priors can be considered an appropriate choice: we recall from Figures 8.1 and 8.2 that sampling the parameters from the default prior generates variation that can be considered to be compatible with typical real communities. This is specifically the case after scaling the **X** and **T** matrices, which is done to avoid the bimodal predictions illustrated in the lowest row of panels in Figure 8.2. In the case of the Poisson model, the expected count is given by the exponential of the linear predictor. Thus, as the linear predictor varies e.g. from −10 to 10, the expected count varies between 0.00005 and 22,000. As the lognormal Poisson model adds the residual variation before the exponential transformation, with the default priors it can predict counts in the order of millions. Thus, applications of the default prior (with the scaling of the **X** and **T** matrices) can be considered to be compatible also with typical count data.

Scaling the **Y** matrix might be needed in the case of normally distributed data. With such data, there is usually no natural unit, as is the case with presence–absence data or count data. If, for example, y_{ij} is the log-transformed biomass of species j in the sampling unit i, the value of y_{ij} will vary with the unit used to measure biomass. To make the default priors compatible with any normally distributed data, one possibility is to scale each column of the community data matrix **Y** to zero mean and unit variance. Unfortunately, the effect of such a scaling to parameter estimates cannot be back-transformed in the same way as the scaling of the **X** and **T** matrices can be. Thus, if applying this scaling to the **Y** matrix, the parameter estimates will be in units compatible with the scaled data, not with the original data. For this reason, even if scaling the **Y** matrix for normally distributed data is implemented in Hmsc, it is not the default option. However, if the user selects the scaling option, the model predictions are back-transformed to the original scale as, unlike the parameters, the predictions are easy to back-transform.

8.4 The Prior Distribution of Species Associations

The species associations are modelled in HMSC through the random-effect part of the model with the help of latent variables. We recall from Chapter 7 that in its basic form the latent variable model reads as $L^R_{ij} = \sum_{h=1}^{n_f} \eta_{ih}\lambda_{hj}$, where η_{ih} are the site loadings and λ_{hj} are the species loadings. As both of these are estimated, we need to define a prior distribution for both. The priors for the site and species loadings are independent of each other, and thus the prior density related to the random effects decomposes as:

$$p(\mathbf{H}, \boldsymbol{\alpha}, \boldsymbol{\Lambda}, \boldsymbol{\Phi}, \boldsymbol{\delta}) = p(\mathbf{H}, \boldsymbol{\alpha})p(\boldsymbol{\Lambda}, \boldsymbol{\Phi}, \boldsymbol{\delta}) \tag{8.9}$$

8.4.1 The Prior for Site Loadings

We have already discussed the prior density $p(\mathbf{H}, \boldsymbol{\alpha})$ in Chapter 7. Namely, for non-spatial latent variables, we assumed that $\eta_{ih} \sim N(0, 1)$ independently for each i and h. For spatial latent variables, we assumed the Gaussian process prior $\boldsymbol{\eta}_{\cdot h} \sim N(0, \boldsymbol{\Sigma}_h)$, where the variance-covariance matrix $\boldsymbol{\Sigma}_h$ equals the correlation matrix with elements $\exp(-d/\alpha_h)$, where d is the distance between sampling units, and α_h is the spatial (or temporal) scale at which the dependency among the site loadings decays for factor h. For spatial models, we need a prior for the scale parameters α_h. The default prior of Hmsc for each α_h parameter is similar to the default prior defined for the ρ parameter in Equation 8.6:

$$\begin{cases} \text{with probability } 0.5, & \alpha_h = 0, \\ \text{with probability } 0.5, & \alpha_h \sim \text{Uniform}(0, \alpha^*) \end{cases} \tag{8.10}$$

In Equation 8.10, $\alpha_h = 0$ should be interpreted as the limit $\alpha_h \to 0$, by which we refer to independent site loadings, corresponding to $\boldsymbol{\Sigma}_h$ being set to the identity matrix. Thus, the default prior in Hmsc assumes that with probability of 0.5, there is no spatial or temporal signal. With the remaining probability of 0.5, the prior assumes the scale of the spatial or temporal decay to be uniformly distributed between 0 and α^*.

Theoretically, the maximal value of α^* should be selected based on the spatial scale at which one expects the relevant ecological processes to operate. However, in practice, its choice is determined by the availability of the data. To understand why, let us assume that the ecologically relevant spatial scale would be $\alpha_h = 1$ km; this could be, for example, the scale at which the species would move and interact, or the scale at which the missing environmental covariate that is captured by the site loadings would vary. Let us further assume that the data originate from a large-scale study where the distance between nearest sampling units is of the order of 10 km. In this case, it is not possible to estimate small values of α_h such as $\alpha_h = 1$ km, because with this value of α_h the distance of 10 km would correspond to essentially independent site loadings. Thus, the model would fit the data equally well whether $\alpha_h = 1$ or $\alpha_h = 0$, and thus it would not be possible to see any spatial signal at the scale of $\alpha_h = 1$ km. Conversely, if the data come from a small-scale study where the largest distance between any two sampling units is 100 m, it is not possible to estimate such large values as $\alpha_h = 1$ km, because this would make the site loadings highly correlated among all sampling units. This would mean that the site loadings would be essentially the same for all sampling units, and hence they would become confounded with the intercept of the model.

Based on the above considerations, the range $(0, \alpha^*)$ should be selected so that it is compatible with the distances between the sampling units in the data. To do so, the default prior of Hmsc sets α^* to the maximal distance within the enclosing rectangle of all sampling units. With this choice, the default prior for α_h is independent of the spatial units, and thus it is equally appropriate whether the spatial coordinates have their units as metres or kilometres. Let us finally note that the HMSC model can include multiple random effects (indexed by r in Equation 7.10) and each of them can involve multiple factors h. In such a case, the prior of Equation 8.10 is applied independently for each α_h^r.

8.4.2 The Multiplicative Gamma Process Shrinking Prior for Species Loadings

We next discuss the prior of the species loadings $\mathbf{\Lambda}$, which in turn implicitly defines the prior for the species association matrix $\mathbf{\Omega}$ through $\mathbf{\Omega} = \mathbf{\Lambda}^{\mathrm{T}}\mathbf{\Lambda}$. This is probably the most important part of this chapter, as for $\mathbf{\Lambda}$ it is more difficult to define a generally applicable default prior than for all the other parameters. Furthermore, while the results of HMSC analyses are usually not very sensitive to the other prior choices (assuming a high availability of data), they are usually sensitive to the prior chosen for the species loadings $\mathbf{\Lambda}$. The prior for the species loadings is the multiplicative gamma process shrinking prior that Bhattacharya and Dunson (2011) proposed for modelling of high-dimensional covariance matrices, such as the $\mathbf{\Omega}$ matrix for a large number of species. The density of the multiplicative gamma process shrinking prior decomposes as:

$$p(\mathbf{\Lambda}, \mathbf{\Phi}, \boldsymbol{\delta}) = p(\mathbf{\Lambda} \mid \mathbf{\Phi}, \boldsymbol{\delta})p(\mathbf{\Phi})p(\boldsymbol{\delta}) \tag{8.11}$$

The matrix $\mathbf{\Phi}$ with elements ϕ_{hj} has the same dimensions $n_f \times n_s$ as the matrix $\mathbf{\Lambda}$, and it models local shrinkage of species loadings. The parameter $\boldsymbol{\delta}$ with elements δ_h is a vector of length n_f, and it models global shrinkage of species loadings. The three components of Equation 8.11 are defined as follows:

$$\lambda_{hj} \mid \phi_{hj}, \delta \sim N\left(0, \phi_{hj}^{-1}\tau_h^{-1}\right), \tau_h = \Pi_{l=1}^{h}\ \delta_l \tag{8.12}$$

$$\phi_{hj} \mid \upsilon \sim \mathrm{Ga}(\upsilon/2, \upsilon/2) \tag{8.13}$$

$$\delta_1 \mid a, b \sim \mathrm{Ga}(a_1, b_1),\ \delta_l \mid a,\ b \sim \mathrm{Ga}(a_2, b_2) \text{ for } l \geq 2 \tag{8.14}$$

Thus, the prior of the species loadings is normally distributed (Equation 8.12), with a mean of zero and precision (inverse of variance) modelled as a product of gamma distributed random variables (Equations 8.13 and 8.14). The choice of this form of the prior may look overly complicated, but there are good reasons for it (Bhattacharya & Dunson 2011), some of which we discuss below.

The user needs to decide about the prior parameters υ, a and b, where υ is a scalar, and a and b are vectors of two numbers, $a=(a_1, a_2)$, and $b=(b_1, b_2)$. As default values, Hmsc assumes $\upsilon=3$, $a=(50, 50)$ and $b=(1, 1)$. Out of these values, we propose that the user pays particular attention to the choice of $a=(50, 50)$, as these two parameters can be used to adjust the level of shrinkage that the prior implies for the matrix $\mathbf{\Omega}$. Let us illustrate this by simulating association matrices from the multiplicative gamma process shrinking prior. This can be done with the script below, which simply implements the above equations.

```
nu = 3
a1 = 50
b1 = 1
a2 = 50
b2 = 1
ns = 50
nf = 10

Delta = matrix(c(rgamma(1, a1, b1), rgamma(nf - 1, a2, b2)))
Psi = matrix(rgamma(nf*ns, nu/2, nu/2), nf, ns)
tau = apply(Delta, 2, cumprod)
tauMat = matrix(tau, nf, ns)
Lambda = matrix(rnorm(nf*ns)*sqrt(Psi*tauMat)^-1, nf, ns)
Omega = t(Lambda) %*% Lambda
isd = diag(1/sqrt(diag(Omega)))
R = isd %*% Omega %*% isd
```

We note that in the last two lines of the script above we have scaled the matrix $\mathbf{\Omega}$ to a correlation matrix $\mathbf{R}$. We further note that we have selected the number of latent factors to be $n_f = 10$; we will return to this point later.

We next visualise the simulated association matrices.

```
plotOrder = corrMatOrder(R, order = "AOE")
corrplot(R[plotOrder,plotOrder], method = "color",
        col = c("grey","white","black"))
```

In Figure 8.3, the association matrices simulated with the choice of $a_2=50$ (panels A and C) show simpler patterns than those simulated by the choice of $a_2=3$ (panels B and D). With $a_2=50$, the species essentially split into two groups that show a positive co-occurrence within groups and negative co-occurrence between groups. With $a_2=3$, the structure of the association network is more complex. The value of a_1 does not actually influence the prior distribution of the correlation matrices $\mathbf{R}$, and

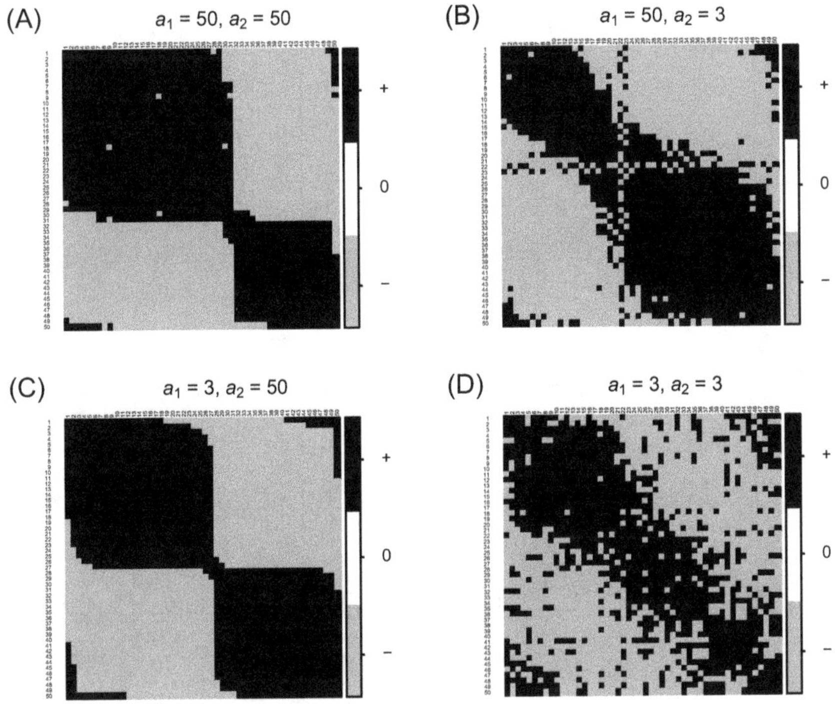

Figure 8.3 Prior samples of species association matrices $\mathbf{\Omega}$ converted to correlation matrices $\mathbf{R}$. The prior parameters a_1 and a_2 vary among the four panels, with the assumed values shown on the top of each panel. In each panel, the fifty species have been ordered in a way that best shows the structure of the association matrix.

thus the fact that the upper and lower panels of Figure 8.3 resemble each other is not a coincidence. What the value of a_1 does influence is the variances of the diagonal elements of the matrix $\mathbf{\Omega}$. To illustrate this, we extract the variances and plot their distributions.

```
s2 = diag(Omega)
```

Figure 8.4 shows the variances not just for a single matrix $\mathbf{\Omega}$, but for 100 matrices $\mathbf{\Omega}$ sampled from the prior. Thus, while Figure 8.3 illustrates individual samples from the prior, Figure 8.4 illustrates the prior density. The y-axis in these panels represents the prior density for a diagonal element Ω_{jj}, or more precisely for $\log_{10}\Omega_{jj}$, as we have log-transformed to be able to show both very small and very large values in the same histograms.

In Figure 8.4, the variances are the smallest with $a_1 = 50$ and $a_2 = 50$ (panel A), and the largest for $a_1 = 3$ and $a_2 = 3$ (panel D). While the

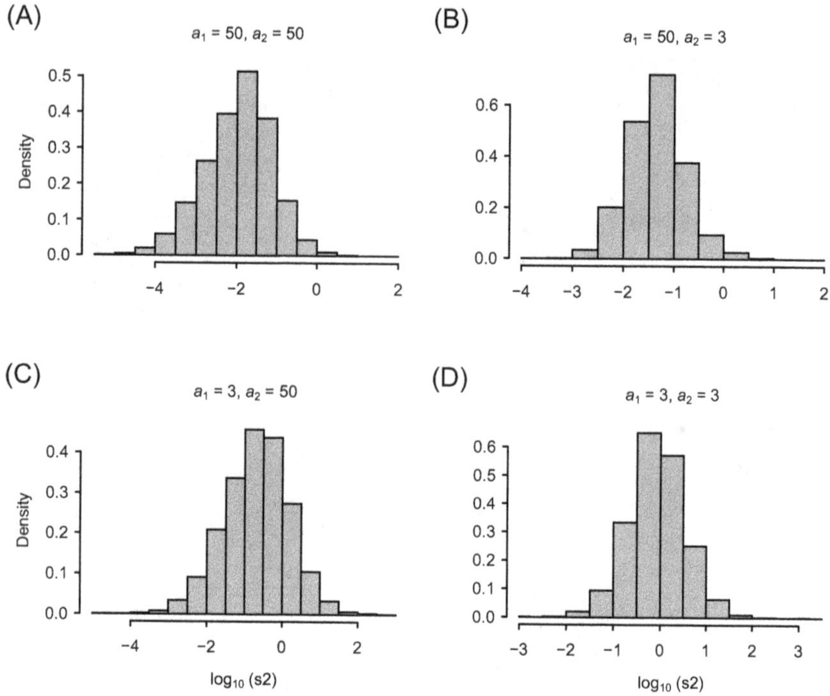

Figure 8.4 Prior distributions of diagonal elements of the covariance matrices **Ω**. The prior parameters a_1 and a_2 vary among the four panels, with the assumed values shown on the top of each panel. Each panel is based on 100 prior samples for a community of 50 species, and hence there are 5,000 values behind each histogram. Note that the histograms are shown for $\log_{10}$ transformed values.

differences between these two extremes may look small in Figure 8.4, the plotting is done at the log-scale, thus the differences are in fact quite substantial. The reason why the default prior has the values of $a_1 = 50$ and $a_2 = 50$ is that these shrink the variances of the random effect close to zero, and thus a strong signal in the data is required to overcome the shrinkage. In contrast, the prior with $a_1 = 3$ and $a_2 = 3$ has much less shrinkage, and consequently a much higher risk of overfitting and thus capturing noise rather than signal. With the choice of these shrinkage parameters, there is always a trade-off: if one assigns very large values to a_1 and a_2, then shrinkage becomes so high that the model becomes essentially identical to a model without any random effects.

To disentangle the roles of a_1 and a_2, let us further examine another aspect of these prior simulations. Namely, we will look at the prior

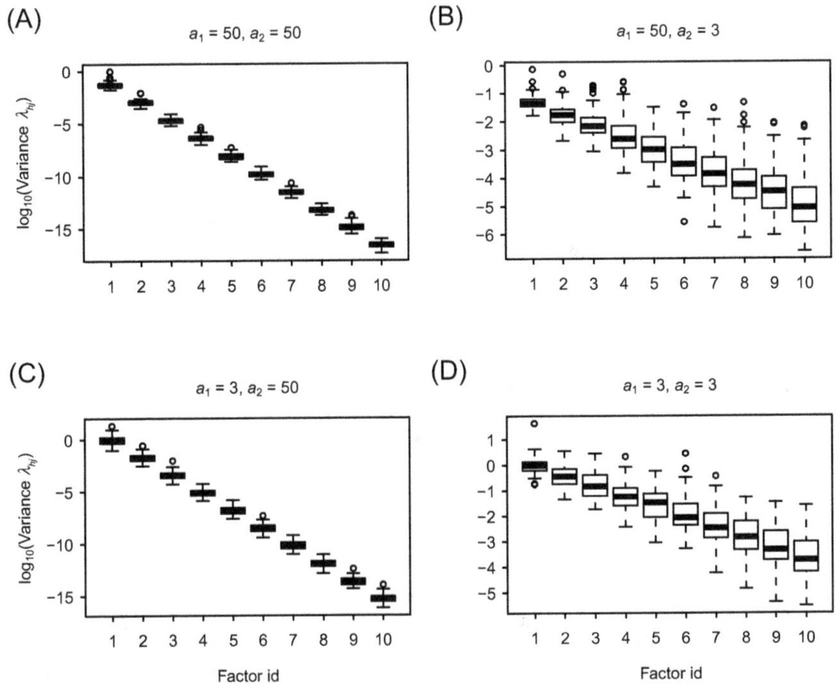

Figure 8.5 Prior distributions of variances of species loadings $\mathbf{\Lambda}$. The prior parameters a_1 and a_2 vary among the four panels (A–D), with the assumed values shown on the top of each panel. Each panel is based on 100 prior samples of a community of 50 species, and boxplots show the variances of the lambda parameters for each of the 10 factors included in the model. Note that the histograms are shown for $\log_{10}$ transformed values.

variances of the species loadings λ_{hj}. To do so, we repeated the prior simulation that was used to generate Figures 8.3 and 8.4, but this time recording the variance among the fifty λ_{hj} values (one for each species j), separately for the n_f= 10 factors indexed by h:

```
lambda.var = apply(Lambda, 1, var)
```

The prior distributions of the variances of species loading are illustrated in Figure 8.5, using the same combinations of the prior parameters a_1 and a_2 that were used in Figures 8.3 and 8.4. We observe that in all cases, the variances decrease with increasing factor h. This is due to the global shrinkage. The decreasing variance implies that the first factor is the most important, as it carries the largest amount of variation, whereas the contribution of the forthcoming factors decreases gradually.

Figure 8.5 illustrates how the two prior parameters a_1 and a_2 control the shrinkage, as defined by Equations 8.12 and 8.14. Namely, the parameter a_2 sets how fast the species loadings shrink to zero as a function of factor identity. This explains why in Figure 8.3 we observed much simpler species association networks with $a_2 = 50$ than with $a_2 = 3$. Namely, with $a_2 = 50$ they are generated with a lower effective number of factors than with $a_2 = 3$. The parameter a_1 plays a somewhat smaller role in Figure 8.5, but is important as well, as it sets the basic level of shrinkage, i.e. the prior variance of the very first factor.

In Hmsc the prior parameters can be changed with the setPriors function. For example, if a random effect is included in the object rL, the default prior parameter choices $a_1 = 50$ and $a_2 = 50$ can be modified to the alternative choices $a_1 = 3$ and $a_2 = 3$ by

```
rL = setPriors(rL, a1 = 3, a2 = 3)
```

8.4.3 How Many Factors Are Needed?

Let us finally return to the question of how many factors are needed and how their number should be selected. In the context of the multiplicative gamma process shrinking prior, the answer is provided by the title of the paper by Bhattacharya and Dunson (2011), which is 'sparse Bayesian infinite factor models'. Thus, theoretically, the number of factors is infinite. This is theoretically a feasible choice because the increasing shrinkage with respect to factor identity (illustrated in Figure 8.5) makes the 'tail' of the factors so small that their summed variance is finite even if the number of factors would be infinite (for a more technical description, see Bhattacharya & Dunson 2011). While there may well be an infinite number of factors in the theoretical model, the number of factors should necessarily be finite in the software implementation. With the default option of Hmsc, the number of factors is adjusted so that the negligible ones are truncated away. Thus, while with $a_1 = a_2 = 50$ only a few factors might be sufficient, with $a_1 = a_2 = 3$ more factors might be needed. As a default option in Hmsc, the adjustment of the number of factors is conducted during the burn-in iterations, the number of which is set by the transient input parameter of the sampleMcmc function. If the user wishes to change this, the number of burn-in iterations (during which the number of factors is adjusted) can be changed by the adaptNf input parameter of the sampleMcmc function. For example, if setting adaptNf = 0, then the number of factors will not be adjusted, but rather fixed to its initial value.

Another question that relates to the number of factors is the identifiability of the factors. In statistics, the term 'parameter identifiability' relates to the issue of whether the parameter could be estimated accurately even if there would be an infinite amount of data. In non-Bayesian factor models, identifiability is ensured, for example, by constraining the loading matrix to be a lower triangular with positive diagonal entries (Geweke & Zhou 1996). In the multiplicative gamma process shrinking prior that is assumed in HMSC, there are no such constraints, and actually neither the species nor the site loadings are identifiable. To see why this is the case, we note that e.g. the pair ($\mathbf{\Lambda}$, $\mathbf{H}$) yields exactly the same linear predictor as the pair ($-\mathbf{\Lambda}$, $-\mathbf{H}$). This also means that the posterior mean of both the species and the site loadings is zero, because their posterior distributions are symmetric around zero.

Whether the non-identifiability of the species and site loadings is a problem or not depends on the application. If one is interested in the association matrix $\mathbf{\Omega}$ instead of the primary parameters $\mathbf{\Lambda}$ and $\mathbf{H}$, it is not a problem, because the matrix $\mathbf{\Omega}$ is identifiable (Bhattacharya & Dunson 2011). But if one is interested in conducting a model-based ordination (see Hui et al. 2015), then constructing a biplot in terms of the posterior means of ($\mathbf{\Lambda}$, $\mathbf{H}$) can clearly be problematic. For this reason, Hmsc by default aligns the posterior samples by switching some of the ($\mathbf{\Lambda}$, $\mathbf{H}$) pairs into ($-\mathbf{\Lambda}$, $-\mathbf{H}$) if that improves their correlation with the posterior mean of ($\mathbf{\Lambda}$, $\mathbf{H}$).

Let us then discuss the question of why species associations are implemented in HMSC through the latent variable approach. The reason here is that in species-rich communities (meaning that there are tens or hundreds of species), the matrix $\mathbf{\Omega}$ will be high dimensional. For example, with $n_s = 100$ species, there are 10,000 numbers in the matrix $\mathbf{\Omega}$. The requirements that a covariance matrix must be symmetric and positive definite will make the degrees of freedom smaller than the number of matrix elements, yet there are still far more degrees of freedom than can be estimated in practice. With the latent variable approach, the dimensionality of the problem is greatly reduced, as the number of parameters to be estimated scales as $n_s n_f$ instead of n_s^2. This is beneficial if there are many fewer factors than there are species. In an ancestral version of HMSC (Ovaskainen et al. 2010), species associations were not implemented through a latent variable matrix, but by assuming an inverse-Wishart prior for the matrix $\mathbf{\Omega}$. Because of this, the method was applicable only for small species communities consisting of a maximum of few tens of species.

As a related consideration, we note that the approach $\mathbf{\Omega} = \mathbf{\Lambda}^{\mathrm{T}}\mathbf{\Lambda}$ is able to produce a full rank association matrix $\mathbf{\Omega}$ if the number of factors is at least equal to the number of species. For this reason, it is hard to imagine why one would like to choose having more factors than species, even if this is theoretically possible. Hence, by default Hmsc truncates the number of factors to be at most the number of species. Thus, in the case of the univariate models considered in Chapter 5, the number of factors was fixed to one.

8.5 The Prior Distribution of Data Models

Above, we have discussed the prior distribution assumed for the fixed (Section 8.3) and random (Section 8.4) effects of the model; by summing these up, we have defined the prior distribution of the matrix of linear predictors **L**. What thus remains to be described is the link from the linear predictor to the data **Y**. For the probit and Poisson models, there are no additional parameters involved, as in the probit model y_{ij}~Bernoulli($\Phi(L_{ij})$) and in the Poisson model y_{ij}~Poisson(exp(L_{ij})). But the normal and lognormal Poisson models have the additional variance parameter $\mathbf{\Sigma}$, as in the normal model $y_{ij} = L_{ij} + \varepsilon_{ij}$ and in the lognormal Poisson model y_{ij}~Poisson(exp($L_{ij} + \varepsilon_{ij}$)), where in both models $\varepsilon_{ij} \sim N\left(0, \sigma_j^2\right)$. The residual variance parameters σ_j^2 are collected as the diagonal elements of the matrix $\mathbf{\Sigma}$.

In HMSC, the prior distribution for each σ_j^2 is the inverse gamma distribution with a shape parameter $a_{\sigma j}$ and rate parameter $b_{\sigma j}$, where the prior parameters $a_{\sigma j}$ and $b_{\sigma j}$ can be adjusted by the user. As a default, Hmsc assumes for the normal model that $a_{\sigma j} = 1$ and $b_{\sigma j} = 0.01$ for each species j. Let us illustrate this choice of prior by sampling 1,000 random deviates.

```
sigmaj = 1/rgamma(1000, shape = 1, rate = 0.01)
```

As shown in Figure 8.6, the prior assumes the residual variance σ_j^2 to be typically between 0.01 and 1. This is a reasonable assumption if, in the case of the normal model, the variance of the data is of the order of one, either originally or because the scaling option of the **Y** matrix has been used. However, if the variance of the data is much lower or higher (e.g. 10^{-6} or 10^6), then the default prior can be problematic. In such a case, the user should either change the prior parameters $a_{\sigma j}$ and $b_{\sigma j}$ or scale the data.

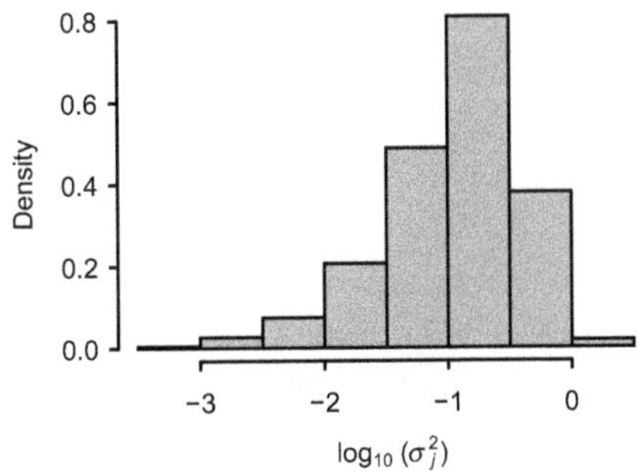

Figure 8.6 Prior density of residual variance σ_j^2. Note that the histogram is shown for $\log_{10}$ transformed values.

8.6 What HMSC Users Need and Do Not Need to Know about Posterior Sampling

Sections 8.3–8.5 included detailed discussions about the prior distribution assumed by HMSC. In this section we discuss posterior sampling, i.e. how the parameters are estimated. While the definition of the prior distribution was technically quite involved, this topic is technically even more involved, and thus discussing it in full detail is probably not of major interest for an ecologically orientated reader. Further, while an ecologist applying HMSC should be aware about the prior choices, it is actually not necessary to know how posterior sampling is technically conducted for adequate interpretation of the results. This is because the way the posterior distribution is sampled does not influence the results – as long as it is done correctly. However, it is necessary to understand some of the basic principles of posterior sampling, at least to the extent of being able to check MCMC convergence and hence that the posterior is sampled appropriately.

In HMSC, the posterior is sampled with a MCMC method. MCMC is a very broad class of methods that enable posterior sampling in almost any kind of Bayesian model (Gelman et al. 2013). The basic principle of this approach is that the posterior sampling proceeds as a Markov chain. The chain proceeds through a number of iterations that we index here by $s = 1, 2, 3, \ldots$ The state variable of the Markov chain is the parameter vector $\boldsymbol{\theta}$, which, in the case of HMSC, contains all the model parameters as defined by Equation 8.2. For iteration round s, the values of the parameters are denoted by $\boldsymbol{\theta}_s$. When moving to the next iteration round $s+1$, the parameter vector $\boldsymbol{\theta}_s$ changes to $\boldsymbol{\theta}_{s+1}$. We will later return to how this exactly happens, but for now the main issue is that, given the current value of the parameter vector $\boldsymbol{\theta}_s$, there is some way to sample $\boldsymbol{\theta}_{s+1}$. This is why the method is called a Markov chain method: to be able to sample $\boldsymbol{\theta}_{s+1}$, it is sufficient to know only $\boldsymbol{\theta}_s$ – one does not need to know, for example, the previous value $\boldsymbol{\theta}_{s-1}$.

The rule of how to sample $\boldsymbol{\theta}_{s+1}$ given $\boldsymbol{\theta}_s$ is selected so that the stationary state of the Markov chain is the posterior distribution. This means that if one iterates the Markov chain for long enough, samples $\boldsymbol{\theta}_s$ obtained from the Markov chain will be samples from the posterior distribution. In other words, for large enough s it holds that $p(\boldsymbol{\theta} \mid \boldsymbol{y}) = p(\boldsymbol{\theta}_s)$. Thus, if one runs the method for long enough, the sampled values represent the posterior distribution. This is why we have cut out a transient (or burn-in) in our examples. Initially, the values of $\boldsymbol{\theta}_s$ may not represent the posterior distribution yet, but simply the value $\boldsymbol{\theta}_1$ at which the chain was initialised.

Even if the value of the chain $\boldsymbol{\theta}_s$ for any large enough s can be considered as a sample from the posterior, it is just one sample. To be able to assess, for example, the posterior mean and 95 per cent credible interval, one needs many samples. In our examples, we have typically obtained 1,000 samples from each Markov chain. Furthermore, it is desirable that these samples are independent of each other – in other words, that they are not correlated with each other. If the samples are correlated, they will not provide an accurate representation of the posterior distribution. To clarify this, let us make a simple toy example. We sample the initial value of a single parameter as $\theta_1 \sim N(0, 1)$, and then set $\theta_{s+1} = \theta_s$, so that we freeze the chain to its initial value. In this case, the chain is still technically a Markov chain, and each θ_s will be an unbiased sample from the distribution $N(0, 1)$. But the collection of the values of θ_s for $s = 1, \ldots, 1{,}000$ are not a good representation of the distribution $N(0, 1)$, as they are not independent samples from it. If $N(0, 1)$ were the posterior distribution from which one would consider the θ_s as samples, one would get a very biased assessment of parameter uncertainty, as with this toy example the full posterior distribution would be concentrated into a single value.

Typically, consecutive posterior samples $\boldsymbol{\theta}_s$ and $\boldsymbol{\theta}_{s+1}$ obtained through MCMC are not independent of each other. Even if consecutive samples typically just resemble each other rather than being identical, the above toy example illustrates why this is a problem. To obtain independent samples, one option is to not record all the samples $\boldsymbol{\theta}_1, \boldsymbol{\theta}_2, \boldsymbol{\theta}_3, \ldots$, but only the samples $\boldsymbol{\theta}_1, \boldsymbol{\theta}_{\Delta s+1}, \boldsymbol{\theta}_{2\Delta s+1}, \ldots$, where Δs is called the thinning interval. Increasing Δs reduces the correlation between consecutive samples $\boldsymbol{\theta}_s$ and $\boldsymbol{\theta}_{\Delta s+s}$, and thus can solve the problem. But at the same time it introduces another problem, as now one needs to run the MCMC Δs times longer. For example, obtaining 1,000 samples with thinning $\Delta s = 1{,}000$ requires one million iterations, which can take a long time. That is why we recommend the user to first run the MCMC with $\Delta s = 1$, and examine if MCMC convergence diagnostics are satisfactory. If not, the MCMC can be run next with $\Delta s = 10$. If this does not lead to satisfactory convergence, one can go on as $\Delta s = 100, 1{,}000, \ldots$, continuing as long as needed to get satisfactory convergence, or to exceed the time available for doing the computation. If the latter occurs, it means that the parameter estimates obtained from the model cannot be fully trusted, as they may represent only a corner of the full posterior distribution. While the thinning Δs is increased, the burn-in period should also be increased. For example, we have set the burn-in period as 500 Δs, so if the MCMC chain is run for a total of 1,500 Δs iterations, the first third of which is discarded as potentially transient.

As illustrated in the examples of Chapters 5–7, MCMC convergence can be checked using different diagnostics. In these examples, we have checked MCMC convergence in terms of the potential scale reduction factor and the effective number of samples, both implemented through the coda package (Plummer et al. 2018). The potential scale reduction factor, also called the Gelman-Rubin convergence diagnostic, compares different independently run MCMC chains to each other (Brooks & Gelman 1998; Gelman & Rubin 1992). This is important to do, because if different MCMC chains give different results, one can be sure that the posterior distribution is not well sampled. Instead, if the different chains do show similar results, this increases the trust that the posterior is well sampled. The effective number of samples evaluates the autocorrelation structure within MCMC chains. If consecutive samples are independent, then the effective number of samples equals the actual number of samples. But if the consecutive samples are correlated, then the effective number of samples will be lower than the actual number of samples. For example, in the toy example where we froze the MCMC chain by setting $\theta_{s+1} = \theta_s$ for $s = 1, \ldots, 1{,}000$, the actual number of samples is 1,000 but their effective number is one.

Let us then return to the issue of how posterior sampling is conducted specifically in Hmsc, i.e. how $\boldsymbol{\theta}_{s+1}$ is sampled given $\boldsymbol{\theta}_s$. As there are many parameters included in $\boldsymbol{\theta}$, a Gibbs sampling strategy is assumed. This means that the parameters in $\boldsymbol{\theta}$ are divided into groups, and the parameters from each group are sampled one after the other, conditional on the values of the parameters in the other groups. The parameters in each group are updated by sampling them from their full conditional distributions. We note this sampling strategy is different from the widely used Metropolis-Hastings algorithm (Gelman et al. 2013), in which new values are proposed and then accepted or rejected. Instead, sampling from the full conditional distribution means that the focal parameters are sampled directly from the distribution that would be their posterior distribution if the other parameters are fixed to their current values. While the Metropolis-Hastings algorithm can be applied with basically any kind of prior distribution, sampling directly from the full conditional distributions is possible only if the prior distributions are chosen in a particular way. For example, the prior for the **V** matrix has been chosen as the inverse-Wishart distribution (Equation 8.7) because this distribution is conjugate to the multivariate normal distribution, which enables it to be sampled from its full conditional distribution (Gelman et al. 2013). As another example, the prior for the phylogenetic signal parameter ρ has been implemented through discrete grid (see discussion after Equation 8.6), because in this case the full conditional on ρ can be computed numerically and thus the parameter can then be sampled directly from it (Gelman et al. 2013). The technical details on how each parameter is sampled in turn from its full conditional distribution are described in Tikhonov et al. (2019).

Let us finally note that even if much effort has been spent to make the posterior sampling in Hmsc computationally efficient, MCMC sampling often takes a lot of time, and MCMC convergence can be problematic. This is especially the case when the model is complex, when the data are large or when the data are not normally distributed. Thus, reaching satisfactory MCMC convergence is expected to take a considerable time if the model involves many sampling units and many species, if there are multiple random effects, and in particular if the probit, Poisson or lognormal Poisson models are assumed. As we will discuss in Section 12.2.3, implementing alternative methods that would speed up computations and improve MCMC convergence for such cases is a continuous challenge, not only for HMSC but for the entire field of Bayesian statistics in general.

8.7 Sampling from the Prior with HMSC

In this section, we illustrate how the prior distribution can be sampled in Hmsc. The user might want to sample the prior distribution because this is the easiest way to understand the consequences of the prior parameter choices explained in Sections 8.3–8.5. Furthermore, being able to sample not only from the posterior but also from the prior makes it possible to compare the prior and posterior distributions for evaluating how informative the data are about the parameters or model predictions.

As our example, we use the Community A that we simulated in Section 6.6 and illustrated in Figure 6.5A.

Thus, we continue from that example, assuming that we have constructed the model as usual.

```
m = Hmsc(Y = Y, XData = XData, XFormula = ~x, TrData = TrData,
    TrFormula = ~trait, phyloTree = phy, distr = "probit")
```

We next call the same function sampleMcmc that is used for model fitting, but now we set the option fromPrior = TRUE.

```
nChains = 2
thin = 1
samples = 1000
transient = 0
verbose = 0
m = sampleMcmc(m, thin = thin, samples = samples,
  transient = transient, nChains = nChains, verbose = verbose,
  fromPrior = TRUE)
```

When the option from Prior = TRUE is chosen, the function sample Mcmc samples the parameters directly from the prior, using the equations described in Sections 8.3–8.5. This means that the samples are independent of each other, and thus there is no reason for doing any thinning or cutting of a transient.

Let us look at the ρ parameter as an example. We may apply the usual tools to study MCMC convergence, and to view its posterior summary.

```
mprior = convertToCodaObject(m)
effectiveSize(mprior$Rho)

## var1
## 1912.254

gelman.diag(mprior$Rho,multivariate = FALSE)$psrf

##      Point est. Upper C.I.
## [1,] 0.9993746   1.000067

summary(mprior$Rho)
```

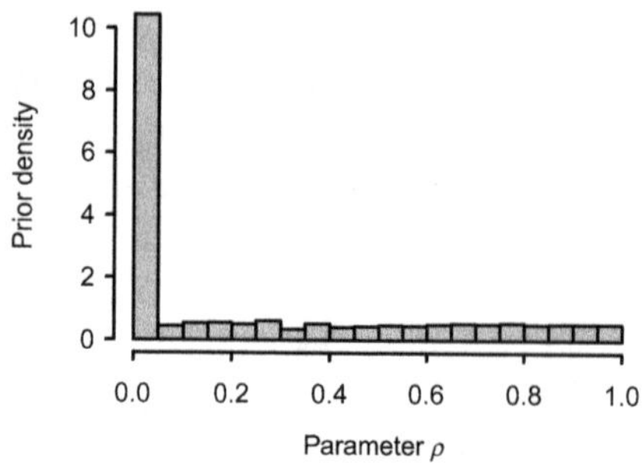

Figure 8.7 Prior density of the phylogenetic signal parameter, based on 1,000 samples obtained directly from the prior distribution.

```
##
## Iterations = 1:1000
## Thinning interval = 1
## Number of chains = 2
## Sample size per chain = 1000
##
## 1. Empirical mean and standard deviation for each variable,
##    plus standard error of the mean:
##
##          Mean       SD       Naive SE Time-series SE
##      0.257265 0.329012     0.007357        0.007536
##
## 2. Quantiles for each variable:
##
## 2.5%  25%  50%  75% 97.5%
## 0.00 0.00 0.02 0.52  0.96
```

We observe that the MCMC convergence statistics are very good. This is to be expected, because we have independent samples from the prior. The posterior summary of the ρ parameter is fully compatible with the prior assumed for this parameter (Equation 8.6), covering the full range from zero to one, and with 50 per cent of the posterior mass given for the value $\rho = 0$. To plot the sampled prior distribution, we may pool the samples from the two chains into one.

```
prior.rho = c(mprior$Rho[[1]], mprior$Rho[[2]])
```

The histogram of the pooled samples shown in Figure 8.7 matches with how the default prior was defined (Equation 8.6). The peak corresponds to the probability mass of 0.5 assigned to $\rho = 0$, whereas the remaining probability mass is distributed uniformly from zero to one.

8.7.1 If There Are No Data, the Posterior Distribution Equals the Prior Distribution

Let us recall from Section 8.2 that the difference between the posterior and the prior distribution tells what we have learned from the data. Thus, if there are no data to learn from, these two should be equal. Let us test this by declaring all data as missing, which can be done by setting all values of the **Y** matrix to NA. In contrast, we will leave the data on environmental covariates (**X**), traits (**T**) and phylogeny (**C**) as they are, because they will not inform us about the parameter estimates if the community data **Y** are missing. We note that while Hmsc allows any pattern of missing data for **Y**, it does not allow any missing data for the other data matrices.

```
Y[ ,] = NA
m = Hmsc(Y = Y, XData = XData, XFormula = ~x, TrData = TrData,
    TrFormula = ~trait, phyloTree = phy, distr = "probit")
m = sampleMcmc(m, thin = thin, samples = samples, transient
    = transient, nChains = nChains, verbose = verbose)
```

Note that we have not applied the option fromPrior = TRUE, and thus we are technically sampling the posterior distribution, not the prior distribution. But because all of the data in the **Y** matrix were declared missing (Y[,] = NA), the posterior should be identical to the prior. Now the samples are not independent of each other, as they are obtained by the usual MCMC approach of posterior sampling:

```
mpost = convertToCodaObject(m)
effectiveSize(mpost$Rho)

##   var1
## 46.808

gelman.diag(mpost$Rho,multivariate=FALSE)$psrf

##      Point est. Upper C.I.
## [1,]  1.121479   1.330756
```

Consequently, the posterior distribution shown Figure 8.8A is not perfectly in line with the prior distribution shown in Figure 8.7.

If we change thin = 100 in the MCMC sampling parameters, MCMC convergence improves.

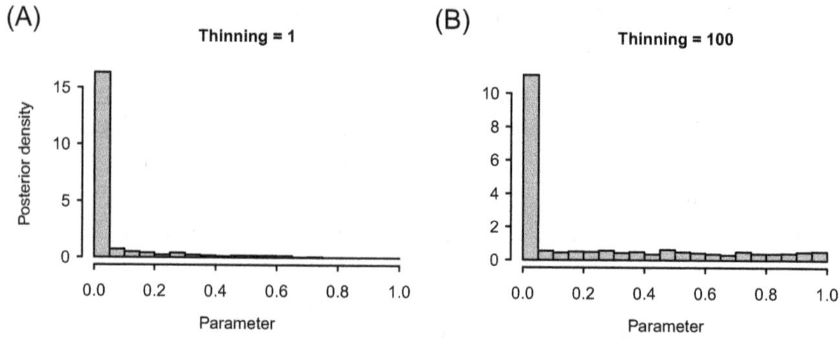

Figure 8.8 Posterior density of the phylogenetic signal parameter based on 1,000 MCMC samples with either thinning = 1 (panel A) or thinning = 100 (panel B). In both panels, all data have been set as missing.

```
mpost = convertToCodaObject(m)
effectiveSize(mpost$Rho)

##      var1
## 135.9322

gelman.diag(mpost$Rho, multivariate = FALSE)$psrf

##       Point est. Upper C.I.
## [1,]   1.007947    1.03725
```

As expected, this makes the sampled posterior a better approximation of the prior distribution (Figure 8.8B).

Sampling the posterior distribution with all data declared missing is a powerful tool for testing the implementation of an MCMC sampling scheme. This is because with all data missing, the true posterior distribution equals that assumed for the prior. Thus, if the sampled distribution does not converge to the prior distribution, there is a bug somewhere in how the MCMC is implemented. In addition to running these tests, Hmsc has been tested by simulating data with known parameter values, and then comparing the estimated values to the true ones, as we did in Section 6.6. Conducting both kinds of tests is necessary when implementing MCMC algorithms for posterior sampling, as MCMC algorithms for sampling complex models are often quite complex, and thus prone to error.

8.8 How Long Does It Take to Fit an HMSC Model?

Performing posterior sampling through MCMC is computationally intensive for complex models and big data comprising many species and/or sampling units. In such cases, we do not recommend trying to fit the 'final' model as the first step, but to first develop the model using a small subset of the data. By doing so, the entire analysis pipeline (from importing data to outputting the result tables and figures) can be developed without the need to wait very long for the computations to finish. Only after the entire pipeline is set up and tested will it be meaningful to fit the model with the full data.

The computational time depends on how long it takes to perform one MCMC iteration, and how many MCMC iterations are needed to achieve satisfactory MCMC convergence. While the exact answer of how long the MCMC sampling takes is case-specific, it is informative to examine how the computational time scales with the size of the data. This is done in Figure 8.9, where we have fitted different HMSC models for ten MCMC iterations, and recorded the total time for model fitting. We note that the time per iteration is not exactly the time for model fitting divided by the number of iterations, because the model fitting also includes a preparatory phase where, for example, spatial covariance matrices are pre-computed. Figure 8.9 includes the computational time for this preparatory phase as well.

As expected, the computational time increases with increasing amount of data, measured either by the number of sampling units (Figure 8.9A), the number of species included (Figure 8.9B) or the number of covariates included (thick versus thin continuous lines in Figure 8.9A). Note that both the x- and the y-axes in Figure 8.9 are shown in logarithmic scale, so each additional unit means that the computational time becomes ten times longer. The main message of Figure 8.9 is that the computational time depends very drastically on the structural assumptions of the model. In Figure 8.9A, the computational times for the models without random effects are essentially negligible compared to those with a random effect, especially compared to the case where the random effect has a spatial structure. Likewise, in Figure 8.9B, the computational times for models without a phylogeny are essentially negligible compared to those with a phylogeny. Furthermore, including a spatially structured random effect not only increases the computational time by a constant factor, but also causes the computational time to increase with the number of sampling units much faster than linearly. Likewise, including a phylogeny increases the computational time with respect to the number of species units much faster than linearly.

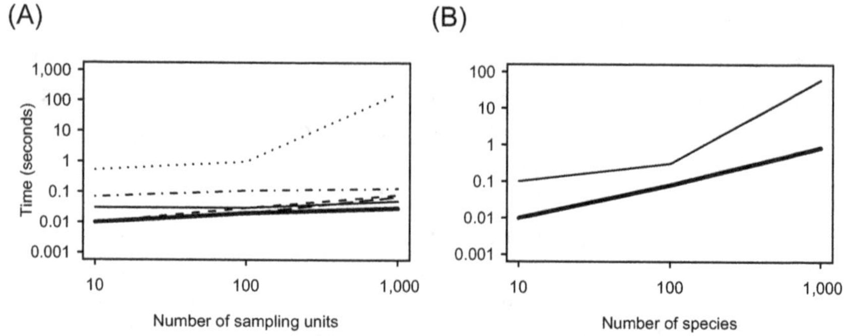

Figure 8.9 Time required for fitting an HMSC model for ten MCMC iterations for different numbers of sampling units (panel A) and species (panel B). In panel A, the baseline case (thick continuous black line) is a normal model fitted for ten species. The baseline includes intercept only ($n_c = 1$), and has no traits, phylogeny or random effects. The other cases deviate from the baseline by having environmental covariates ($n_c = 10$; thin continuous line, or red line in the version shown in the Colour Plate), being based on a probit (dashed line with long dash or green line in the version shown in the Colour Plate) or Poisson (dashed line with short dash, or yellow line in the version shown in the Colour Plate) model, including non-structured random effects (dashdotted line, or blue line in the version shown in the Colour Plate), or spatial random effects (dotted line, or purple line in the version shown in the Colour Plate). In panel B, the baseline case (thick continuous line) is the baseline case of panel A, fitted for ten sampling units. The thin continuous line (or red line in the version shown in the Colour Plate) deviates from the baseline by including a phylogeny. The models were fitted with an HP EliteBook 840 G3 laptop, with two 2.40 GHz cores and 16.0 GB RAM.

The above results mean that with the baseline implementation of Hmsc, spatial models are practically infeasible to fit for very large data, and that models with a phylogeny are practically infeasible to fit for very large species communities. Thus, improved computational algorithms or other special solutions are needed to overcome these limitations. The case of big spatial data was recently addressed by Tikhonov et al. (2020), who approximated the full Gaussian process prior either by the Gaussian predictive process or by the nearest neighbour Gaussian process. This approximation makes it possible to fit spatial models with tens or hundreds of thousands of sampling units, which would not be possible with the baseline implementation of the spatial models due to the adverse scaling of computational time illustrated in Figure 8.9A.

9 · *Evaluating Model Fit and Selecting among Multiple Models*

As we have illustrated in many examples, a community ecologist may apply the HMSC model for many kinds of tasks. These include addressing questions related to ecological inference, such as asking which environmental and spatial drivers influence community assembly processes and how, or questions related to predictions, such as what kind of communities are expected to be found in non-surveyed sampling units. When using HMSC – or any kind of statistical model – for addressing these kinds of questions, it is crucial that the model is as good as possible. Essentially, we are trying to find a model that is as close to the underlying reality as possible. At this point, let us recall the famous quote by Box: 'all models are wrong but some are useful' (1976; 1979), which makes the point that models simplify reality rather than replicate it. A model can never be expected to be perfect, and thus the issue that one should evaluate is whether a model is 'good enough', or which among a set of candidate models is the 'best' one. This is a complex issue, partially because there are many ways of determining how 'good' a model is. The reason is that what the best model is depends on the purpose of the modelling. For example, if the motivation for the modelling is to perform inference, then the best model will be the one that is most in line with the processes and mechanisms underlying the data (Burnham & Anderson 2002; Dunson 2018). However, if the aim is merely to make as good predictions as possible, then a model that makes the best predictions is preferred, even its assumptions would be hard to match with the scientific understanding of the system (Dunson 2018).

In the previous chapters, we have already addressed the topic of model evaluation by measuring both the explanatory and predictive powers in many of the examples. In this chapter, we will discuss model evaluation and selection more thoroughly. In Section 9.1, we start by noting that even if there are automated procedures for model selection, the most important step is actually done by the ecologist when deciding what kind of models will be fitted and what the model will be used for. In Section

9.2, we discuss different ways of measuring model fit based on contrasting the model predictions with the observed data. In Section 9.3, we discuss the use of information criteria as a method for evaluating model fit. The methods discussed in these sections can be used to select among models that differ in their structure. For example, we may compare models with different sets of environmental covariates, models with and without spatial random effects, models with and without traits or phylogenetic information or models that differ in their prior distributions. In Sections 9.4 and 9.5 we present additional methods that are specific for variable selection, i.e. for comparing models that are structurally identical but differ in the included environmental covariates. Among the many methods that can be used for this purpose, HMSC implements variable selection by the spike and slab prior approach (Section 9.4). HMSC also implements reduced rank regression (RRR), which is not for selecting individual predictors but for combining them to reduce their dimensionality (Section 9.5).

9.1 Preselection of Candidate Models

While we present several methods that can be used to select the 'best model' based on quantitative criteria, we do not recommend applying the blind strategy of running all possible model variants through an automated model selection pipeline. Instead, we emphasise that an important task for the ecologist applying HMSC – or any statistical modelling approach – is to use their prior knowledge of the system to restrict the model structure and the sets of candidate variables to be included. The first reason for this is to ensure that the models make sense in light of ecological theory, which is especially important if the focus is on model inference (Burnham & Anderson 2002). The second reason for preselecting some candidate models is pragmatic, as the comparison of all possible models may become computationally intractable. For example, if there are five candidate environmental predictors, there are 243 model alternatives considering all possible combinations of linear and quadratic effects (without even considering the interactions among the predictors). If the possibility of including or excluding the linear or quadratic effects of five candidate species traits (but again excluding interactions) is included, the number of candidate models increases to 59,049; adding different structural options related to spatial or hierarchical structures can literally lead to millions of possible models. The third reason for preselecting some candidate models in light of ecological knowledge is avoiding so-called data fishing. If we would fit millions of candidate models, the model that would best fit independent test data would not necessarily be the best model, as with that many candidate models one might fit not only the training data but also the test data just by coincidence.

Concerning variable selection, there are typically several predictor variables that the researcher hypothesises to influence the occurrence or abundance of the species. Among these, some might not be influential, or some might be collinear with other predictor variables and thus carry redundant information. Increasing the number of variables in a model increases the risk of over-parameterisation, and thereby increase the risk of spurious influence. Thus, it is generally recommended to preselect the least number of variables possible which are as informative and independent of each other as possible. As the very first step, before moving on to fitting any model, predictor variables should be examined for correlation or multicollinearity. In doing so, the researcher may already exclude predictor variables that are strongly correlated with other predictors, either by selecting the one that makes the most ecological sense, or summarising the joint variation among the correlated variables, for example through PCA (see Section 9.5).

Beyond variable selection, one needs also to make many kinds of decisions about the structure of the model, i.e. whether to include random effects, traits and/or phylogeny, and whether to take a hurdle approach to abundance data or to model the zeros and non-zeros together, for example with the lognormal Poisson model. These kinds of decisions can greatly influence how well the model fits the data or how accurately it makes predictions. Making good choices about model structure is typically more difficult than merely selecting among candidate variables.

After preselecting and fitting the candidate models, the next step is to compare among the fitted models more formally. This can be done by evaluating various aspects of predictive performance (Section 9.2) with a cross-validation approach, or by applying an information-theoretic approach (Section 9.3). Alternatively, variable selection can be performed in the model-fitting phase (Sections 9.4 and 9.5).

9.2 The Many Ways of Measuring Model Fit

Model fit can be evaluated in many ways, the choice of which depends on the purpose for which the model will be used. We discuss below

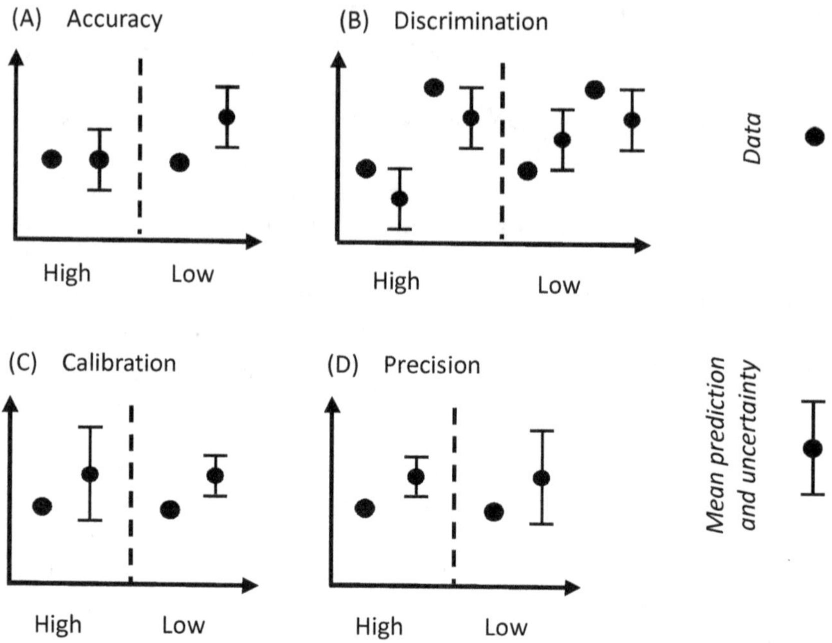

Figure 9.1 Illustration of four aspects of model fit: accuracy, discrimination, calibration and precision. In each panel the dots without an error bar show the true values, and the dots with an error bar show the mean model predictions and their confidence intervals.

three aspects by which model fit can be evaluated. First, model fit can be measured in terms of accuracy, discrimination power, calibration, or precision. Second, model fit can be evaluated for each of the individual species, or for the entire community. Third, model fit can be evaluated in terms of explanatory or predictive performance.

9.2.1 Accuracy, Discrimination Power, Calibration and Precision

Figure 9.1 illustrates how the match between model predictions and the truth can be evaluated in terms of accuracy, discrimination power, calibration, or precision (Norberg et al. 2019). In statistical terminology, accuracy is the opposite of bias – it measures the degree of proximity between the predicted and the true value. In Figure 9.1A, the left-hand case shows that the mean prediction is closer to the true value, while the right-hand case shows that it is farther away from the true value. Hence, the left-hand case has a higher accuracy than the right-hand case. Discrimination power does not examine the absolute match between predicted and true values, but how well predictive values can discern different types of true values. In Figure 9.1B, the left-hand case has higher discrimination power than the right-hand case. This is because in the left-hand case, low and high predictions discriminate well between low and high true values, whereas in the right-hand case, the two predictions overlap and thus discriminate the low and high true values less clearly. Note that in Figure 9.1B, the left- and right-hand predictions are equally accurate, as in both cases the mean predictions are equally far away from the true values. Calibration refers to statistical consistency between distributional predictions and the true values; that is, in calibrated predictions the relative frequency of test values with predictive probability p should be p (Gneiting & Raftery 2007). In Figure 9.1C, the prediction in the left-hand case is better calibrated than the prediction in the right-hand case, since the true data point for the latter case is outside the confidence interval while it is within the confidence interval for the former case. However, well-calibrated predictions do not always contain the true values within their confidence intervals. For well-calibrated predictions, the true value will be inside the 50 per cent confidence interval for 50 per cent of the cases, and hence the true value will be outside the 50 per cent confidence interval for 50 per cent of the cases. Similarly, the true value will be inside the 95 per cent confidence interval for 95 per cent of the cases, and hence the true value will be outside the 95 per cent confidence interval for 5 per cent of the cases. Precision (also referred to as sharpness) measures the width of the predictive distribution and thus its information content. Hence, in Figure 9.1D the prediction in the left-hand case has a higher precision than that of the right-hand case, because the confidence interval of the prediction is smaller for the left-hand case.

While the quality of predictions unambiguously increases with increasing accuracy, increasing discrimination power and increasing calibration, the interpretation of precision depends on the accuracy of the predictions (Gneiting & Raftery 2007). If the predictions are accurate, their quality increases with precision. However, if the predictions are not accurate, the true value will increasingly fall outside the prediction interval with increasing precision, meaning that a high value of precision actually decreases the calibration of predictive distributions. This is illustrated in panels C and D of Figure 9.1, where the well-calibrated prediction is less precise than the poorly calibrated prediction.

Whether to measure model fit in terms of accuracy, discrimination power, calibration or precision, or with a combination of these, depends on the aspect of model performance that is especially critical given the aim of the modelling. For example, if the goal is to predict the probability that a focal species is present in a site, or the expected species richness in a site, or the expected level of beta-diversity between a pair of sites, then measures of accuracy are likely to be the most relevant criterion. If instead the goal is to prioritise sites in terms of their species occurrences, species richness, or community composition, then measures of discrimination are likely to be the most relevant. If the goal is to make statements about prediction uncertainty, e.g. whether the predicted species occurrence probabilities are reliable, or whether the uncertainty estimates involved in predictions of species richness or community composition are valid, then measures of calibration are likely to be important. In theory, measures of precision would be relevant if one wishes to minimise uncertainty. However, since models that involve little uncertainty in their predictions may behave badly with respect to the other measures of performance, we do not recommend using precision as the sole measure of predictive performance.

The way that different aspects of model fit are measured in practice depends on the nature of the data. For any kind of data (presence–absence, count or continuously distributed data), HMSC implements the root mean squared error (RMSE), defined for species j as:

$$RMSE_j = \sqrt{\frac{1}{n}\sum_{i=1}^{n}\left(y_{ij} - p_{ij}\right)^2} \qquad (9.1)$$

where y_{ij} is the observed data and p_{ij} is the model prediction for species j and sampling unit i. In the case of the probit and normal models, the posterior distribution of predictive data is summarised into a single model prediction by taking the posterior mean, whereas for Poisson and log-normal Poisson models the posterior median is applied. As RMSE measures how close the best predictions are to the data, it is a measure of accuracy.

For presence–absence data analysed with probit regression, HMSC implements the Area Under the Curve (AUC) (Pearce & Ferrier 2000) and Tjur R^2 (Tjur 2009). Both of these are measures of discrimination, asking how well the occurrence probabilities discriminate sampling units as either occupied or empty. The units of AUC and Tjur R^2 are different; a model that behaves ‘as well as by chance’ will yield an AUC of 0.5 and a Tjur R^2 of 0, whereas a model that perfectly discriminates empty and occupied sampling units will have an AUC of 1 and a Tjur R^2 of 1. While we consider both equally valid, and both generally give consistent results, it is important to report which one has been applied. An advantage of AUC is that it is often used in ecology and thus many ecologists are familiar with it. An advantage of Tjur R^2 is that it has a very simple definition and thus is easy to interpret: it is the average predicted occurrence probability over those sampling units where the species does occur, minus that where the species does not occur.

For continuously valued data analysed with linear regression, HMSC implements the usual coefficient of determination R^2, defined for each species as the square of the Pearson correlation coefficient between the observed and predicted data. When computing R^2 related to explanatory power, i.e. making predictions for the same data used to fit the model, it can be expected that the model does not predict worse than by chance. However, when making predictions for new data, e.g. when performing cross-validation, it can happen that the model has basically no predictive power for some of the species, in which case it is expected to predict (just by chance) equally often worse than by random than better than by random. Predictions that are worse than those by chance will have a negative Pearson correlation coefficient, but this will be invisible for R^2 as it involves squaring the number. To be able to determine if the predictions are worse or better than by chance, HMSC reverts the sign of R^2 if the correlation coefficient is negative. Thus, if the model has no predictive power, the species-specific R^2 values can be expected to be symmetrically distributed around zero. For count data modelled with Poisson or lognormal Poisson regression, HMSC implements a pseudo-R^2, defined otherwise as the R^2 for normally distributed data, but applying the Spearman rank correlation instead of the Pearson correlation. As both R^2 and pseudo-R^2 compare the correlation between the predicted and observed values, they can be viewed as measures of discrimination. Additionally, when measuring R^2 for the data used to fit the model, it can be interpreted as the proportion of variance explained by the model.

As discussed above, the measures of model fit that are implemented in the current version of Hmsc are related to accuracy and discrimination. While measures of calibration or precision are not currently implemented in Hmsc, such measures can be defined for species distribution models (Norberg et al. 2019), and they can be computed based on the full posterior distribution of model predictions, which is the output of the predict function of Hmsc.

HMSC returns measures of model fit separately for each species. To get an overall assessment of model fit, the species-specific values of model fit can simply be averaged over all species or over focal subsets of species. Model fit can also be evaluated in terms of diversity metrics. For example, Norberg et al. (2019) compared stacked and joint SMDs not only in terms of species-specific performance, but also in terms of the accuracy, discrimination, calibration and precision of predicted species richness and beta-diversity.

9.2.2 Evaluating Model Fit for Different Types of Predictions

All the measures of model fit can be applied to measure explanatory power, i.e. how well the model predicts the data that were used to fit it, as well as to measure predictive power, i.e. how well the model predicts data not used for model fitting. With respect to the latter, we recall that one common approach is to apply cross-validation, where the data are split into a number of folds, out of which one is considered test data and the other ones are training data. We recall from the hierarchical study design example of Section 5.6.8 that this splitting of the data can be done in multiple ways, for example by randomly selecting subsets of the sampling units or some higher-level hierarchical units such as plots. We further recall from Section 7.7.2 that both the sampling units and the species may be split into different folds. In this case, the data for species in the non-focal species fold are assumed to be known, which allows potentially better predictions through utilising information on species-to-species residual associations.

As cross-validation approaches are based on splitting the data randomly into different folds, they generally deal with interpolation, i.e. making predictions for sampling units that have similar environmental conditions to the training data, and for sampling units that are close in space and time to the training data. A more difficult task is that of extrapolation, i.e. making predictions for sampling units that are dissimilar in their environmental or spatio-temporal predictors to those seen in the training data (Norberg et al. 2019; Pearson et al. 2006; Thuiller et al. 2004). For example, SDMs are commonly used to predict how species will respond to climate change. For such predictions to be trusted, one should critically evaluate not only the predictive power of the model in terms of the present-day data, but also its ability to predict already happened changes in species distributions based on fitting the model to past data (e.g. Algar et al. 2009; Johnston et al. 2013).

9.3 The Widely Applicable Information Criterion (WAIC)

Information-theoretic approaches calculate statistical metrics that describe the level of model parsimony, compensating for the level of goodness-of-fit and punishing for model complexity (Burnham & Anderson 2002). In single-species distribution modelling, the Akaike Information Criterion (Akaike 1974, AIC) is the most widely used metric (Grueber et al. 2011; Johnson & Omland 2004). AIC has been developed for the maximum likelihood inference framework, and its analogue in the Bayesian inference framework is the Bayesian Information Criterion (BIC, Schwarz 1978). The AIC and the BIC are defined by:

$$\mathrm{AIC} = L_n(\hat{\theta}) + \frac{d}{n} \tag{9.2}$$

$$\mathrm{BIC} = nL_n(\hat{\theta}) + \frac{d}{2}\log n \tag{9.3}$$

Here θ denotes the model parameters, n the number of data points used to fit the model, $\hat{\theta}$ the maximum likelihood estimate, d the number of model parameters and

$$L_n(\theta) = -\frac{1}{n}\sum_{i=1}^{n}\log p(Y_i|\theta) \tag{9.4}$$

is called the log loss function or the minus log likelihood function. The most parsimonious models are selected based on the lowest AIC or BIC values. The difference between these two is that AIC corresponds (asymptotically) to the model that would be selected based on leave-one-out cross-validation, whereas BIC is designed to identify the true model that generated the data. In practice, BIC selects simpler models, as it penalises more for model complexity.

Neither the AIC nor BIC apply as such to HMSC. To see one reason for this, we note that it is already difficult to define the number of parameters d. For example, the latent variable structure behind species associations involves a theoretically infinite number of factors and thus an infinite number of variables. Assume that the first three factors are sufficient to explain the variation in the data, and thus that the remaining factors are shrunk so close to zero they are in practice negligible. Then the model – and hence the likelihood of the data – would essentially be the same whether the number of factors is truncated to 3 or 100, but the number of model parameters (and hence AIC) will critically depend on this choice.

Furthermore, AIC and BIC only apply to regular statistical models, but not hierarchical models (Spiegelhalter et al. 2002), which in the statistical terminology belong to the family of singular models (Watanabe 2013). HMSC is a singular model because it contains hierarchical layers and latent variables. To make AIC applicable to hierarchical models, Spiegelhalter et al. (2002) introduced the deviance information criterion (DIC) that generalises AIC for hierarchical models, and replaces the actual number of parameters by the so-called effective number of parameters. The effective number of parameters depends on the prior distribution, e.g. correlations among the parameters in the prior distribution reducing their effective number. Spiegelhalter et al. (2002) noted that with hierarchical models, both the likelihood and model complexity depend on the 'focus' of the modelling exercise. For example, in the fixed-effects part of HMSC, the species-specific β parameters are conditional on the community-level γ parameters. If applying DIC, one should choose whether the focus is on β or γ.

Watanabe (2010; 2013) overcame the limitation of the AIC and BIC approaches by developing 'widely applicable' versions of them: the WAIC (Watanabe 2010), and the Widely Applicable Bayesian Information Criterion (WBIC; Watanabe 2013). WAIC and WBIC can be applied to both singular and regular models, and they generalise AIC and BIC for regular models; for regular models, WAIC is equivalent to AIC and WBIC to BIC. Whether the model is regular or singular, WAIC has the same asymptotic behaviour as the Bayes cross-validation loss and the Bayes generalisation loss, whereas the same is not true for DIC (Watanabe 2010). For this reason, WAIC is considered to be better justified for singular models than DIC; this is why WAIC has been chosen for HMSC.

Following Watanabe (2010), WAIC is defined as:

$$\text{WAIC} = B_t L(n) + \frac{1}{n} V(n) \tag{9.5}$$

where $B_tL(n)$ is the so-called Bayesian-training loss defined by:

$$B_t L(n) = -\frac{1}{n} \sum\nolimits_{i=1}^{n} \log p^*(Y_i) \tag{9.6}$$

and $V(n)$ is the functional variance defined by:

$$V(n) = \sum_{i=1}^{n} \{ \mathrm{E}_{\theta|y}[(\log p(Y_i|\theta))^2] - \mathrm{E}_{\theta|y}[\log p(Y_i|\theta)]^2 \} \tag{9.7}$$

In Equations 9.6 and 9.7, $\mathrm{E}_{\theta|y}[\cdot]$ denotes the posterior mean, and $p^*(Y_i)$ denotes the Bayes predictive distribution $p^*(y) = \mathrm{E}_{\theta|y}[p(y|\theta)]$. In the context of the multivariate HMSC model, each datapoint Y_i is the vector of species abundances or occurrences in the sampling unit i. Equations 9.6 and 9.7 are actually a special case of those presented by Watanabe (2010); we have set the so-called inverse temperature to one, as that choice corresponds to the standard Bayes estimation.

As mentioned above, choosing the model with the lowest WAIC value maximises its predictive power, in the sense that WAIC coincides asymptotically (i.e. as the number of data points $n \rightarrow \infty$) both with the Bayes cross-validation loss and the Bayes generalisation loss (Watanabe 2010). The Bayes generalisation loss $B_gL(n)$ is defined by $B_gL(n) = -\mathrm{E}_Y[\log p^*(Y)]$, and thus it measures how accurately the model is able to predict previously unseen data. The cross-validation loss $C_vL(n)$ is defined by $C_vL(n) = -\frac{1}{n}\sum_{i=1}^{n} \log \mathrm{E}_{\theta|y}^{(i)}[p(Y_i|\theta)]$, where $\mathrm{E}_{\theta|y}^{(i)}[\cdot]$ denotes the posterior mean for a model fitted without the data point i, and thus it asks how accurately the model is able to predict the data used for model fitting when applying leave-one-out cross-validation.

One advantage of WAIC is that it can be computed quickly, as all the components needed for WAIC computation can be recorded at the same time as the posterior is sampled. In contrast, performing leave-one-out cross-validation requires fitting the model to the data as many times as there are sampling units, which is computationally highly demanding. However, the access to WAIC values does not mean that the cross-validation procedure implemented in HMSC is redundant. For instance, one does not necessarily wish to specifically perform leave-one-out cross-validation. As one example, in a hierarchical sampling design where several sampling units have been surveyed in each plot, one may wish to train the model by leaving out all sampling units from the focal plot in order to test how well it predicts communities in unseen plots (see the hierarchical case study in Section 5.6.8). In the same line, one may wish to test the predictive power for spatial, temporal or environmental extrapolation rather than interpolation. This can be done by splitting the data into training and test data in a suitable way, but not through evaluating WAIC. Furthermore, WAIC uses specifically log posterior predictive score to evaluate models, but sometimes one may wish to evaluate predictive power from another point of view, such as accuracy, discrimination, calibration, or precision (Section 9.2.1). This can be done by applying cross-validation, but not through WAIC. Therefore, WAIC and cross-validation approaches may provide complementary information for selecting the best model. We will illustrate the use of WAIC in the context of HMSC with simulated data in Sections 9.4 and 9.5, and with real data in Section 11.1.3.

9.4 Variable Selection by a Spike and Slab Prior

In the Bayesian context, one way to implement variable selection for regression models is to assume the so-called spike and slab prior (Mitchell & Beauchamp 1988; O'Hara & Sillanpää 2009). To describe what the spike and slab stand for, we return to the most basic case, namely the univariate linear model described in Section 5.2. We recall that the linear predictor is written as $L_i = \sum_{k=1}^{n_c} x_{ik}\beta_k$, where n_c denotes the number of columns of the design matrix $\mathbf{X}$, and hence the number of regression parameters β_k to be estimated. We recall from Section 5.2 that the intercept is included as the first 'covariate' so that $x_{i1} = 1$ for all i, and that categorical variables with m levels expand to $m - 1$ dummy variables.

Through variable selection we will include a subset of variables that we initially consider. To illustrate how variable selection by the spike and slab prior procedure works, let us start by assuming that we initially consider two environmental variables, in addition to the intercept. That is, $n_c = 3$, and the full model reads $L_i = \beta_1 + x_{i2}\beta_2 + x_{i3}\beta_3$. Here x_{i2} and x_{i3} are the values of the two environmental covariates in the sampling unit i. As these two covariates might be included or excluded, there are four model variants to assess, the simplest being the intercept-only model $L_i = \beta_1$, the intermediate cases $L_i = \beta_1 + x_{i2}\beta_2$ and $L_i = \beta_1 + x_{i3}\beta_3$, and the most complex is the full model $L_i = \beta_1 + x_{i2}\beta_2 + x_{i3}\beta_3$. As discussed above, the question of which model performs the best can be addressed by applying cross-validation or information criteria. In the spike and slab prior approach, the variable selection is already done in the model-fitting phase. To explain how this is done, we first note that the simpler models can be obtained as special cases of the full model, by setting the different regression parameters to zero. For example, the model $L_i = \beta_1 + x_{i3}\beta_3$, which does not include the covariate x_{i2}, can be obtained as a special case of the full model if the regression parameter β_2 is estimated to be exactly zero, as in that case the value of the covariate x_{i2} will not influence the linear predictor. In the Bayesian context, the possibility that a regression parameter is exactly zero can be included by setting some of the prior mass of the regression parameter to zero. This is the 'spike' part of the spike and slab prior.

In the basic univariate linear model, each β_k is typically given a prior such as $\beta_k \sim N(0, \sigma^2)$, where the choice of the prior variance parameter σ^2 may depend for example on how the covariates have been scaled and on the type of the response variable (see Section 8.3). With the baseline prior $\beta_k \sim N(0, \sigma^2)$, the prior probability that β_k is exactly zero is zero, because the normal distribution is a continuous distribution and thus sampling any specific value from it takes place with a probability of zero. To incorporate the prior assumption that the covariate x_{ik} may not influence the response variable at all, the spike and slab prior is constructed as $\beta_k = p_k \widetilde{\beta}_k$, where $p_k \sim \text{Bernoulli}(q_k)$ and $\widetilde{\beta}_k$ follows the baseline prior of $\widetilde{\beta}_k \sim N(0, \sigma^2)$. Here q_k is the prior probability by which the covariate x_{ik} will be selected to the model, and hence the regression parameter β_k will be non-zero. Thus, $1 - q_k$ is the probability that β_k is zero, modelling the 'spike' part of the prior. If the covariate x_{ik} is selected, its value $\widetilde{\beta}_k$ is assumed to follow the prior distribution $N(0, \sigma^2)$, which is the 'slab' part of the prior. Special cases of the spike and slab prior are obtained by setting the prior parameter $q_k = 0$, in which case the covariate is never selected, or by setting it to $q_k = 1$, in which case the covariate is always selected. These special cases are redundant, as $q_k = 0$ corresponds simply to not including the covariate in the model, and $q_k = 1$ corresponds simply to including it and not applying variable selection. Thus, if applying variable selection, the prior parameter q_k is chosen to have some value with $0 < q_k < 1$, the choice of q_k reflecting the prior belief of how likely the environmental covariate should be included in the model. Figure 9.2 illustrates the spike and slab prior for two choices of the prior parameter q_k. If applying the prior of Figure 9.2A, the covariate is more likely to be selected in the model than if applying the prior of Figure 9.2B.

In HMSC, variable selection through the spike and slab prior can be applied to the covariates included in the matrix $\mathbf{X}$. To do so in the full model containing traits and phylogeny, we recall from Equation 8.1 that in HMSC the species niches are modelled as $\text{vec}(\mathbf{B}) \sim N(\text{vec}(\mathbf{\Gamma T}^{\text{T}}), [\rho\mathbf{C} + (1-\rho)\mathbf{I}] \otimes \mathbf{V})$. This equation defines the prior for the regression parameters, and thus it plays the role of the baseline prior $\beta_k \sim N(0, \sigma^2)$ of the univariate linear model discussed above. Variable selection through spike and slab is incorporated as $\beta_{kj} = p_{kj} \widetilde{\beta}_{kj}$, where $\text{vec}(\widetilde{\mathbf{B}}) \sim N(\text{vec}(\mathbf{\Gamma T}^{\text{T}}), [\rho\mathbf{C} + (1-\rho)\mathbf{I}] \otimes \mathbf{V})$, and p_{kj} describes whether variable k is included ($p_{kj} = 1$) or excluded ($p_{kj} = 0$) for species j. As we will discuss next, this can be done jointly for all species or individually for each species.

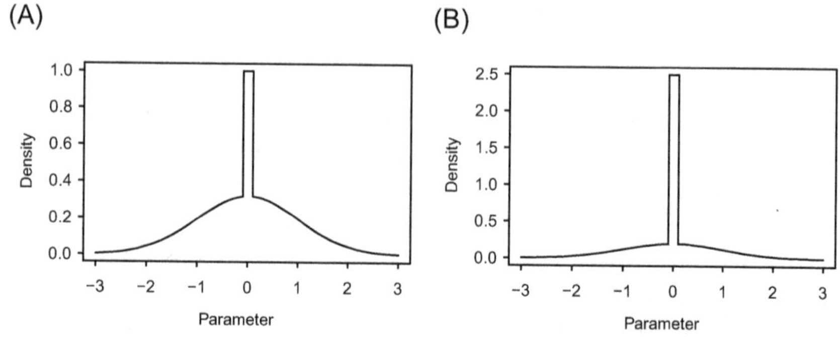

Figure 9.2 Illustration of spike and slab priors. In both panels, the slab follows the standard normal distribution with mean of 0 and variance of 1. The prior probability q of the variable being selected is $q = 0.8$ in panel A and $q = 0.5$ in panel B. For illustrative purposes, the prior mass related to the spike is shown as being uniformly distributed in $[-0.1, 0.1]$, whereas the prior mass of the actual spike is concentrated to the single value of zero.

9.4.1 Selecting Variables Jointly for All Species or Individually for Each Species

In the multivariate context of HMSC, variables can be selected separately for each species, jointly for all species or jointly for groups of species. Applying variable selection for the covariate x_{ik} separately for each species corresponds to the prior $\rho_{kj} \sim \text{Bernoulli}(q_{kj})$ assumed independently for each species. In this case, the covariate may be selected for one species (say, if $\rho_{k1} = 1$, then covariate k is selected for species 1) but not selected for another species (say, if $\rho_{k2} = 0$, then the same covariate k is not selected for species 2). Selecting the variables jointly for all species means that the covariate is selected or not selected simultaneously for all species. This corresponds to the prior $\rho_{kj} = \rho_k$, where $\rho_k \sim \text{Bernoulli}(q_k)$. One may hypothesise that species with different trait categories (e.g. species from different guilds) or from different taxonomical groups respond to different sets of variables. In this case, it is possible to apply species group-specific variable selection. The species are divided into groups $g_s = 1, \ldots, n_{g_s}$, and the covariates are selected or not selected simultaneously for all the species within groups, but independently among groups. This corresponds to the prior $\rho_{kj} = \rho_{kg_s(j)}$, where $g_s(j)$ is the group to which species j belongs, and the prior $\rho_{kg} \sim \text{Bernoulli}(q_{kg_s})$ is assumed independently for each species group g_s.

In addition to the choice of whether the covariates are selected separately or jointly among the species, different sets of covariates can also be selected separately or jointly. One example where joint selection of regression coefficients is needed is the case of categorical explanatory variables. To illustrate this, let us return to the example from Section 5.2.1, where we considered the categorical explanatory variable of habitat type, which included the categories of coniferous forests, broadleaved forests, and mixed forests. In Section 5.2.1, this variable was included in a regression model by setting $x_{i1} = 1$ as the intercept, x_{i2} as an indicator variable for broadleaved forests, and x_{i3} as an indicator variable for mixed forests. In this case, variable selection should clearly be applied simultaneously to both of the regression parameters β_{2j} and β_{3j}, as these two parameters model the effect of the habitat type. To enable joint selection of multiple covariates, the covariates are divided into groups $g_c = 1, \ldots, n_{g_c}$. The prior for p_{kj} is set up as $p_{kj} = p_{g_c(k)g_s(j)}$, where $g_c(k)$ is the covariate group of covariate k, $g_s(j)$ is the species group of species j, and the assumption of $p_{g_c g_s} \sim \text{Bernoulli}(q_{g_c g_s})$ is applied independently for each pair of covariate group g_c and species group g_s.

9.4.2 Simulated Case Study with HMSC

To examine the performance of the slab and spike prior in variable selection, we simulate a case study where variable selection can be expected to be required: a dataset with many covariates but a small number of sampling units. We consider a community of $n_s = 10$ species sampled in $n_y = 20$ sampling units, and we focus on examining which of the $n_c = 10$ potential environmental covariates are most influential and thus should be kept in the model. To keep the example as simple as possible, we assume that the data are normally distributed, and we do not include species traits, phylogenies or random effects.

We will generate three datasets that differ from each other in the set of covariates that influence the community data. We start by simulating variation in ten environmental covariates over the n_y sampling units. In the script below, we assume that each of the covariates follows the standard normal distribution, and we store the covariates in the dataframe XData.

```
ny = 20
nc = 10
ns = 10
XData = data.frame(matrix(rnorm(ny* nc), ncol = nc, nrow = ny))
X = cbind(rep(1,ny), as.matrix(XData))
nc = nc + 1
```

In the script above, we have also constructed an **X** matrix that includes the intercept in addition to the covariates. As the inclusion of the intercept adds one column to the **X** matrix, we have $n_c = 11$, and thus there are eleven parameters to be estimated for each species. We next simulate the responses of the $n_s = 10$ species to the environmental covariates.

```
beta = matrix(rnorm(ns*nc), ncol = ns, nrow = nc)
eps = matrix(rnorm(ns*ny), ncol = ns, nrow = ny)
```

In the script above, we sample both the species-specific regression parameters β_{kj} and the residuals ε_{ij} from the standard normal distribution. To make the three datasets as comparable as possible, we will include the same residuals to all three datasets.

In the first dataset (called henceforth Data FULL), we assume that all covariates do matter. Thus, in the script below, we construct the linear predictor as usual, and add the residuals to form the species data.

```
Beta.FULL = beta
L.FULL = (X %*% beta.FULL)
Y.FULL = L.FULL + eps
```

In the second dataset (called henceforth Data JOINT), we assume that the first environmental covariate ($k = 2$) matters for all species, but that the remaining nine environmental covariates do not influence any of the species. In the name Data JOINT, the word JOINT stands for the covariate that jointly influences all species. Additionally, we include a regression parameter related to the intercept ($k = 1$) for all species. To follow the notation introduced above, we make these choices by constructing the matrix **P**, with $p_{kj} = 1$ if covariate k is selected for species j, and $p_{kj} = 0$ if covariate k is not selected for species j.

```
p.JOINT = matrix(1, ncol = ns, nrow = nc)
for (k in 3:nc){
 p.JOINT[k,] = 0
}
beta.JOINT = p.JOINT * beta
L.JOINT = (X %*% beta.JOINT)
Y.JOINT = L.JOINT + eps
```

In the third dataset (called henceforth Data SEPARATE), we assume that different species respond to different sets of environmental covariates. More precisely, in the script below we assume that the first species

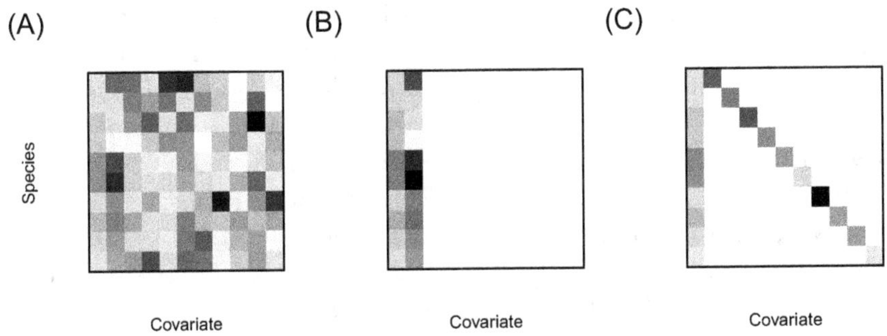

Figure 9.3 Illustration of true matrices of species-specific regression parameters β_{kj} assumed in the three datasets. The white elements correspond to zero values, and the shaded grey area shows the absolute value for non-zero elements. The panels correspond to Data FULL (A), Data JOINT (B) and Data SEPARATE (C).

responds to the first environmental covariate, the second species to the second environmental covariate, and so on until the tenth species responds to the tenth environmental covariate. We additionally include the intercept for all species. In the name Data SEPARATE, the word SEPARATE stands for the covariates that separately influence the different species.

```
p.SEPARATE = matrix(1, ncol = ns, nrow = nc)
for (j in 1:ns){
   for (k in 2:nc){
       if ((k-1)!= j){
          p.SEPARATE[ k,j] = 0
       }
   }
}
beta.SEPARATE = p.SEPARATE * beta
L.SEPARATE = (X %*% beta.SEPARATE)
Y.SEPARATE = L.SEPARATE + eps
```

Figure 9.3 illustrates the simulated variation in the true regression parameters underlying each of the three datasets.

To run the HMSC analyses in a compact notation, we combine the three datasets and the three sets of true values of the regression parameters into lists:

```
Y = list(Y.FULL, Y.JOINT, Y.SEPARATE)
beta = list(beta.FULL, beta.JOINT, beta.SEPARATE)
```

Variable selection is implemented in Hmsc through the input argument XSelect of the Hmsc function. We will apply variable selection in three different ways. In Model FULL, we will force the inclusion of all covariates, and thus not perform variable selection at all. In Model JOINT, we will select each variable jointly for all species. In Model SEPARATE, we will select each variable separately for each species. In this way, Data FULL is compatible with Model FULL, Data JOINT is compatible with Model JOINT and Data SEPARATE is compatible with Model SEPARATE. Fitting each of the three models to each of the three datasets will show whether the models that are best compatible with the data are able to select the right covariates. It will also show what happens when one fits a model where the variable selection setup is not compatible with the data.

The script below defines the object XSelect for the three model variants.

```
qq = 0.1 #prior probability for a covariate to be included

#1: No variable selection
XSelect.FULL = NULL

#2: Variable selection jointly for all species
XSelect.JOINT = list()
for (k in 2:nc){
 covGroup = k
 spGroup = rep(1, ns)
 q = rep(qq, max(spGroup))
 XSelect.JOINT[[k-1]] = list(covGroup = covGroup,
  spGroup = spGroup, q = q)
}

#3: Variable selection separately for each species
XSelect.SEPARATE = list()
for (k in 2:nc){
 covGroup = k
 spGroup = 1:ns
 q = rep(qq, max(spGroup))
 XSelect.SEPARATE[[k-1]] = list(covGroup = covGroup,
  spGroup = spGroup,q = q)
}
XSelect = list(XSelect.FULL, XSelect.JOINT, XSelect.SEPARATE)
```

In the script above, XSelect.FULL is set to NULL, corresponding to no covariate selection and thus the inclusion of all covariates. XSelect.JOINT and XSelect.SEPARATE are both lists of length ten, with each element of the list describing how each of the ten environmental covariates is to be selected.

To illustrate how variable selection is set up, let us look at the third element of the list XSelect.JOINT.

```
XSelect.JOINT[[3]]

## $covGroup
## [1] 4
##
## $spGroup
## [1] 1 1 1 1 1 1 1 1 1 1
##
## $q
## [1] 0.1
```

This element indicates that variable selection is to be applied for the covariate that is found from column four of the X matrix. We recall that column four corresponds to the third environmental covariate, as the first column of X corresponds to the intercept (which one typically wishes to include in any model). All species belong to species group 1, and thus the covariate is selected for all species jointly. *A priori*, we have assumed that this covariate should be included in the model with a probability of 0.1.

As another example, let us look at the third element of the list XSelect.SEPARATE.

```
XSelect.SEPARATE[[3]]

## $covGroup
## [1] 4
##
## $spGroup
## [1] 1 2 3 4 5 6 7 8 9 10
##
## $q
## [1] 0.1 0.1 0.1 0.1 0.1 0.1 0.1 0.1 0.1 0.1
```

This element also concerns the selection of the covariate found from column four of the **X** matrix. However, now each species belongs to its own group, and thus the covariate is selected independently for each species. The prior probability by which the covariate is to be selected is now a vector, with each value giving the prior probability by which the covariate is to be included for each species.

We are now ready to set up and fit the HMSC models.

```
models = list()
for(dataset in 1:3){
 tmp = list()
 for (model in 1:3){
    m = Hmsc(Y = Y[[dataset]], XData = XData, XFormula = ~.,
        XSelect = XSelect[[model]], distr = "normal")
    m = sampleMcmc(m, thin = thin, samples = samples,
        transient = transient, nChains = nChains,
        verbose = verbose)
    tmp[[model]] = m
    }
models[[dataset]] = tmp
}
```

With the script above, we have included all nine cases (three models fitted to each of the three datasets) into the object models, so that the HMSC object of a given dataset and model is found from models [[dataset]][[model]]. Note that we have included the linear effects of all ten covariates included in the dataframe XData with XFormula = ~.

Figure 9.4 evaluates the MCMC convergence of the β parameters in terms of the potential scale reduction factors. MCMC convergence is better when variable selection is not applied (panels A, D and G). This can be expected, as variable selection requires additional MCMC steps for sampling the indicator parameters p_{kj} describing which variables are to be selected and which are not.

Let us next examine whether variable selection with spike and slab was successful in selecting the right covariates. To do so, we compute for each β_{kj} parameter the posterior probability by which the parameter was selected. We note that the posterior probability for the β_{kj} parameter being selected – and hence being non-zero – equals the posterior mean of the indicator variable p_{kj}, i.e. $\Pr(\beta_{kj} \neq 0) = \mathrm{E}[p_{kj}]$. With the script below, we construct Figure 9.5, which illustrates the true value of each β_{kj} parameter in the x-axis, and the posterior probabilities by which the covariate k is selected for species j (i.e. β_{kj} is estimated to be non-zero) in the y-axis.

```
par(mfrow = c(3,3))
for(dataset in 1:3){
 for (model in 1:3){
   postBeta = getPostEstimate(models[[dataset]][[model]],
     parName = "Beta")
   Ep = as.vector(postBeta$support + postBeta$supportNeg)
```

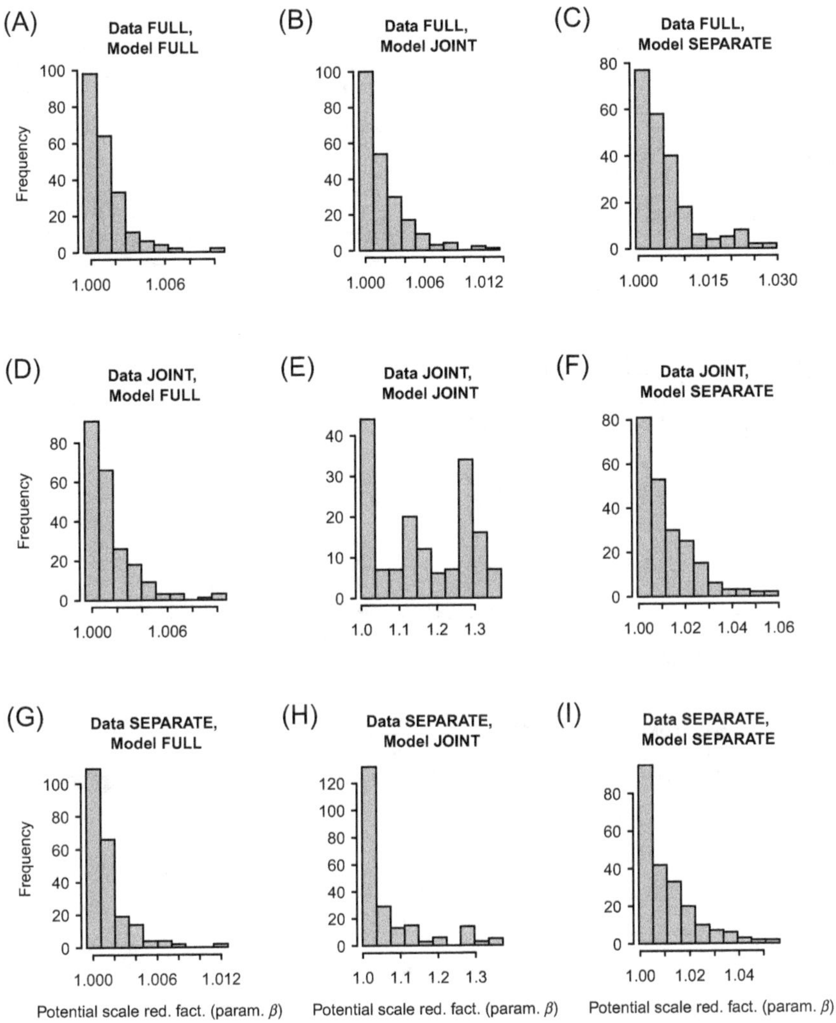

Figure 9.4 MCMC convergence diagnostics for the β_{jk} parameters measured in terms of the potential scale reduction factor. The rows of panels correspond to the three datasets (panels A, B and C correspond to Data FULL, panels D, E and F to Data JOINT and panels G, H and I to Data SEPARATE), and the columns of panels to the three models fitted to the data (panels A, D and G correspond to Model FULL, panels B, E and H to Model JOINT and panels C, F and I to Model SEPARATE).

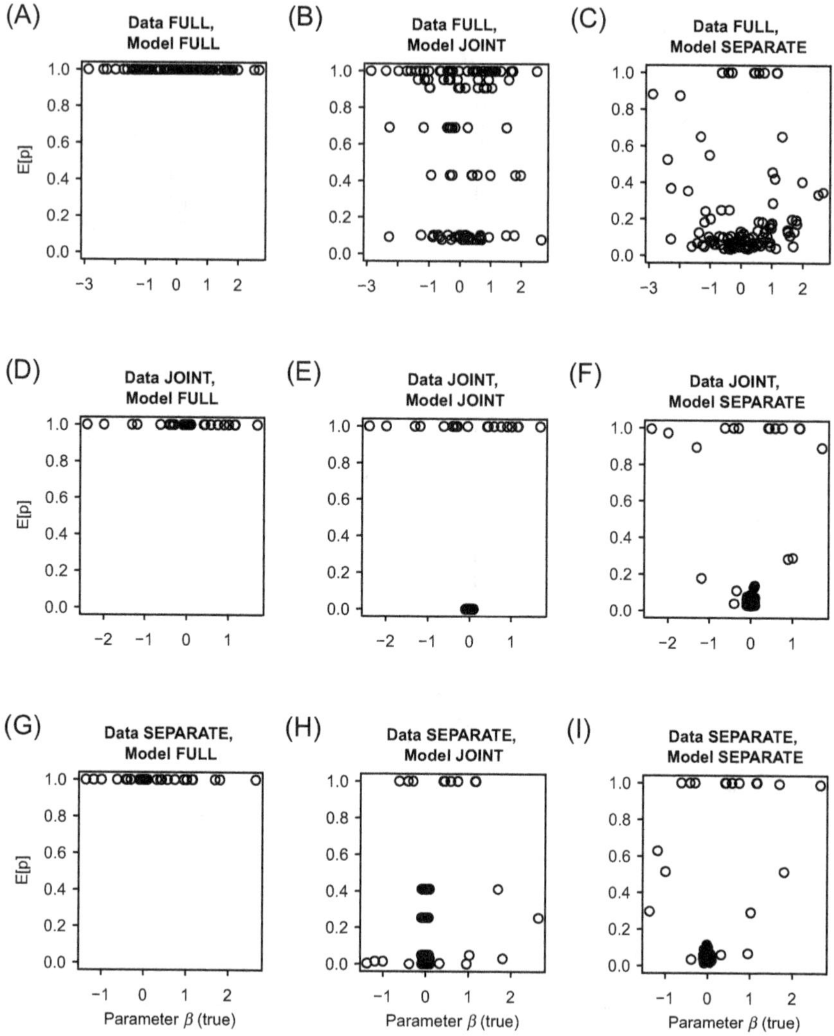

Figure 9.5 Results of variable selection. In each panel, the x-axis shows the true β_{kj} parameter assumed when generating the data, and the y-axis shows the posterior probability by which the parameter is estimated to be non-zero, and hence the covariate is selected. The open dots refer to cases where the true parameter is non-zero, and the filled dots to cases where the true parameter is zero. The vertical locations of the filled dots have been jiggered to make overlapping dots visible. The rows of panels correspond to the three datasets (panels A, B and C correspond to Data FULL, panels D, E and F to Data JOINT, and panels G, H and I to Data SEPARATE) and the columns of panels to the three models fitted to the data (panels A, D and G correspond to Model FULL, panels B, E and H to Model JOINT, and panels C, F and I to Model SEPARATE).

```
    true.beta = as.vector(beta[[dataset]])
    plot(true.beta[!true.beta==0], Ep[!true.beta==0],
      ylim = c(0,1), xlab = "true beta", ylab = "E[p]",
      main = paste0("Data ", as.character(dataset),
      ", Model ", as.character(model)))
    points(runif(sum(true.beta==0), min = -0.1, max = 0.1),
      Ep[true.beta==0],pch = 19)
  }
}
```

If variable selection worked perfectly, the open dots in Figure 9.5 would be located at E[p] = 1 and the filled dots would be located at E[p] = 0. In the uppermost row of panels corresponding to Data FULL, the data generating the model assumed that all covariates matter, and thus we would have liked to see all covariates to be selected for all species. This is trivially the case of Model FULL (panel A), for which model selection is not applied at all. For Model JOINT (panel B) and Model SEPARATE (panel C), only a subset of the regression parameters is selected with a probability of one. Especially in Model SEPARATE, many of the regression parameters have E[p] close to zero and thus are not selected. However, we note that is not entirely terrible; this is especially the case for those β_{kj} for which the value is close to zero, and thus the true effect of is small, hence it may not make a big difference if the covariate is excluded.

In the middle row of the panels corresponding to Data JOINT, the data generating model assumed that one of the covariates influences all species, and the remaining nine covariates do not influence any of the species. Model JOINT, which selects the covariates jointly for all species, does a perfect job in finding the relevant covariate: it includes the relevant covariate with a probability of one, and the irrelevant covariates with a probability of zero (Figure 9.5E). Model SEPARATE, which selects the covariates individually for each species, also does relatively well in selecting the relevant covariates (Figure 9.5F), though not quite as well as Model JOINT. The superior performance of Model JOINT compared to Model SEPARATE is to be expected, as Model JOINT utilises the additional information that each covariate matters either for all or none of the species, making the task of variable selection easier.

In the lowermost row of the panels corresponding to Data SEPARATE, the data generating model assumed that each species is influenced by one covariate that differs for each species. As expected, now Model SEPARATE is most in line with the data generating model, i.e. it works the best for dropping out the irrelevant covariates (Figure 9.5I). As

Model JOINT selects the covariates jointly for all species, it cannot replicate the assumptions of the data generating model. Model JOINT has dropped some of the covariates with a probability of one, perhaps because they had little effect, even for the species that they did influence. For other covariates, Model JOINT shows uncertainty in whether they should be included or not.

Let us next examine how variable selection influences the explanatory and predictive powers of the models. We start by computing explanatory power in terms of R^2.

```
meR2 = matrix(NA, nrow = 3, ncol = 3)
for(dataset in 1:3){
   for (model in 1:3){
      m = models[[dataset]][[model]]
      predY = computePredictedValues(m)
      MF = evaluateModelFit(m, predY = predY)
      meR2[dataset, model] = mean(MF$R2)
   }
}
##                     Model FULL Model JOINT Model SEPARATE
## Data FULL                0.929       0.822          0.729
## Data JOINT               0.712       0.470          0.603
## Data SEPARATE            0.713       0.286          0.587
```

We observe that for all three datasets, the mean explanatory power (averaged over the species) is the highest for Model FULL. This is to be expected, as explanatory power generally increases with the number of covariates included in the model. However, part of the success in explaining the data may be due to overfitting, i.e. explaining noise rather than signal. To study the extent to which this is the case, we evaluate predictive power in terms of two-fold cross-validation. We note that as we have data on twenty sampling units, the predictions will be based on models fitted to ten sampling units only.

```
partition = createPartition(models[[1]][[1]],
  nfolds = 2)
meR2CV = matrix(NA, nrow = 3, ncol = 3)
for(dataset in 1:3){
   for (model in 1:3){
      m = models[[dataset]][[model]]
      predY = computePredictedValues(m, partition
        = partition)
```

```
    MFCV = evaluateModelFit(m, predY = predY)
    meR2CV[dataset, model] = mean(MFCV$R2)
  }
}
##                 Model FULL Model JOINT Model SEPARATE
## Data FULL            0.375       0.228          0.263
## Data JOINT           0.154       0.284          0.074
## Data SEPARATE        0.099       0.067          0.106
```

As is generally the case, the predictive powers are lower than the explanatory powers. Model FULL has the best predictive power for Data FULL, Model JOINT has the best predictive power for Data JOINT, and Model SEPARATE has the best predictive power for Data SEPARATE. This is reassuring, as it means that the models that were structurally in line with the data yielded the best predictions. In particular, in the cases of Data JOINT and Data SEPARATE, variable selection not only identified the covariates that actually did matter, but also helped to make better predictions. In the case of Data SEPARATE, the predictive power of Model FULL is almost as good as that of Model SEPARATE, showing that for the sake of predictive performance it may not be so critical whether the regression coefficients are estimated to be close to zero or exactly zero.

Let us then compute the WAIC for these same cases.

```
WAIC = matrix(NA, nrow = 3, ncol = 3)
for(dataset in 1:3){
  for (model in 1:3){
    WAIC[dataset, model] =
      computeWAIC(models[[dataset]][[model]])
  }
}
##                 Model FULL Model JOINT Model SEPARATE
## Data FULL           18.516      20.860         25.515
## Data JOINT          16.669      16.074         17.188
## Data SEPARATE       17.255      18.834         17.480
```

Model FULL has the lowest WAIC for the Data FULL, and Model JOINT has the lowest WAIC for the Data JOINT. The lowest WAIC for the Data SEPARATE is however not obtained by Model SEPARATE, but by Model FULL, although the difference between these two is small. This reflects the above results that Models FULL and SEPARATE performed essentially equally well also in terms of cross-validation.

9.5 Reduced Rank Regression (RRR)

As discussed in Section 9.4, one way to tackle the case of many explanatory variables is to apply a variable selection procedure to decide which variables should be kept in the model. Another commonly applied strategy is to compress the number of predictors into the main axes of variation through, for example PCA, and then use the first few principal components as the predictors. This can be straightforwardly conducted in HMSC or any other SDM, as it only requires preprocessing the data before the model-fitting step takes place. However, it might be more efficient to combine the two steps of reducing predictor dimensionality and fitting the model into a single step. This can be done by RRR, introduced by Anderson (1951) and Izenmann (1975), and formulated in the Bayesian context by Geweke (1996). The RRR approach that we present in this section is a generalisation of the method developed in Ovaskainen et al. (2017a) specifically for time-series community data. The main difference between the two-step approach of using principal components as predictors and the one-step approach of RRR is that the former identifies the main axes of environmental variation independently of the species data, whereas the latter identifies those axes of environmental variation that best explain the species data. That is, the one-step RRR approach connects environmental variation to community variation when selecting the axes.

For example, when assessing how climatic conditions influence a given community, there are typically 'too many' measurements of the climatic conditions to choose from, such as the minimum, maximum and mean temperatures measured at a daily resolution. In such a case, it might be difficult to identify the effects of each of the variables separately. Further, variable selection may not be the most appropriate approach, as the community is likely to be influenced by the combined climatic conditions rather than by the conditions at a specific day or month. Thus, one may wish to either summarise the climatic variables by the first few principal components, or apply the RRR approach. Another example that motivated the methodological developments of Ovaskainen et al. (2017a) is community-level time-series modelling. In time-series models, the abundances or occurrences of the species in the previous time step are used as predictors of those in the next time step. While it might be possible to assess how each species influences each other species in small species communities, it is highly challenging in large communities consisting of many species. With RRR, the individual predictors (i.e. the abundances of the individual species) are summarised into those main axes of variation in community structure to which the species respond most strongly in their time-series dynamics. In Ovaskainen et al. (2017a), these main axes of variation were called 'community-level drivers'.

To implement RRR into HMSC, we recall that the linear predictor related to the fixed effects is written as $L_{ij}^{F} = \sum_{k=1}^{n_c} x_{ik}\beta_{kj}$. Here we decompose the n_c covariates to two sets: $n_c = n_c^* + n_c^{RRR}$. The covariates $k = 1, \ldots, n_c^*$ are those for which RRR is not applied, and they are thus treated in the regression model as usual. The intercept is also included as usual for this set of covariates. The covariates $k = \left(n_c^* + 1\right), \ldots, \left(n_c^* + n_c^{RRR}\right)$ are those for which RRR is applied. This latter set typically consists of many covariates that are expected to influence the community data, but the effects of which are difficult to model separately.

We denote by n_c^{RRR} the number of covariates after the dimension reduction has been applied, whereas $n_c^{O,RRR}$ denotes the original number of covariates for which RRR is to be applied. We denote the original covariates for which RRR is applied by $\widetilde{x}_{il}$, where $l = 1, \ldots, n_c^{O,RRR}$. The reduced covariates are linear combinations of the original covariates, so that

$$x_{i\left(n_c^* + k\right)} = \sum_{l=1}^{n_c^{O,RRR}} w_{kl}\widetilde{x}_{il} \tag{9.8}$$

for $k = 1, \ldots, n_c^{RRR}$, where the weighting w_{kl} determines the contribution of the original covariate l to the reduced covariate k.

The weights w_{kl} and the regression coefficients $\beta_{\left(n_c^* + k\right)j}$ are estimated. Thus, for each of these, a prior needs to be defined. Concerning the regression coefficients, the usual HMSC prior of $\text{vec}(\mathbf{B}) \sim N(\text{vec}(\mathbf{\Gamma T}^{\text{T}}), [\rho\mathbf{C} + (1-\rho)\mathbf{I}] \otimes \mathbf{V})$ is assumed whether the regression coefficients model the effect of a covariate that does ($k = n_c^* + 1, \ldots, n_c^* + n_c^{RRR}$) or does not ($k = 1, \ldots, n_c^*$) relate to dimension reduction. Concerning the weights w_{kl}, the first prior choice concerns the number n_c^{RRR} of reduced dimensions to be included. While n_c^{RRR} could be estimated in the Bayesian context by setting it as a prior, in Hmsc it is a fixed number that is set by the user. The RRR approach can be expected to be beneficial when the reduced number of covariates is much smaller than the original number of covariates, i.e. if $n_c^{RRR} \ll n_c^{O,RRR}$, as in such a case the reduced model has much fewer parameters to be estimated than the original model. Typically, n_c^{RRR} is chosen to be a small number, such as one or two.

HMSC applies the same multiplicative shrinkage prior to the weights w_{kl} that applies for the species loadings of the latent variable approach related to the random effects (see Section 8.4.2). Thus, the first reduced covariate undergoes the least shrinking and is thus generally the one that will be the most important; the importance of the remaining reduced covariates decreases with their number k. Analogously to Equations 8.12–8.14, the prior is defined as:

$$w_{kl} \mid \phi_{kl}^{RRR}, \delta \sim N\left(0, \left(\phi_{kl}^{RRR}\right)^{-1}\left(\tau_k^{RRR}\right)^{-1}\right), \tau_k^{RRR} = \Pi_{h=1}^{k} \delta_h^{RRR} \quad (9.9)$$

$$\phi_{kl} \mid \upsilon \sim \mathrm{Ga}(\upsilon^{RRR}/2, \upsilon^{RRR}/2) \quad (9.10)$$

$$\delta_1 \sim \mathrm{Ga}\left(a_1^{RRR}, b_1^{RRR}\right), \delta_h \sim \mathrm{Ga}\left(a_2^{RRR}, b_2^{RRR}\right) \text{ for } h \geq 2 \quad (9.11)$$

As default values of the prior parameters, Hmsc assumes $\upsilon^{RRR} = 3$, $a^{RRR} = (1, 50)$ and $b^{RRR} = (1, 1)$.

9.5.1 Simulated Case Study with HMSC

To examine the performance of RRR, we consider a similar case study that we used to illustrate variable selection with the spike and slab prior. We thus consider again a community of $n_s = 10$ species sampled in $n_y = 20$ sampling units. While in the case of the spike and slab prior our aim was to determine which of the ten candidate environmental variables influence the community data, our aim is now to determine which linear combinations of the ten candidate environmental variables influence the community data. To keep the example as simple as possible, we again assume that the data are normally distributed, and we do not include species traits, phylogenies or random effects in the model.

We will generate three datasets that differ from each other in how the covariates influence the species data. We start by simulating variation in the ten environmental covariates over the n_y sampling units.

```
ny = 20
nc = 10
ns = 10
XData = data.frame(matrix(rnorm(ny*nc), ncol = nc, nrow = ny))
```

We then simulate the responses of the $n_s = 10$ species to the environmental covariates. Our first dataset (called Data FULL) assumes that the species respond individually to all the ten environmental covariates.

```
X.FULL = as.matrix(XData)
IX.FULL = cbind(rep(1, ny), X.FULL)
nc.FULL = nc + 1
beta.FULL = matrix(rnorm(ns*nc.FULL), ncol = ns,
  nrow = nc.FULL)
eps = matrix(rnorm(ns*ny), ncol = ns, nrow = ny)
L.FULL = (IX.FULL %*% beta.FULL)
Y.FULL = L.FULL + eps
```

In the second dataset (called Data PC, where PC stands for principal components), we assume that the species do not respond individually to each environmental covariate, but to a univariate gradient of environmental conditions. We further assume that the univariate gradient corresponds to the main axis of variation in the environmental data, as given by the first principal component of the ten environmental covariates.

```
pc = princomp(X.FULL)
X.PC = pc$scores[ , 1]
IX.PC = cbind(rep(1, ny), X.PC)
nc.PC = 2
beta.PC = matrix(rnorm(ns*nc.PC), ncol = ns,
  nrow = nc.PC)
L.PC = (IX.PC %*% beta.PC)
Y.PC = L.PC + eps
```

In the third dataset (called Data RRR, where RRR stands for reduced-rank regression), we also assume that the species do not respond individually to each of the ten environmental covariates, but to a univariate gradient of environmental conditions. In this case, the univariate gradient that is assumed to be relevant for the species does not, however, coincide with the main axis of variation in the environmental data. Instead, it is given by another linear combination of the environmental covariates. Arbitrarily, we assume that the relevant linear combination weights every second environmental covariate by +1 and every second environmental covariate by −1.

```
wRRR = matrix(c(1,-1,1,-1,1,-1,1,-1,1,-1))
X.RRR = X.FULL %*% wRRR
IX.RRR = cbind(rep(1, ny), X.RRR)
```

```
nc.RRR = 2
beta.RRR = matrix(rnorm(ns*nc.RRR), ncol = ns,
  nrow = nc.RRR)
L.RRR = (IX.RRR %*% beta.RRR)
Y.RRR = L.RRR + eps
```

To enable running the HMSC analyses in a compact notation, we combine the three datasets and the three sets of true values of the regression parameters into lists.

```
Y = list(Y.FULL, Y.PC, Y.RRR)
beta = list(beta.FULL, Y.PC, beta.RRR)
```

We next fit three different models to each of these three datasets. The models are set in so that Data FULL is compatible with Model FULL, Data PC is compatible with Model PC, and Data RRR is compatible with Model RRR. Thus, in Model FULL, we include all covariates as predictors as usual. In Model PC, we first compute the principal components of the environmental data, and use the first of these as the sole environmental covariate. In Model RRR, we use the reduced-rank regression approach described above. To do so, we utilise the XRRRData, XRRRFormula, and ncRRR input arguments when constructing the HMSC object with the function Hmsc.

```
models = list()
for(dataset in 1:3){
   tmp = list()
   for (model in 1:3){
      switch(model,{
        m = Hmsc(Y = Y[[dataset]], XData = XData,
          XFormula = ~., distr = "normal")
      },
      {
        pc = princomp(XData)
        XData.PC = data.frame(pc$scores[ ,1])
        m = Hmsc(Y = Y[[dataset]], XData = XData.PC,
          XFormula = ~., distr = "normal")
      },
      {
        m = Hmsc(Y = Y[[dataset]], XData = XData, XFormula = ~1,
          XRRRData = XData, XRRRFormula = ~.-1, ncRRR=1,
            distr = "normal")
      }
      )
```

```
        m = sampleMcmc(m, thin = thin, samples = samples,
          transient = transient, nChains = nChains, verbose
          = verbose)
        tmp[[model]] = m
    }
    models[[dataset]] = tmp
}
```

With the script above, we have included all nine cases (three models times three datasets) in the object models, so that the HMSC object for a particular dataset and a particular model is found from models[[dataset]][[model]]. To see how the model with reduced-rank regression is set up, let us look at the respective model object, for example in the case of Data FULL.

```
head(models[[1]][[3]]$X)

##   (Intercept)
## 1           1
## 2           1
## 3           1
## 4           1
## 5           1
## 6           1

head(models[[1]][[3]]$XRRR)

##           X1          X2         X3          X4         X5
## 1 -0.6264538  0.91897737 -0.1645236  2.40161776 -0.5686687
## 2  0.1836433  0.78213630 -0.2533617 -0.03924000 -0.1351786
## 3 -0.8356286  0.07456498  0.6969634  0.68973936  1.1780870
## 4  1.5952808 -1.98935170  0.5566632  0.02800216 -1.5235668
## 5  0.3295078  0.61982575 -0.6887557 -0.74327321  0.5939462
## 6 -0.8204684 -0.05612874 -0.7074952  0.18879230  0.3329504
##            X6         X7         X8         X9        X10
## 1 -0.62036668 -0.5059575 -1.9143594  0.4251004 -1.2313234
## 2  0.04211587  1.3430388  1.1765833 -0.2386471  0.9838956
## 3 -0.91092165 -0.2145794 -1.6649724  1.0584830  0.2199248
## 4  0.15802877 -0.1795565 -0.4635304  0.8864227 -1.4672500
## 5 -0.65458464 -0.1001907 -1.1159201 -0.6192430  0.5210227
## 6  1.76728727  0.7126663 -0.7508190  2.2061025 -0.1587546

models[[1]][[3]]$ncRRR

## [1] 1
```

We observe that the usual design matrix X contains the intercept only, which choice we made by setting XFormula = ~1. In addition, there is another design matrix XRRR (where RRR stands for reduced-rank regression) that contains the linear effects of the ten environmental covariates. This design matrix does not contain an intercept, which choice is made by setting XRRRFormula = ~.-1. By including the intercept in the 'regular' part of the HMSC model, we have chosen to estimate the regression coefficient related to the intercept as usual. By including the covariates in the XRRR part of the model, we have chosen to estimate their effects through the RRR approach. Thus, while fitting the model, HMSC is seeking for the linear combinations of the covariates that best explain the data. The number of linear combinations to be included in the model is controlled by ncRRR, which we have set to one, to correspond to the data generating model where we assumed that a single linear combination of the explanatory variables influences the community data. With real data applications, the correct number of the dimensions is not known, and thus one may wish, for example, to compare models in which ncRRR equals one or two.

Figure 9.6 evaluates the MCMC convergence of the β parameters in terms of the potential scale reduction factors for the nine fitted models. We observe that MCMC convergence is better for the cases where RRR is not applied (panels A, D and G and panels B, E and H) than for the cases where RRR is applied (panels C, F and I). This can be expected, as RRR requires additional MCMC steps for sampling the weights w describing the linear combinations of explanatory variables that best explain the data.

We will next examine how variable selection influences the explanatory and predictive powers of the models. We start by computing explanatory power in terms of R^2.

```
meR2 = matrix(NA, nrow = 3, ncol = 3)
for(dataset in 1:3){
   for (model in 1:3){
      m = models[[dataset]][[model]]
      predY = computePredictedValues(m)
      MF = evaluateModelFit(m, predY = predY)
      meR2[dataset, model] = mean(MFs[[dataset]]
        [[model]]$R2)
   }
}
##                  Model FULL   Model PC   Model RRR
## Data FULL          0.929       0.108       0.431
## Data PC            0.667       0.374       0.392
## Data RRR           0.797       0.070       0.607
```

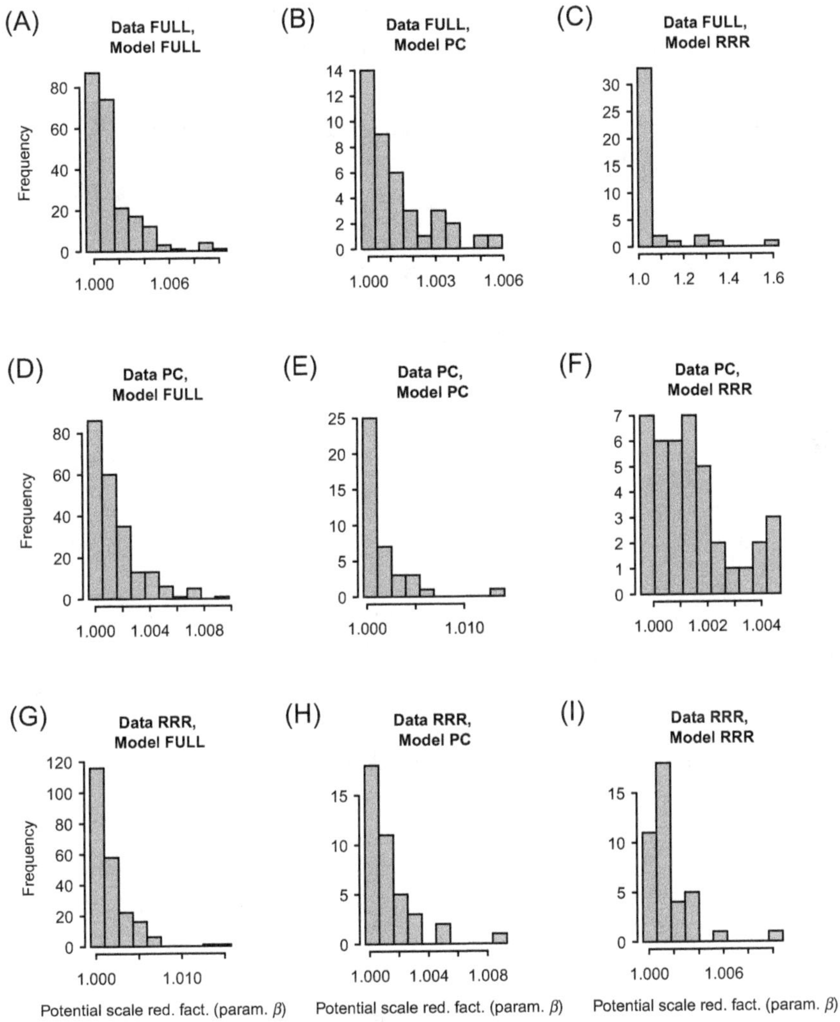

Figure 9.6 MCMC convergence diagnostics for the β parameters measured by the potential scale reduction factor. The rows of panels correspond to the three datasets (panels A, B and C correspond to Data FULL, panels D, E and F to Data PC, and panels G, H and I to Data RRR), and the columns of panels to the three models fitted to the data (panels A, D and G correspond to Model FULL, panels B, E and H to Model PC and panels C, F and I to Model RRR).

We observe that for all three datasets, the mean explanatory power (averaged over the species) is the highest for the full model that individually estimates the influence of each covariate on each species. This is to be expected, as explanatory power generally increases with model complexity – but this can also arise from overfitting. To examine the level of overfitting, we next evaluate predictive power based on two-fold cross-validation.

```
partition = createPartition(models[[1]][[1]], nfolds = 2)
meR2CV = matrix(NA, nrow = 3, ncol = 3)
for(dataset in 1:3){
    for (model in 1:3){
        m = models[[dataset]][[model]]
        predY = computePredictedValues(m, partition = partition)
        MFCV = evaluateModelFit(m, predY = predY)
        meR2CV[dataset, model] = mean(MFs[[dataset]][[model]]
          $R2)
    }
}
##                    Model FULL Model PC Model RRR
## Data FULL               0.491   -0.113     0.093
## Data PC                 0.155    0.259     0.228
## Data RRR                0.300   -0.011     0.416
```

As is generally the case, the predictive powers are lower than the explanatory powers. Reassuringly, the Model FULL has the best predictive power for the Data FULL, the Model PC has the best predictive power for the Data PC, and the Model RRR has the best predictive power for the Data RRR.

Let us then compute the WAIC for these same models.

```
WAIC = matrix(NA, nrow = 3, ncol = 3)
for(dataset in 1:3){
    for (model in 1:3){
      WAIC[dataset, model] = computeWAIC(models[[dataset]]
        [[model]])
    }
}
##                    Model FULL  Model PC  Model RRR
## Data FULL              18.539    26.436     24.828
## Data PC                16.679    16.113     16.221
## Data RRR               16.781    22.058     16.076
```

From this, we can see Model FULL has the lowest WAIC for the Data FULL, the Model PC has the lowest WAIC for the Data PC, and the Model RRR has the lowest WAIC for the Data RRR. Thus, model selection based on WAIC is consistent with the model selection based on cross-validation.

Figure 9.7 visualises the estimated β parameters for the model that involves RRR (Model RRR applied to Data RRR). As we included only one axis of variation by setting ncRRR = 1, the responses of the

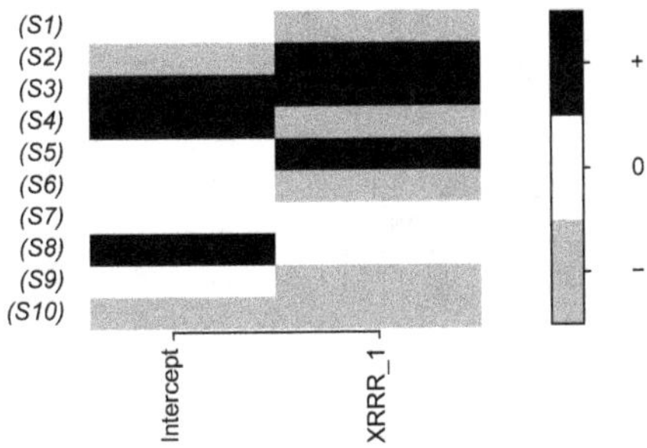

Figure 9.7 Heatmap of estimated species niches for Model RRR applied to Data RRR. Black and grey colours show parameters that are estimated to be positive and negative, respectively, with at least 0.95 posterior probability.

species to the reduced covariates are represented in Figure 9.7 by a single column entitled 'XRRR_1'. Figure 9.7 shows that each species has its individual response to the reduced covariates, some responding to it positively and some others negatively.

We may also examine the parameters w_{kl} with which the original covariates are weighted to construct the reduced covariates (Equation 9.8). The script below prints the posterior mean estimates as well as the posterior probability by which the values are estimated to be positive.

```
postwRRR = getPostEstimate(m, parName = "wRRR")
postwRRR$mean

##          [,1]       [,2]       [,3]       [,4]       [,5]
##[1,] -1.147676 0.8014653 -0.9198365 0.9685467 -1.395205
##          [,6]        [,7]       [,8]       [,9]     [,10]
## [1,] 0.8678293 -0.6780748 1.026627 -1.009054 1.228237

postwRRR$support

##   [,1] [,2] [,3]    [,4] [,5] [,6]    [,7] [,8] [,9] [,10]
## [1,] 0    1    0 0.99975    0    1 0.00025    1    0    1
```

We observe that the posterior mean values follow roughly a pattern where -1 and 1 alternate. When generating the data, we assumed the opposite order, i.e. that 1 and -1 alternate. This reflects the fact that the parameters w_{kl} and $\beta_{(n_c^*+k)j}$ are not identifiable in the HMSC model. This is because their influence on the linear predictor has a product structure, and thus multiplying both by -1 leads to an identical model.

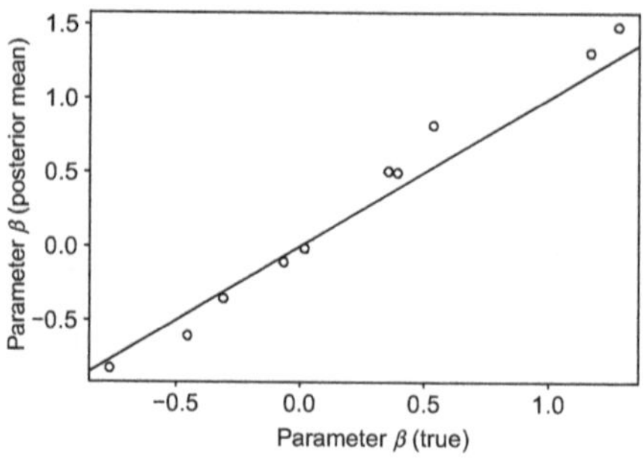

Figure 9.8 Match between assumed (x-axis) and estimated (y-axis) regression parameters related to reduced-rank regression. The figure is based on Model RRR fitted to Data RRR. The line shows the identity $y = x$.

This is fully analogous to the species and site loadings in the random effect part of the HMSC model being not identifiable (see Section 8.4).

As the estimated w_{kl} parameters need to be multiplied by -1 to make them match with those used to generate the data, the estimated β parameters also need to be multiplied by -1 to make them match with those used to generate the data. The script below generates Figure 9.8, showing that after this transformation there is a good match between the assumed and estimated β parameters.

```
plot(beta.RRR[2,], -postBeta$mean[2,])
abline(0, 1)
```

Part III

Applications and Perspectives

10 · *Linking HMSC Back to Community Assembly Processes*

Although HMSC was developed specifically for linking community data to the underlying assembly processes (Ovaskainen et al. 2017b), it does not include an explicit description of those processes. As discussed in Chapters 4–9, HMSC is a multivariate hierarchical GLMM implemented in Bayesian inference or, very simply put, a correlative statistical model. The fact that correlation does not imply causation is a general and long-known issue in statistical ecology (Cale et al. 1989), and this is not a problem exclusive to HMSC. Therefore, as when applying any other statistical method of correlative nature, the ecologist should interpret the results cautiously and in the light of ecological knowledge. While HMSC analyses thus allow for correlative and not causal inference, the framework has been developed to facilitate the formulation of data-driven hypotheses regarding the processes that structure communities. This is because different parts of the model are conceptually linked to theory on assembly processes, as described in Chapters 4–7.

In this chapter, we examine the links between HMSC outputs and the underlying community ecological processes, with the help of simulations. More precisely, we will apply HMSC to data generated from a mechanistic model with known underlying assembly processes, and then assess how those processes are captured from the patterns in the data. The reader may wonder what is then different from the many case studies with simulated data from Chapters 5–9. Unlike in those previous chapters, we will now simulate the data with an agent-based model that explicitly involves some of the community assembly processes, instead of simulating the data with HMSC. This means that while the simulated data in Chapters 5–9 were structurally fully in line with the assumptions of HMSC, the data that we now generate will violate some of the underlying assumptions of GLMMs. We note that data on real communities are likely to violate those assumptions to an even greater extent than the data that we will simulate here. Thus, while the main motivation of this chapter is to assess how community assembly processes translate into HMSC outputs, another motivation is to examine the robustness of HSMC to violations against structural model assumptions. After simulating data with the spatial agent-based model (Section 10.1), we simulate two 'virtual ecologists' who sample data from the simulations, one applying a spatial study design (Section 10.2) and the other a temporal study design (Section 10.3). The virtual ecologist approach is a widely applied method in ecology for testing the capability of sampling and statistical methods in finding the signal from the data (Zurell et al. 2010; Ovaskainen et al. 2019). In Section 10.4 we finish this chapter by summarising what the virtual ecologists learned by applying HMSC to their data, particularly in light of the assembly processes that were used to simulate the data.

10.1 Simulating an Agent-Based Model of a Competitive Metacommunity

We generate the simulated data with an agent-based resource-consumer model of a competitive metacommunity. In this section we first describe the environmental context of our simulation and the properties of the simulated species, then the underlying metacommunity dynamics, then finally how the virtual ecologists sample data from the simulated communities.

10.1.1 Environmental Variation

The environmental context of the study is depicted in Figure 10.1. The landscape shown in Figure 10.1A consists of two kinds of habitats – open and forest. The study area also involves a climatic gradient, with northern areas having a colder climate than southern areas (Figure 10.1B). Note that the climatic gradient occurs at the spatial scale of 20 km, and thus it is not generated by large-scale variation such as latitude, but by small-scale variation such as altitude. We may view the study area as part of an island, where a mountain slope starts at the sea level where the y-coordinate is 0 and continues along the y-coordinate.

10.1.2 Species Traits

The environment depicted in Figure 10.1 is inhabited by a community of $n_s = 100$ species. While the species are virtual and our model represents a simplistic caricature of any real community, the reader may visualise the

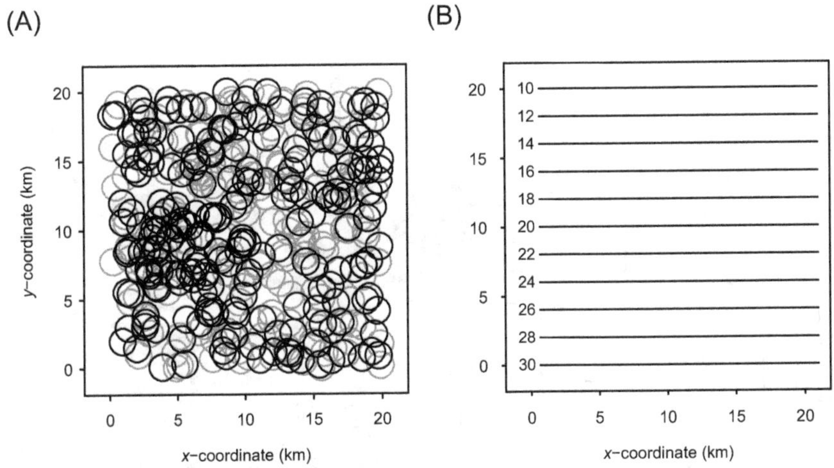

Figure 10.1 Environmental context in which metacommunity dynamics are simulated. In Panel A, the grey circles represent open habitat resource patches, and the black circles represent forest habitat resource patches. Panel B shows a contour plot of the mean annual temperature over the study area.

species as dung beetles, which compete for resources. We assume that dung resources used by the species are produced by some herbivore species dwelling in open habitats, and by other herbivore species dwelling in forest habitats. We assume that all of the 100 species have undergone an adaptive radiation from a single ancestral species. To do so, we recycle the phylogenetic tree from Figure 6.3 where the 100 species are represented as the tips of that tree.

We next simulate a set of traits for each species. These traits will determine metacommunity dynamics through their influence on demographic dynamics, i.e. the birth and death rates of the individuals. First, we characterise each species j with its specialisation level S_j to forest habitats, relative to open habitats. With $S_j > 0$, the species will be better adapted to forest habitat resources than to open habitat resources, whereas with $S_j < 0$, the species will be better adapted to open habitat resources than to forest habitat resources. If $S_j = 0$, the species will be a generalist in the sense that it is equally well-adapted to both forest and open habitat resources. We assume that the specialisation level has evolved according to the diffusion model of trait evolution (Beaulieu et al. 2012), and thus shows a high phylogenetic signal (Figure 10.2).

We next assume a trade-off between fecundity and tolerance to cold conditions, so that species with a higher thermal optimum (denoted by

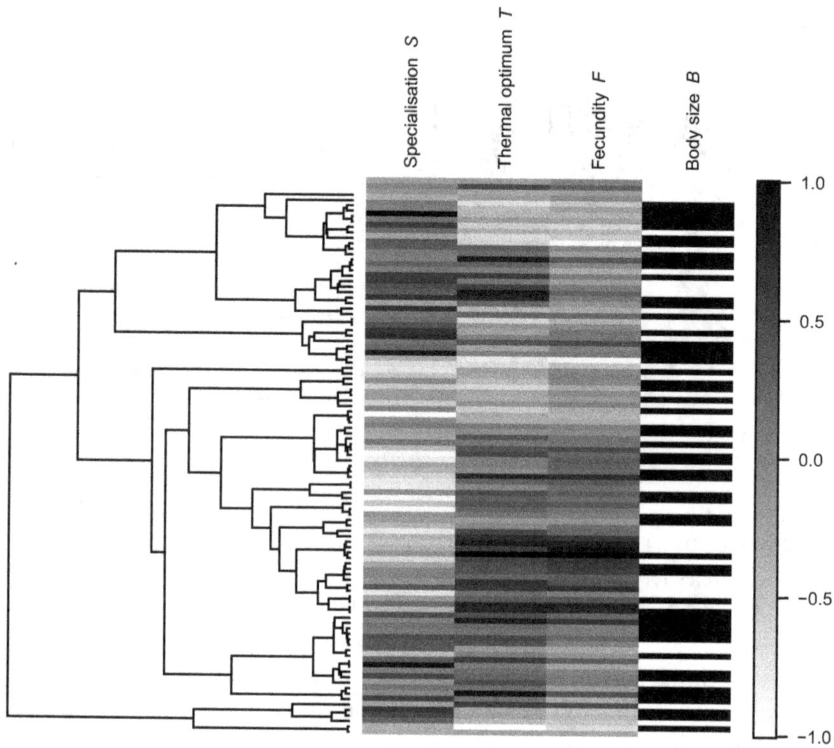

Figure 10.2 Phylogenetic relationships and species traits assumed in the community of 100 species for which metacommunity dynamics are simulated. The shades of grey show the trait values, which are scaled to range from −1 to +1 for each trait.

the trait T_j) have a higher fecundity (denoted by the trait F_j). As the specialisation level S, also both thermal optimum T and fecundity F have evolved according to a diffusion model of evolution. The high phylogenetic signal in all of these traits is visible in Figure 10.2, in the sense that the values of the traits are not randomly distributed across the phylogeny (it is possible to see a pattern, as the shades of grey are grouped according to different clades).

We further characterise the species by their body size B_j, which for simplicity we classify into two classes of small and large. We assume that body size influences competition among the species, so that large-bodied species compete for dung produced by large herbivores, whereas small-bodied species compete for dung produced by small herbivores. We assume that body sizes are not phylogenetically constrained, and we thus randomly assign each species to large (+1) or small (−1) body size.

Consequently, the distribution of body size lacks any correlation with phylogeny (Figure 10.2). We note that, from the evolutionary point of view, this may be the case if there is strong selection on body size, if the species-specific trait optima for body size are unrelated to the locations of the species in the phylogeny and if the genetic architecture underlying body size allows for its rapid evolution.

While our metacommunity model will be a mechanistic agent-based model, the model that we have used to generate the phylogeny and species traits is just a phenomenological statistical model. It is actually the same model that we used to simulate the traits in Section 6.4. For the traits S, T and F we have assumed the strength of the phylogenetic signal to be $\rho = 0.8$, whereas for the trait B it is assumed to be zero. Further, we have assumed a correlation of 0.9 between the traits T and F, and no correlation between the other pairs of traits. To put all traits into the same scale, we have scaled them so that their smallest value is -1 and largest value is $+1$. In an even more mechanistic approach, we could have simulated evolutionary processes over time, including processes such as speciation and local adaptation, rather than simply simulating the final outcome of these processes. But for our purposes it is more important to take a mechanistic approach to simulating the ecological rather than the evolutionary processes. Thus, we will next place a community with the trait values illustrated in Figure 10.2 into the environmental context of Figure 10.1, and focus on their ecological dynamics.

10.1.3 The Metacommunity Model

The agent-based model that we will next describe is technically a spatio-temporal point-process in continuous space and continuous time (Ovaskainen et al. 2014). When generating the landscape of Figure 10.1A, we have assumed that the density of both the forest and open habitat patches is 0.5. We have further assumed that the locations of the patches are completely random, so that we have generated them as a realisation of a spatial Poisson process. Thus, in the landscape of spatial dimension 20×20, the expected total number of patches is 400, and the actual number is Poisson distributed with this expectation. All resource patches are assumed to be circular with a radius of one. We assume that each patch produces two types of resources: those consumed by small-bodied species and those consumed by large-bodied species. We assume that each patch produces resources with a rate of one, which is divided

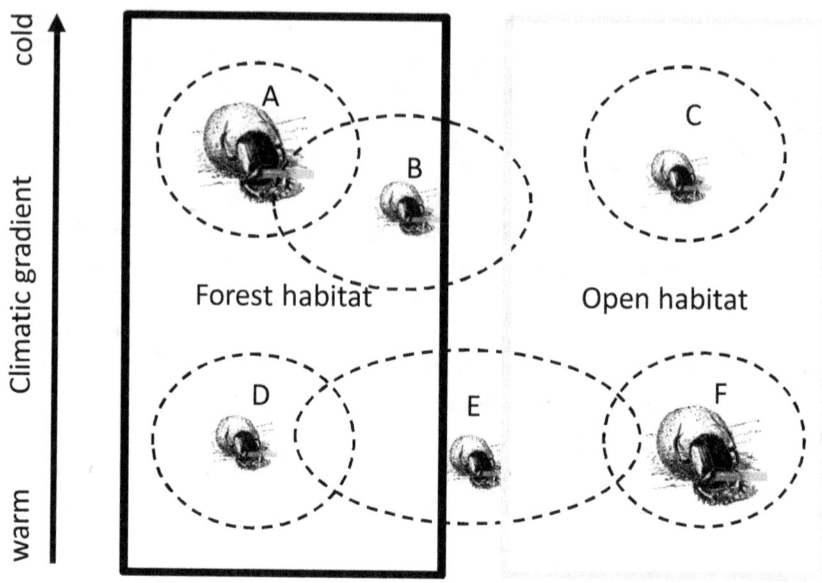

Figure 10.3 Conceptual illustration of variation in species niches in the competitive metacommunity model. Species A and F represent large-bodied species, whereas species B, C, D and E represent small-bodied species. Species A, B and C are adapted to cold climates and species D, E and F to warm climates. Species A and D are specialised to forest habitat resources, species C and F to open habitat resources, species B is a partial generalist, preferring forest habitat resources, and species E is a full generalist. Species D and E compete partially for the same resources, whereas species A and B do not, as the latter belong to different size classes and thus to a different ecological guild in terms of resource use. Note that this graph illustrates species niches only for six species, but the entire metacommunity consists of 100 species.

equally between these two types of resources. As the resources are different in forest and open habitats, in total there are four types of resources. All resources are assumed to decay and thus disappear at a per capita rate of 0.1 unless they are consumed before that.

Due to the variation in their traits, the 100 species comprising the metacommunity fill in the niche space in terms of habitat availability as well as the availability of the climatic niches (Figure 10.3). The extent to which two species will compete with each other will depend on the overlap of their niche. We next describe how this happens through individual-level processes.

The individuals belonging to each of the 100 species need resources for their survival and reproduction. Each individual can be in two states:

resource-deprived or *resource-satiated*. When at the satiated state, the individual gradually loses its energy and thus changes back to the deprived state at a rate of 0.25. A deprived individual may die, which takes place at a per individual mortality rate of one. Alternatively, the individual can change to the satiated state by consuming a resource unit from its proximity. This takes place at a rate that depends on the ability of the individual to utilise the available resource type. More specifically, the rate at which individuals of species j consume forest resources that match with their size (dung produced by small or large herbivores) is

$$r_j^F = \frac{\exp(10\ S_j)}{1 + \exp(10\ S_j)} \tag{10.1}$$

assuming that the resource particle is in the proximity, i.e. located at a maximum distance of 0.25 from the individual. If the resource particle is farther away, it is assumed that the individual does not have access to it. The rate at which individuals of species j consume open resource particles from their proximity is assumed to be $r_j^O = 1 - r_j^F$, so that the rate of use of forest and open habitat resources sums to one for each individual. This means that all species are equally efficient in using resources, but they vary in the extent at which they are specialised to each type of resource. We have used the logistic function of Equation 10.1 to map the specialisation parameter S_j to the scale from 0 to 1, so that a large value of S_j corresponds to r_j^F being close to one. This means that a species with a high S_j will be efficient at consuming forest habitat resources but inefficient at consuming open habitat resources, and vice versa.

The rate at which the satiated individuals produce new offspring depends on their fecundity trait F_j and the match between their thermal optimum T_j and the climatic conditions at their location $T(x, y)$. More precisely, the rate at which satiated individuals of species j produce propagules is:

$$\exp(1.5\ F_j)\ \exp\left(-2\ d[T(x,y), T_j]^2\right) \tag{10.2}$$

In Equation 10.2, the distance $d[T(x, y), T_j]$ between the temperature at the individual's location and its thermal optimum is measured as:

$$d[T(x,y), T_j] = \left| T_j - \frac{T(x,y) - 20}{10} \right| \tag{10.3}$$

In Equation 10.3, we have scaled the temperature $T(x, y)$ to range from -1 to $+1$, so that it matches with the range of values assumed for the thermal optimum trait T_j. With this definition, the smallest possible distance $d(T(x, y), T_j)$ is zero and the largest possible distance is two. The distance zero is obtained if the individual is located in a climate that perfectly matches its thermal optimum, and the distance is two if, for example, the individual is located in the coldest corner of the climatic gradient but its thermal optimum corresponds to the warmest corner.

In Equation 10.2, we have then assumed a Gaussian fitness function, where the fecundity of the species decreases with increasing distance from its thermal optimum. For example, if species j has the maximal thermal optimum of $T_j = 1$, then it will reach its maximal fecundity exp (1.5 F_j) at the warmest corner of the climatic gradient where $T(x, y) = 30$. The fecundity of an individual of this species will drop to the proportion $\exp(-2) = 0.13$ of its maximum if the prevailing climatic conditions are intermediate with $T(x, y) = 20$, whereas it will drop to the proportion $\exp(-8) = 0.0003$ of its maximum in the coldest part of the climatic gradient where $T(x, y) = 10$.

The newly born propagules are assumed to disperse and to be deposited anywhere within a distance of one from their mother with an equal likelihood. The deposited propagules are assumed to emerge without time delay as new individuals at the deprived state. Moreover, to avoid all species from eventually going extinct, we assume that for each species, immigration of deprived individuals outside the simulation area occurs at a rate of 0.01 per unit area.

10.1.4 Simulated Metacommunity Dynamics: The Underlying Reality

We initiated the metacommunity by assuming that there are no individuals, so that the species arrive to the simulation domain by random immigration, as would be the case if they arrived to an empty island. We simulated the dynamics of the metacommunity for 100 time units, so that the species have time to colonise the parts of the island that are suitable for them. During the course of the simulation, the species first increase rapidly in their population size due to the availability of unused resources. After the initial phase, they show more stable dynamics due to density dependence acting through resource competition (Figure 10.4).

To be familiarised with the simulated patterns, Figure 10.5 shows the locations of the individuals of two species at the end of the simulation at $T = 100$. Out of these, the species shown by black dots is similar to species A in Figure 10.3. Namely, this species is a forest habitat resource

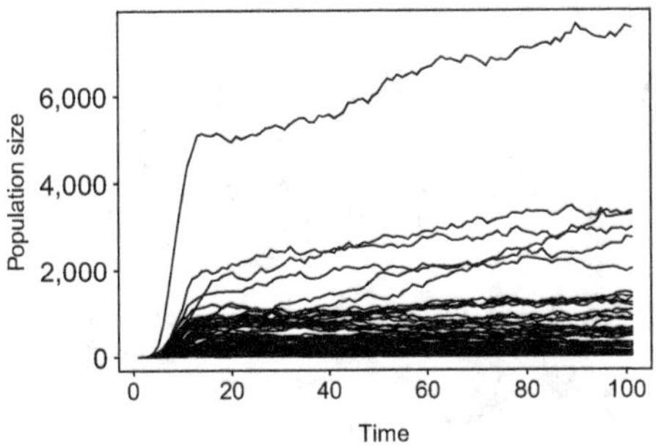

Figure 10.4 The time-evolution of population sizes in the simulation of the metacommunity model. Each line represents one of the 100 species constituting the simulated metacommunity.

specialist with a low thermal optimum and intermediate fecundity. In contrast, the species shown by grey dots is similar to species E in Figure 10.3. Namely, this species is essentially a generalist with a high thermal optimum and high fecundity. As seen from Figure 10.5, the distributions of these two species (i.e. their realised niches) partially correspond to their trait values (i.e. their fundamental niches), illustrating the relevance of environmental filtering, and hence the species sorting paradigm in our simulated metacommunity (Holyoak et al. 2005).

While Figure 10.5 exemplifies the distributions of two species, Figure 10.6 illustrates the final state of the simulation for all species. As expected, there is a positive relationship between the fecundity parameter F_j and the population size n_j of the species (Figure 10.6A). The climatic conditions averaged over the locations of the individuals correspond well to the thermal optima of the species (Figure 10.6B), reflecting the importance of climate for species sorting in the simulated metacommunity. There is a lot of variation around the trends shown in Figure 10.6, reflecting the fact that population dynamics are also influenced by other factors, such as competitive interactions among the species.

10.1.5 Sampling Data: Observations Made by Virtual Ecologists

With simulated data, we know the full state of the system, i.e. the exact locations of all individuals at all times. With real data, this is obviously not the case. Rather, species data are typically available from only a subset of

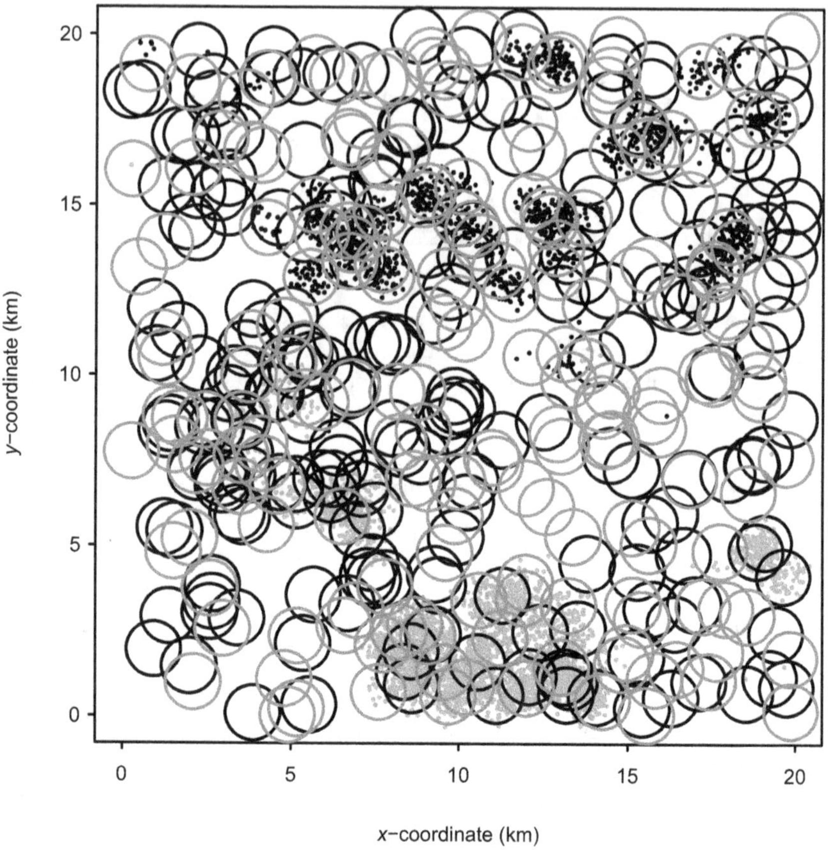

Figure 10.5 The locations of individuals of two species (out of the simulated 100 species) at the end of the simulation. The black and grey dots depict the individuals of the two species. The trait values of species shown by black dots are $S_1 = 0.65$, $T_1 = -0.34$ and $F_1 = 0.07$, whereas the trait values of the species shown by grey dots are $S_2 = -0.14$, $T_2 = 1.00$ and $F_2 = 1.00$.

all possible spatial locations and time points, and instead of counts of all individuals, the data may be of e.g. presence–absence nature. To convert the underlying dynamics of the metacommunity into data, we assumed that the system has captured the interest of two virtual ecologists (Zurell et al. 2010). The first, called V. E. Space, has acquired snapshot data from multiple locations (Figure 10.7A). The second, called V. E. Time, has acquired time-series data from a single location (Figure 10.7B).

The virtual ecologist V. E. Space made her survey at time $T = 100$ at the end of the simulation. She placed a 10×10 regular grid consisting of

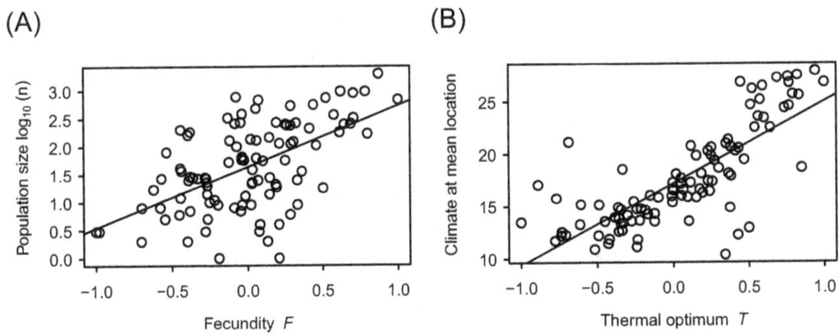

Figure 10.6 Relationships between species traits and realised species distributions in the simulated metacommunity, as recorded from the final state of the simulation. Panel A shows the total population size of each species as a function of its fecundity parameter. Panel B shows the mean climatic conditions over the locations of the individuals as a function of the thermal optimum of the species.

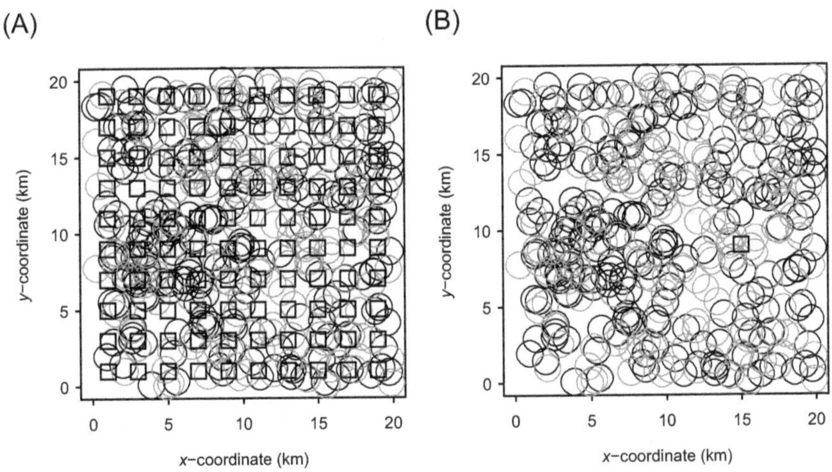

Figure 10.7 Sampling plots used by virtual ecologists V. E. Space (panel A) and V. E. Time (panel B).

$n_y = 100$ sampling units over the simulation area, as illustrated in Figure 10.7A. Each sampling unit is a rectangle of side length 1 km, and the distance between centres of neighbouring sampling units is 2 km. As the baseline case, we assume that V. E. Space conducted a thorough survey in the sampling units and counted all individuals of all species. Thus, the community data in the **Y** matrix consist of counts of the detected species in 100 sampling units. V. E. Space was also aware of variation in habitat quality, i.e. that the landscape varied in the amount of forest and open area habitats. To account for this in her analyses, the habitat quality was scored for each of the sampling units. As illustrated by the figures, the patches may be overlapping. Habitat quality in any location is measured as the number of overlapping patches, as that determines the rate of resource production at that point. V. E. Space measured the habitat quality of a particular sampling unit for each of the two habitat types as the mean habitat quality over the plot.

The sampling by the virtual ecologist V. E. Time was already initiated by his grandfather, who established the permanent sampling plot shown in Figure 10.7B. The survey was continued by V. E. Time's father, and eventually by himself, completing the remarkable 100-year-long time-series dataset. In these time-series data, the matrix **Y** again consists of counts of the detected species in the 100 sampling units, but now the sampling units correspond to years, not spatial locations. As we assumed no temporal variation in environmental conditions or habitat quality, there are no environmental data in this case, just the species data.

10.2 Statistical Analyses of the Spatial Data Collected by a Virtual Ecologist

In this section, we present the analyses conducted by virtual ecologist V. E. Space.

10.2.1 Exploration of the Raw Data

Among the 100 species that we used to simulate the metacommunity, 96 were recorded at least once in the spatial data. Thus, the virtual ecologist V. E. Space might not know that the remaining four species exist. Many of the observed species are very rare. Since species that are observed just a few times contain little information about community assembly processes (e.g. from the species associations point of view) and somewhat complicate the analyses (e.g. increase computational times and are especially problematic for MCMC convergence), V. E. Space decided to include only those fifty-six species that she found in at least five sampling plots.

Let us have a look at the raw data matrices that V. E. Space compiled.

```
head(Y[ ,1:8] )

##            sp_003 sp_008 sp_011 sp_012 sp_013 sp_014 sp_020 sp_021
## sample_001 0           0      0      0      0      0      0      0
## sample_002 0           0      0      0      0      0      0      0
## sample_003 0           0      0      0      0      0      0     52
## sample_004 0           0      1      0      0      1      0      0
## sample_005 0           0     20      0      0      0      1      0
## sample_006 0           0      0      0      0      0      0      0

head(xy)

##            x.coordinate y.coordinate
## sample_001            1            1
## sample_002            1            3
## sample_003            1            5
## sample_004            1            7
## sample_005            1            9
## sample_006            1           11

head(XData)

##               forest      open clim
## sample_001 0.2809917 0.8264463   29
## sample_002 0.0000000 1.3966942   27
## sample_003 1.4545455 1.9752066   25
## sample_004 0.6859504 0.1404959   23
## sample_005 2.0909091 1.9917355   21
## sample_006 2.0000000 2.5041322   19

head(TrData)

##                  S           T           F     B
## sp_003  0.28543765 -0.42731179 -0.53607197 small
## sp_008 -0.07145962  0.94282812  0.41098308 large
## sp_011  0.52355834  0.25049800  0.05692630 large
## sp_012  0.02922536  0.15270451 -0.02451283 large
## sp_013  0.92645749  0.01856742 -0.32948432 small
## sp_014  0.32027511  0.56159859  0.28408241 small
```

As shown above, V. E. Space recorded data on species counts (**Y**), environmental covariates (**X**) and the spatial coordinates of the sampling units (**S**). Further, she compiled from the literature data on species traits (**T**) and their phylogenetic relationships (shown in Figure 10.2).

The species data are quite heavily dominated by zeros, even when the rarest species are excluded. Among the 5,600 numbers (counts of 56 species in 100 sampling units) included in the matrix **Y**, only 662 are non-zero. The total sum over all elements of **Y**, i.e. the total number of individuals observed in the data, is 11,925. Thus, when a typical species was recorded in a given sampling plot, it was relatively abundant, as the average number of individuals conditional on presence is 11,925/662 = 18 individuals.

The distributions of species richness, species prevalence, local abundance and species abundance can be computed as row sums and column means of the **Y** matrix.

```
S = rowSums(Y > 0)
P = colMeans(Y > 0)
LA = rowSums(Y)
SA = colMeans(Y)/P
```

The patterns shown in Figure 10.8 are quite typical for real community datasets. The species richness varies from one to sixteen over the sampling plots, and thus the majority of the fifty-six species were not found from any specific sample. Most species are rare in the sense that they have low prevalence, with the exception of one very common species that occurs in 44 per cent of the sampling plots. The number of individuals recorded from different sampling plots varies greatly, from 1 to 496. The counts of individual species vary from 1 to 58 individuals, when averaged over sampling plots where the species is found.

10.2.2 HMSC Analyses with an Ideal Model

After conducting the above explorations of the raw data, V. E. Space is ready to move to HMSC analyses. We first assume that she is so knowledgeable about the study system that she is able to set up the HMSC model such that it that reflects the underlying community assembly processes in an ideal way.

```
studyDesign = data.frame(sample = row.names(XData))
rL = HmscRandomLevel(sData = xy)
XFormula = ~ forest + open + poly(clim, degree = 2, raw = TRUE)
TrFormula = ~ S + T + F
m = Hmsc(Y = Y, XData = XData, XFormula = XFormula, TrData = TrData,
   TrFormula = TrFormula, phyloTree = phy, studyDesign =
     studyDesign, ranLevels = list(sample = rL),
     distr = "lognormal poisson")
```

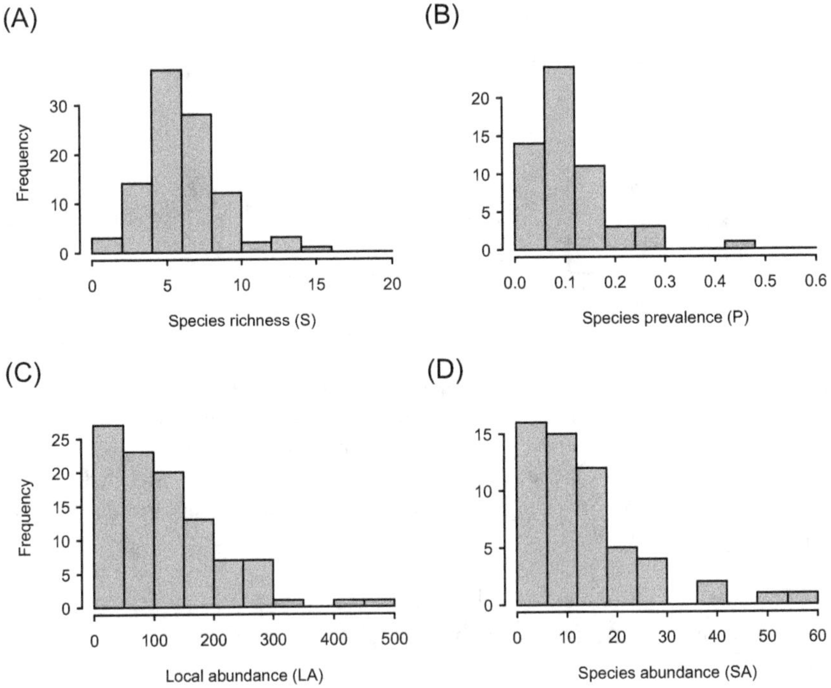

Figure 10.8 Basic summaries of the data collected by V. E. Space. For species richness (panel A) the y-axis (Frequency) corresponds to the number of sampling units, and the variable in the x-axis is the number of species found from each sampling unit. For species prevalence (panel B), the y-axis (Frequency) corresponds to the number of species, and the variable in the x-axis is the fraction of sampling units in which the species is found. For local abundance (panel C), the y-axis (Frequency) corresponds to the number of sampling units, and the variable in the x-axis is the number of individuals found from each sampling unit. For species abundance (panel D), the y-axis (Frequency) corresponds to the number of species, and the variable in the x-axis is the mean number of individuals recorded for that species, averaged over the sampling locations from where it is found.

When defining the HMSC model, V. E. Space has not only included all of the relevant environmental variables, but also assumed a second-order response to climatic conditions in order to allow species abundances to peak at intermediate climatic conditions. She has further included those traits that influence how species sort according to the environmental conditions – but not body size, as this influences ecological interactions among the species rather than their direct responses to

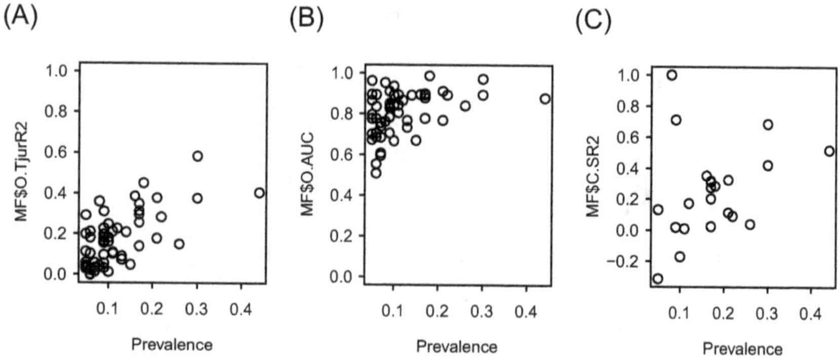

Figure 10.9 An evaluation of the explanatory power of the model. Panels A and B show the ability of the model to discriminate presences (count > 0) from absences (count = 0) in terms of Tjur R^2 (A) and AUC (B). Panel C shows the ability of the model to discriminate high abundances from low abundances in locations where the species is present, measured with pseudo-R^2 conditional on presence (see Section 9.2 on measures of model fit).

environmental variation. She has accounted for phylogenetic relationships among the species by including a phylogenetic tree, and for the spatial nature of the study design by including a spatially structured random effect. Finally, to account for the fact that the count data show a high amount of variation, she has assumed the lognormal Poisson model.

After setting up the model, V. E. Space performs model fitting and evaluates MCMC convergence as usual. We trust that she has ensured that MCMC convergence is sufficiently good and thus we do not further explore it here. As the next step, V. E. Space evaluates the explanatory power of the model.

```
preds = computePredictedValues(m, expected = FALSE)
MF = evaluateModelFit(m, predY = preds)
```

The model is quite successful in explaining both species occurrences (Figure 10.9AB) and their abundances conditional on presence (Figure 10.9C). As expected, the explanatory power generally increases with species prevalence.

To determine which variables contribute to the explained variation, V. E. Space conducts a variance partitioning among the fixed and random effects. To do this, she first looks at the columns of the matrix **X** that Hmsc has constructed based on the objects XData and XFormula.

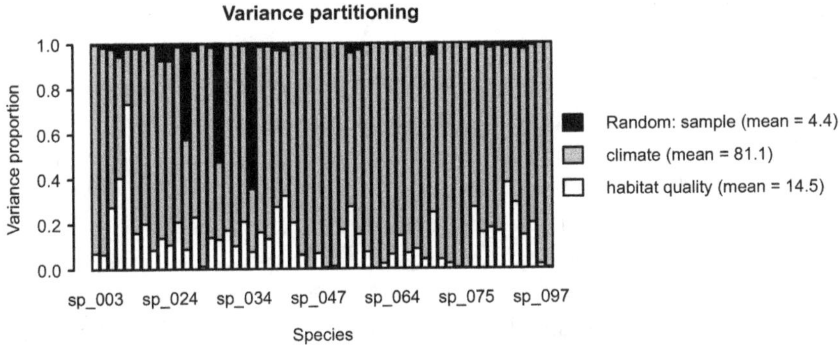

Figure 10.10 Variance partitioning among the fixed and random effects included in the model.

```
colnames(m$X)

## [1] "(Intercept)"
## [2] "forest"
## [3] "open"
## [4] "poly(clim, degree = 2, raw = TRUE)1"
## [5] "poly(clim, degree = 2, raw = TRUE)2"
```

She decides to group the effects of forest and open habitat qualities as 'habitat quality' effects, and the first- and second-order effects of climate as 'climate' effects. She thus assigns columns 1–3 of the matrix **X** to group one (habitat quality), and columns 4 and 5 to group two (climate). We note that while she assigned the intercept to group one, she also could have assigned it to group two.

The variance partitioning (Figure 10.10) shows that variation in climatic conditions is the most important factor that causes variation in species abundances. Habitat quality is also important for some species, but not so important for others. The spatial random effect plays only a very minor role.

The results of the variance partitioning are largely in line with the underlying reality, i.e. the mechanisms that were assumed in the agent-based metacommunity model. However, based on the assumptions of the agent-based model, species abundances can be expected to scale roughly linearly with resource availability, and thus resource availability should make a major difference. Only a relatively small proportion of variance is explained by habitat quality, because the landscape is relatively homogeneous in terms of habitat, compared to the climatic gradient. As quantified by Equation 5.11 – on which the variance partitioning is based – the importance of a particular variable depends both on how strongly the species respond to it, as well as how much the sampling units vary in terms of this variable. The small role of the random effect is explained by the fact that V. E. Space was sufficiently knowledgeable to be able to include all relevant variables in the model, so that the random effects do not capture missing environmental covariates. Thus, in the present model they are expected to capture mainly non-random co-occurrence patterns resulting from ecological interactions among the species. Based on the small proportion of variance attributed to the random effects, density-dependency among species due to resource competition did not leave a strong signal in the data.

Let us then look at the proportions of variance in species niches explained by the traits included in the model, denoted by $R^2_{T\beta}$ in Section 6.3, and by VP$R2T$Beta in Hmsc.

```
VP$R2T$Beta

## (Intercept)        forest        open poly(clim)1 poly(clim)2
## 0.004735835 0.387303512 0.516290321  0.017227887  0.040010670
```

A major part of the variation among species niches related to their responses to habitat quality is explained by the traits. This is to be expected, since we included forest specialisation as species traits. However, traits explain only a minor part of the variation among species niches related to their responses to climate. This is more surprising, since we included their thermal optimum as a trait. We will return to this point later.

The variance partition also includes the proportion of variance that the included traits explain out of species abundances, denoted by R^2_{TY} in Section 6.3, and by VP$R2T$Y in Hmsc.

```
VP$R2T$Y
## [1] 0.114101
```

The traits thus explain 11 percent of the variation in species abundances, measured at the scale of the linear predictor (Section 6.3). This may seem like a surprisingly low proportion considering the fact that basically all the relevant traits were included in the model. Hence, this highlights the difficulty in linking variation in species occurrences with their traits.

V. E. Space then moves to the parameter estimates, first visualising species niches using the plotBeta function (Figure 10.11). This shows that many species respond positively to forest habitat quality, while many

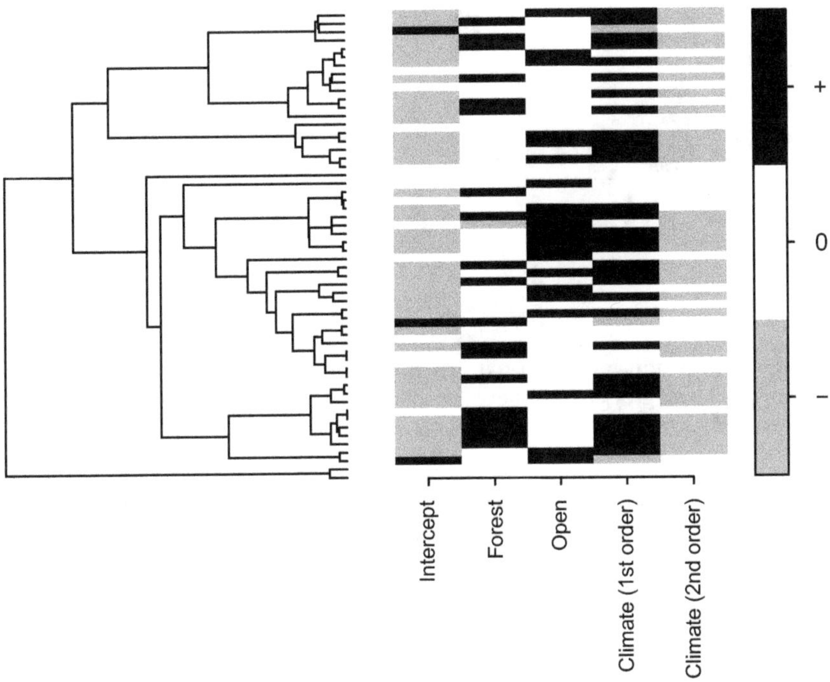

Figure 10.11 Heatmap of estimated β parameters, i.e. species niches. Black and grey colours show the parameters that are estimated to be positive and negative, respectively, with at least 0.95 posterior probability.

others respond positively to open habitat quality. In contrast, almost no species respond negatively to either of these. This is to be expected, as all species need resources for their survival and reproduction. In Figure 10.11, most species show a negative response to the second-order term of climatic conditions. This means that their abundance peaks at some intermediate climatic condition, which is fully in line with what we assumed in the agent-based model. The response to the first-order term of climatic conditions is generally positive. The interpretation of this latter result is slightly more difficult because the model also includes the second-order term of this covariate, and because we did not normalise temperatures to have a zero mean. In this case, a positive response to temperature does not necessarily mean that the species will be more abundant in the warmer part of the climatic gradient than in the cooler part. For this reason, we recommend the user to evaluate the effects of variables with second-order terms or interactions from model predictions rather than from parameter estimates, as we will do soon.

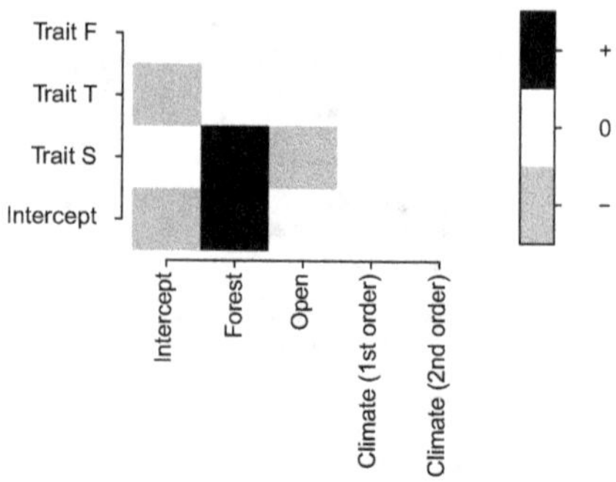

Figure 10.12 Heatmap of estimated γ parameters linking species traits to species niches. Black and grey colours show parameters that are estimated to be positive and negative, respectively, with at least 0.95 posterior probability.

V. E. Space next examines the links between species traits and species niches using the plotGamma function.

```
postGamma = getPostEstimate(m, parName = "Gamma")
plotGamma(m, post = postGamma, param = "Support",
    supportLevel = 0.95)
```

The resulting plot (Figure 10.12) shows that those species with the highest values of the trait *S* respond the most positively to forest habitat quality, but the least positively to open habitat quality. This makes sense, since in the agent-based model a high value of the trait *S* means that the species is highly efficient in utilising forest habitat resources, whereas a low value of the trait *S* means that the species is highly efficient in utilising open habitat resources.

To further explore environmental filtering, V. E. Space evaluates the responses of the community to environmental variation with the help of gradient plots. She starts by examining the influence of climate by normalising the remaining variables (forest and open habitat resource availabilities) to their mean values over the data.

```
Gradient = constructGradient(m, focalVariable = "clim",
 non.focalVariables=list("forest"=list(1),"open"=list(1)))
predY1 = predict(m, Gradient = Gradient, expected = TRUE)
predY2 = predict(m, Gradient = Gradient, expected = FALSE)
```

We note that V. E. Space has computed the model predictions using both expected values (predY1) and realised values (predY2). With non-linear link functions such as the log-link function included in the log-normal model, the posterior median estimate of these two can be quite

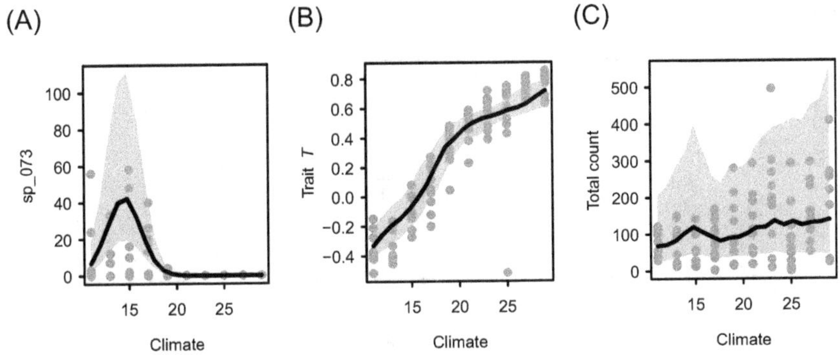

Figure 10.13 Predicted responses to climatic variation. The panels show the expected count of a single focal species (panel A), community-weighted mean value for the trait measuring thermal optimum (panel B), and expected total count over all species (panel C). The lines show posterior median, and the shaded area shows the posterior interquartile range.

different. The one that should be used depends on what exactly one wishes to evaluate.

Figure 10.13 exemplifies some of the many plots that one can create over the climatic gradient with the plotGradient function. The line in panel Figure 10.13A shows the posterior median of the expected count of the species that were illustrated with black dots in Figure 10.5. This species is adapted to cold climates, as seen from both the raw data and the expected response. To her delight, V. E. Space observes a clear positive relationship between the laboratory-measured thermal optimum trait (T) and the field-observed relationship between species occurrences and climatic conditions (Figure 10.13B). Thus, she concludes that the fundamental niche axes of thermal optima and tolerance also play a large role in the realised niche. Figure 10.13C shows that the total abundance over all species is somewhat higher under warm climatic conditions compared to cool conditions. This result also concurs with the assumptions of the agent-based model, as we assumed that fecundity is positively correlated with thermal optimum.

V. E. Space then examines the predicted responses of the community over variation of forest resource availability, setting the other variables to their mean values.

```
Gradient = constructGradient(m, focalVariable = "forest",
 non.focalVariables = list("clim" = list(1), "open" = list(1)))
predY1 = predict(m, Gradient = Gradient, expected = TRUE)
predY2 = predict(m, Gradient = Gradient, expected = FALSE)
```

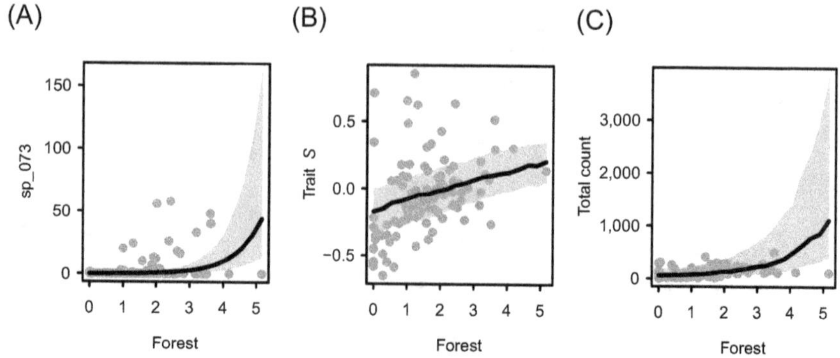

Figure 10.14 Predicted responses to variation in forest resource availability. The panels show the expected count of a single focal species (panel A), community-weighted mean value for the trait measuring forest specialisation S (panel B), and expected total count over all species (panel C). The lines show the posterior median, and the shaded area shows the posterior interquartile range.

The expected count of the species illustrated with black dots in Figure 10.5 increases with forest habitat quality (Figure 10.14A), as does the community-weighted mean trait value of forest specialisation S (Figure 10.14B), and the total count of all species (Figure 10.14C). These predictions were fully expected from the estimates of the β parameters (Figure 10.11) and the γ parameters (Figure 10.12). While Figures 10.11 and 10.12 compactly describe the directions of the responses across all species, covariates, and traits, Figure 10.14 better illustrates the effect sizes – in this case, the impact of variation in forest quality.

Let us then return to the question of why the species traits did not explain variation in their climatic niches, as measured by $R^2_{T\beta}$. To understand why this 'failed', let us start by looking at the relationship between species traits and their niches in relation to habitat quality. In the script below, we extract the posterior mean of the β parameters, and then plot the parameters β_{2j} measuring species responses to forest quality as a function of their trait S_j measuring specialisation to forest resources.

```
postBeta = getPostEstimate(m, parName = "Beta")
plot(TrData$S, postBeta$mean[2,])
mylm = lm(postBeta$mean[2,] ~ TrData$S)
abline(mylm)
```

As expected, the realised niche parameter β_{2j} increases with the fundamental niche parameter S_j (Figure 10.15). The relationship shown in

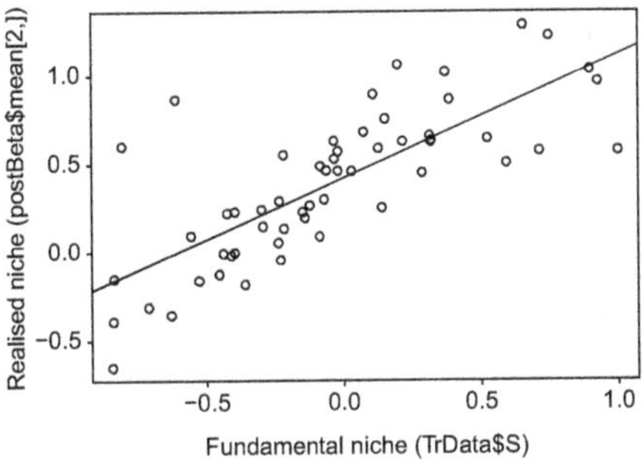

Figure 10.15 Relationship between the realised niche parameter β_{2j} measuring the responses of the species to forest quality and the trait S_j describing the fundamental niche as the extent to which the species is specialised to forest resources. Each species is represented by a dot, and the line shows the linear regression between the parameters.

this figure generated the high value of $R^2_{T\beta} = 0.39$ related to the covariate forest quality.

To plot the analogous relationship concerning the responses of the species to variation in temperature, one needs to account for the fact that the influence of climate was modelled along with a second-order response, as illustrated by the predicted abundance peaking at intermediate climatic values (Figure 10.13A). As the response of species j to temperature T is modelled as $\beta_{4j}T + \beta_{5j}T^2$, the abundance of the species is predicted to be the highest at $T = -\beta_{4j}/(2\beta_{5j})$. To see why this is the case, we recall that the derivative of the function $f(x) = ax^2 + bx$ is $f'(x) = 2ax + b$, and thus $f'(x) = 0$ when $x = -b/(2a)$.

Let us next compute the estimated thermal optima based on the β parameters (the realised niche) with the thermal optima T_j in the measured traits (the fundamental niche).

```
postBeta = getPostEstimate(m, parName = "Beta")
T.opt = -postBeta$mean[4,] / (2*postBeta$mean[5,] )
plot(TrData$T, T.opt)
mylm = lm(T.opt ~ TrData$T)
abline(mylm)
```

Figure 10.16 shows that there is a positive relationship between the fundamental and realised niches related to thermal optima. However, this relationship is blurred in Figure 10.16A by a few species for which the realised niche of thermal optimum obtains very high or very low values. These are primarily species that occur at the ends of the thermal gradient. If the data show e.g. that all the occurrences are found at the coldest part

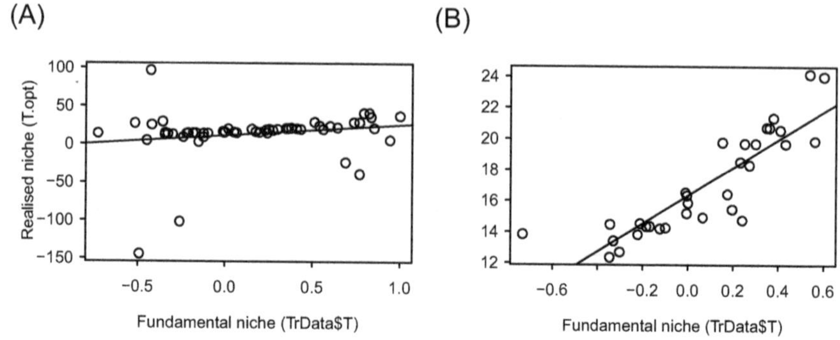

Figure 10.16 Relationship between the thermal optimum based on the realised niche parameters β_{4j} and β_{5j} and the trait T_j describing the fundamental niche of the thermal optimum of the species. Each species is shown by a dot, and the line indicates the linear regression between the parameters. Panel A includes all species, and panel B shows only those species for which the thermal optimum could be estimated with sufficient accuracy (see text).

of the gradient at temperature $T = 10$, a model fitted to these data cannot tell whether the thermal optimum of the species would occur at even colder temperatures, such as $T = -100$. To exclude such cases, in Figure 10.16B V. E. Space has included only those species for which the thermal optimum in the realised niche can be well estimated. To do so, she has written the script:

```
postList = poolMcmcChains(m$postList)
get.T.opt = function(a){ return(-a$Beta[4,] / (2* a$Beta[5,] ))}
post.T.opt = lapply(postList, get.T.opt)
post.T.opt=matrix(unlist(post.T.opt),ncol=m$ns,byrow=TRUE)
T.opt.iqr = apply(post.T.opt, c(2), IQR)
sel.sp = T.opt.iqr < 4
```

In this script, V. E. Space has computed the posterior interquartile range of the realised thermal optima. She has then selected those species for which the range is at most 4, which she considered as reasonably accurate. Figure 10.16B shows the same data as Figure 10.16A, but only for the selected species. This plot most clearly shows the expected positive relationship between the assumed and realised thermal optima.

Based on the R^2 of the linear regression shown Figure 10.16B, $R^2_{T\beta}$ related to thermal optima is very high (0.76), contrary to what the very low estimates in VP$R2T$Beta suggested for both the linear and second-order terms. After the above additional analyses, V. E. Space understands the reason for this apparent contradiction. The relationship between realised and fundamental niches related to thermal optimum was not visible in VP$R2T$Beta because this measure did not properly account for the second-order relationship nor parameter uncertainty. The true relationship was revealed once both of these were accounted for (Figure 10.16B).

We have discussed in length the 'false negative' result derived from VP$R2T$Beta because this example demonstrates the risk of quickly jumping to conclusions, as well as the need to understand the equations building GLMs. While Hmsc includes much ready-made functionality, the most case-specific analyses always need to be implemented by the user. Implementing such analyses requires expertise, as illustrated by the above script that V. E. Space wrote to compute posterior interquartile ranges for the ratio of the two β parameters. This example also demonstrated how posterior uncertainty can be propagated to model predictions, such as the prediction of the realised thermal optima made here.

V. E. Space then returns to more routine HMSC analyses, next examining whether the residual variation in species niches is phylogenetically correlated.

```
mpost = convertToCodaObject(m)
summary(mpost$Rho)[2]

## $quantiles
## 2.5%   25%   50%   75% 97.5%
## 0.00 0.13 0.21 0.29  0.45

mean(c(mpost$Rho[[1]], mpost$Rho[[2]])==0)

## [1] 0.0835
```

The posterior distribution of ρ suggests a modest phylogenetic signal. This was perhaps not so expected, as her model included all traits that influenced species niches. However, now that the data are simulated with the agent-based model, the structural assumptions of the HMSC model are not fully in line with the data-generating process, which may lead to biases in the inference. Furthermore, the posterior probability for no phylogenetic signal is $\Pr(\rho = 0) = 0.08$, which is not negligible, even if most of the posterior mass has positive values of ρ.

As her final analysis step, V. E. Space examines the residual species association networks. First, she examines the data for the presence of a spatial signal.

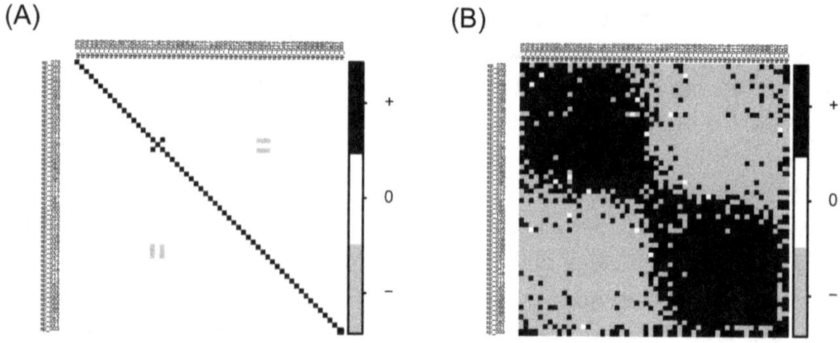

Figure 10.17 Residual species-to-species associations shown in panel A for those species pairs with at least 0.95 support for either a positive or negative association, and in panel B for all species pairs.

```
summary(mpost$Alpha[[1]])[2]$quantiles[1,]

##      2.5%       25%       50%       75%     97.5%
## 2.545584  4.072935  5.091169  6.618519 10.946013
```

The first factor of species loadings indeed shows a spatial signal, as the estimate of the α_1 parameter is bounded away from zero.

Using the threshold level of 0.95, there are basically no species associations (Figure 10.17A). This is perhaps not surprising, considering that in the variance partitioning (Figure 10.10) the random effect contributed very little to the explained variance. However, since V. E. Space is interested in species associations, she decides to visualise them also without any threshold of statistical support. Now the association network suggests that there are two roughly equally large groups of species that show positive associations within groups, and negative associations between groups (Figure 10.17B).

Because of her excellent knowledge of the study system, V. E. Space had hypothesised that the ecological interactions in this species community mainly relate to resource competition, which is structured by body size. To examine if the data produce this signal, she decides to plot the association matrix again so that species are sorted based on their traits. For this, she places species that are similar (both in terms of body size and habitat specialisation) close to each other.

```
small = which(m$TrData$B == "small")
large = which(m$TrData$B == "large")
small = small[order(m$TrData$S[small])]
large = large[order(m$TrData$S[large])]
```

Figure 10.18 Residual species-to-species associations shown for all species pairs. The species are ordered first by their body size (first small, and then large species), and then within the body size class according to their specialisation on forest resources.

```
supportLevel = 0.5
plotOrder = c(small,large)
corrplot(OmegaCor[[1]]$mean[plotOrder,plotOrder],
  method = "color", col = c("grey","white","black"))
```

A visual inspection of the resulting association matrix (Figure 10.18) does not however return the response she was expecting, namely that negative associations would be mostly seen near the diagonal of the matrix, i.e. for species with similar traits. This demonstrates that inferring competitive interactions from negative associations can be very challenging. Even if there are strong competitive interactions in the current case, their influences are perhaps diffused quite evenly through all species, leaving only weak signals in the data. We further note that it is not possible that all species would show a strong negative association with each other, whereas it is possible that all species would show a strong positive association. To see why this is the case, let us assume that there is such a strong negative association between species 1 and species 2 that only one of them is found from any given location. Let us also assume that there is an equally strong negative association between species 2 and species 3. This necessarily means that species 1 and 3 are found from the same locations, and thus they show a positive association.

10.2.3 HMSC Analyses with a Compromised Model

We next assume that V. E. Space has access to only partial data and that she does not have perfect knowledge on the ecology of the study system. Thus, this time she uses presence–absence data only, includes habitat as the only environmental covariate, and the level of forest specialisation as the only trait. Further, instead of the accurate measurements of forest specialisation level (trait S) that she used in the previous subsection, she now has access to the expert opinion-based classification of 'forest species' and 'open area species' (trait S.proxy in the script below).

Because of the above listed shortcomings, we refer to the present model as the compromised model. V. E. Space defines compromised HMSC model as follows:

```
TrData$S.proxy = TrData$S
TrData$S.proxy[TrData$S > 0] = "forest species"
TrData$S.proxy[TrData$S <= 0] = "open area species"
TrData$S.proxy = as.factor(TrData$S.proxy)
XFormula = ~ forest + open
TrFormula = ~ S.proxy
studyDesign = data.frame(sample = row.names(XData))
rL = HmscRandomLevel(sData = xy)
m = Hmsc(Y = 1*(Y > 0), XData = XData, XFormula = XFormula,
 TrData = TrData, TrFormula = TrFormula, phyloTree = phy,
 studyDesign = studyDesign, ranLevels = list(sample = rL),
 distr = "probit")
```

She then moves on as usual to fit the model, check MCMC convergence and evaluate its explanatory power. Figure 10.19 shows that the explanatory power of the compromised model to discriminate presences from absences (Figure 10.19) is very similar to that of the ideal model (Figure 10.9).

However, while the explanatory power of the ideal model was obtained almost solely through the fixed effects (Figure 10.10), in the compromised model roughly half of the explanatory power arises from the random effects (Figure 10.20). Another obvious disadvantage of the compromised model is that it cannot be used to explain variation in abundance, simply because only presence–absence data has been recorded.

The parameter estimates of species niches (Figure 10.21) and how they are related to traits (Figure 10.22) tell essentially the same story as the ideal model. Namely, most species respond to the two kinds of resource availabilities either positively or neutrally (Figure 10.21), and those species that the expert classified as 'open area species' benefit from

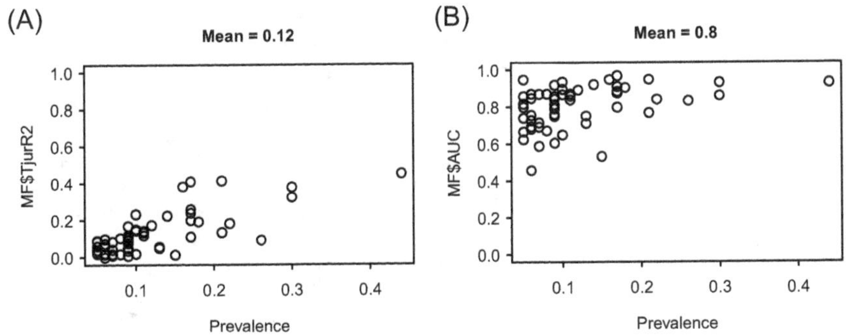

Figure 10.19 The explanatory power of the compromised model measured in terms of Tjur R^2 (panel A) and AUC (panel B).

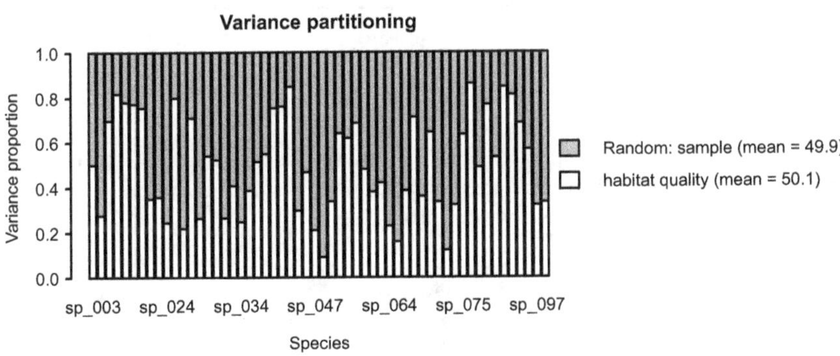

Figure 10.20 Variance partitioning among the fixed and random effects included in the compromised model.

increased forest resource availability less than the average species (Figure 10.22). Yet, unlike in the ideal model, the current parameter estimates do not reveal anything about the climatic niches of the species, for the obvious reason that V. E. Space did not include climatic variation as explanatory variables. Thus, now the random effects might be capturing the climatic effects.

V. E. Space next evaluates the responses of the community to environmental variation with the help of gradient plots:

```
Gradient = constructGradient(m, focalVariable = "forest",
  non.focalVariables = list("open" = list(1)))
predY = predict(m, Gradient = Gradient, expected = TRUE)
```

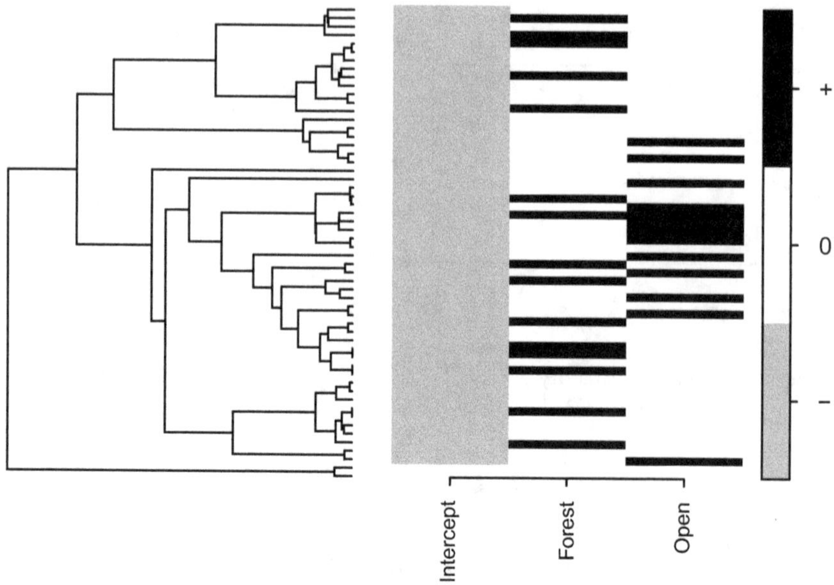

Figure 10.21 Heatmap of estimated β parameters, i.e. species niches. Black and grey colours show the parameters that are estimated to be positive and negative, respectively, with at least 0.95 posterior probability.

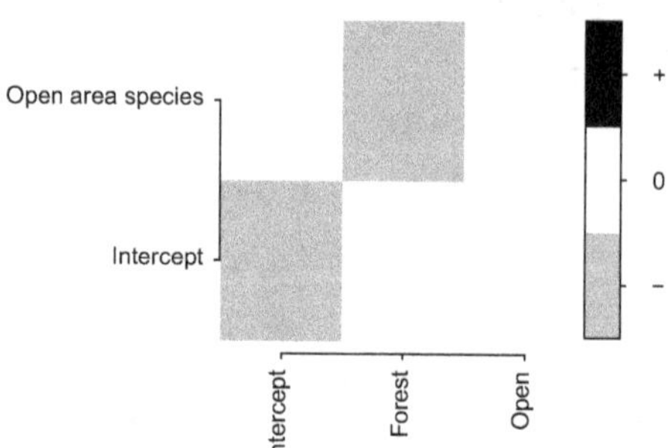

Figure 10.22 Heatmap of estimated γ parameters linking species traits to species niches. Black and grey colours show parameters that are estimated to be positive and negative, respectively, with at least 0.95 posterior probability.

As the compromised model is a presence–absence model, its predictions (plotted with the function plotGradient) are in units of occurrence probabilities and species richness (Figure 10.23), whereas the corresponding predictions made by the ideal model were in terms of counts of individuals (Figure 10.14). In spite of this, the predicted patterns are

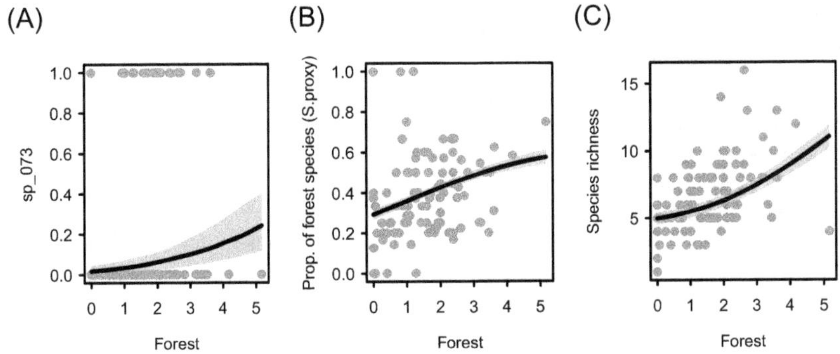

Figure 10.23 Predicted responses to variation in forest resource availability. The panels show the occurrence probability of a single focal species (panel A), proportion of species classified as forest species (panel B), and expected species richness (panel C). The lines show the posterior median, and the shaded area shows the posterior interquartile range.

qualitatively consistent between the two models: in the compromised model, both the occurrence probability of the focal species (Figure 10.23A) and the total species richness (Figure 10.23C) increase with increasing forest resource availability, as does the proportion of species specialised to forests (Figure 10.23B).

V. E. Space then examines whether a phylogenetic signal is present in the residual variation of the species niches.

```
mpost = convertToCodaObject(m)
summary(mpost$Rho)[2]

## $quantiles
##    2.5%    25%     50%     75%   97.5%
## 0.4595 0.7500 0.8400 0.9000 0.9700
```

There is a clear signal, which suggests that she has missed some traits that explain variation in species niches, and that those traits are phylogenetically correlated. We note that the species niche in the compromised model includes only the dependency of the species on resource availability, and that the classification of the species to open and forest area specialists was included in the model. Thus, in the present case the relevant trait was not completely missed, but the proxy used was not fully accurate.

V. E. Space then turns to the random effects, which are of special interest as they explain much of the variation (Figure 10.20). She first examines the presence of a spatial signal:

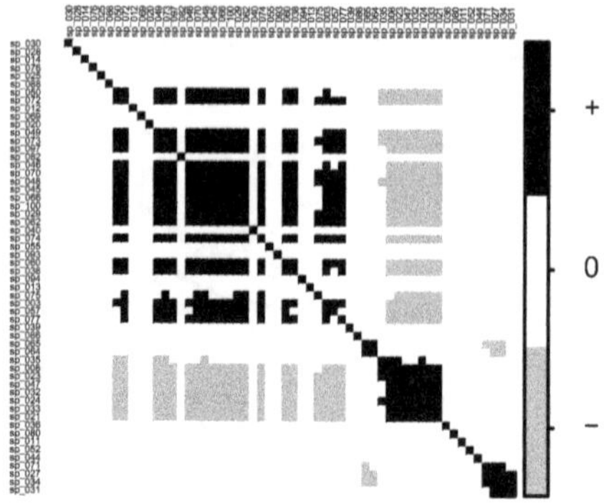

Figure 10.24 Residual species-to-species associations shown for species pairs for which the support for either a positive or negative association is at least 0.95.

```
summary(mpost$Alpha[[1]])[2]
## $quantiles
##                      2.5%     25%      50%      75%    97.5%
## Alpha1[factor1]  6.618   9.164   10.946   13.237   19.85
## Alpha1[factor2]  3.054   5.600    6.618    8.400   15.01
## Alpha1[factor3]  0.000   0.000    0.000    9.673   23.67
## Alpha1[factor4]  0.000   0.000    0.000   12.218   24.43
```

The two leading latent factors (factor1 and factor2) show a clear spatial signal, with the spatial range α_1 of the leading one estimated to be ca. 11 spatial units.

The compromised model also reveals a large number of both positive and negative residual species associations (Figure 10.24). V. E. Space is first somewhat puzzled by these results, as she was expecting to see negative associations rather than positive ones, knowing that the species compete for the same resources. As she does not find the explanation of facilitative interactions plausible, she concludes that she must have missed some relevant environmental variables.

She then realises that the altitude gradient present in her study area might be more important than she originally thought. To test whether altitude is the missing environmental covariate, she uses the mean y-coordinate for each species as a proxy for their location along the altitudinal gradient (note that altitude of the study area is linearly related to the y-coordinate).

```
xy = m$rL[[1]]$s
y.mean = rep(NA, m$ns)
for (j in 1:m$ns){
    y.mean[j] = mean(xy[m$Y[,j]==1, 2])
}
```

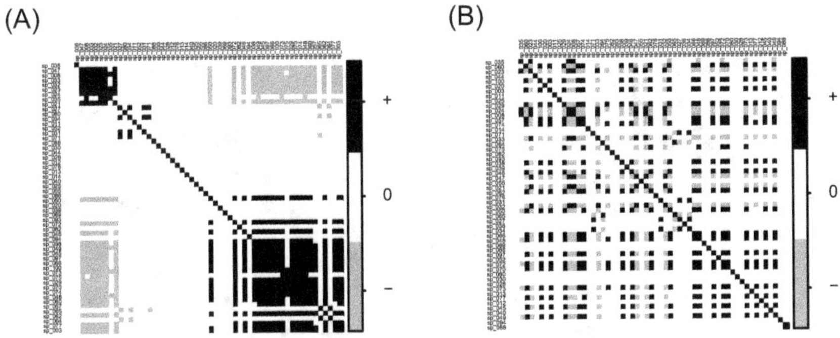

Figure 10.25 Residual species associations shown for species pairs for which the support for either a positive or negative association is at least 0.95. In panel A, the species are ordered according to the mean y-coordinate over their locations, whereas in panel B they are ordered according to the mean x-coordinate.

In the above script, V. E. Space has first pulled the spatial coordinates from the first (and only) random effect (m$rL[[1]]) included in the model. The spatial coordinates are stored in the object $s of the random effect. She then plots the association network again, but now so that the species are sorted according to the mean altitude at which they occur (Figure 10.25A). The results are very clear: species that are found from low altitudes show a positive residual association with each other, and so do the species that are found at high altitudes.

To ensure that the associations relate specifically to altitude and not more generally to spatial variation, she also sorts the species according to the x-coordinate. As this ordering does not reveal almost any pattern (Figure 10.25B), she concludes that the y-coordinate – and hence altitude – is the important missing covariate. To gain further insight into the nature of the 'missing environmental covariate', she illustrates the two leading site loadings by plotting their posterior means over the study area. In the script below, she calls the posterior mean of the leading site loadings η_{i1} as Eta1, and the posterior mean of the second most important site loadings η_{i2} as Eta2.

```
postEta = getPostEstimate(m, parName = "Eta")
Eta1 = postEta$mean[ ,1]
Eta2 = postEta$mean[ ,2]
```

As expected, the spatial variation is much related to the y-coordinate, especially for the leading site loadings (Figure 10.26A). We note that the above analyses do not imply that altitude *per se* would necessarily be causally important (and even less so the y-coordinate), as the underlying

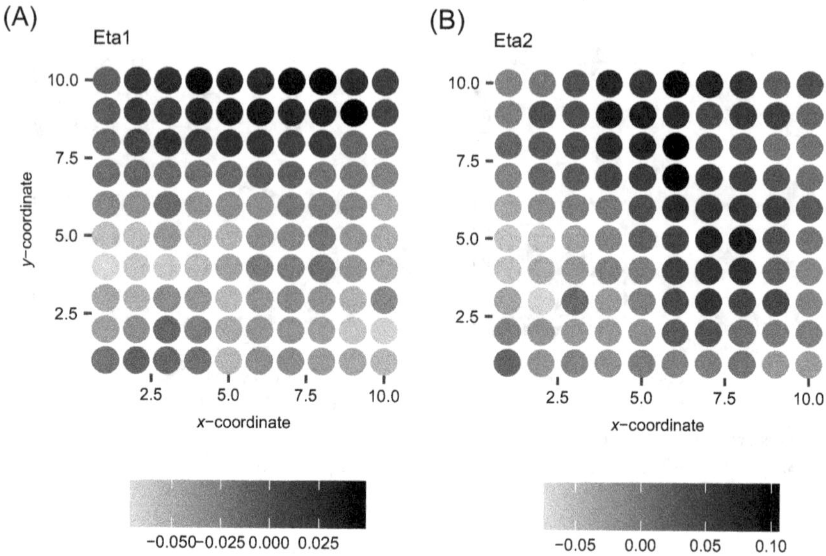

Figure 10.26 Spatial variation in site loadings corresponding to the two leading latent factors.

driver can be any variable that is correlated with altitude. As we created the metacommunity model, we know that the causal/missing variable is actually climatic variation.

As the next step, V. E. Space plans to rerun the analyses with the inclusion of altitude as a fixed effect; in her next fieldwork season, she plans to measure environmental variables that relate to altitude. Furthermore, she aims to conduct an experiment where she exposes a few individuals of each species to environmental conditions representing different parts of the altitudinal gradient, in order to understand which traits underlie how species are sorted along the altitudinal gradient. Finally, to get an even more accurate understanding of the factors influencing the study community, she plans to measure species abundance instead of simply recording them as present or absent. Thus, after next year's fieldwork season, she will be well prepared to conduct analyses that are much closer to those of the ideal model, which is closer to the underlying assembly processes.

10.3 Statistical Analyses of the Time-Series Data Collected by a Virtual Ecologist

We will now turn to the time-series data collected by V. E. Time.

10.3.1 Exploration of the Raw Data

Among the 100 species included in the agent-based metacommunity simulations, 53 were recorded in the time-series data. The number of species detected is much smaller in the time-series study than in the spatial study, as the single study plot that was repeatedly sampled may not represent suitable environment for all the species in the study area. Out of the fifty-three species, nineteen were detected in at least five different years. V. E. Time decides to include only the latter species in his analyses, for the same reasons that V. E. Space dropped the rarest species from her analyses.

Let us have a look at the raw data matrices that V. E. Time has compiled.

```
head(Y[ ,1:6] )

##            sp_005  sp_011  sp_014  sp_017  sp_020  sp_024
## sample_001  0       0       0       0       0       0
## sample_002  0       0       0       0       0       0
## sample_003  0       0       0       0       0       0
## sample_004  0       0       0       0       0       1
## sample_005  0       0       0       0       0       2
## sample_006  0       0       0       0       0       3

head(xy)

##            year
## sample_001  1
## sample_002  2
## sample_003  3
## sample_004  4
## sample_005  5
## sample_006  6

head(TrData)

##                S          T           F     B
## sp_005 0.6640212 -0.3336452 -0.54615770 large
## sp_011 0.5235583  0.2504980  0.05692630 large
## sp_014 0.3202751  0.5615986  0.28408241 small
## sp_017 0.3493696  0.3733680 -0.01550978 small
## sp_020 0.1261791  0.1746892  0.02747405 large
## sp_024 0.1961501  0.7677630  0.78152137 small
```

His data consist of species counts (**Y**) and the temporal coordinates of the sampling units (**S**), i.e. the year when the sampling was conducted.

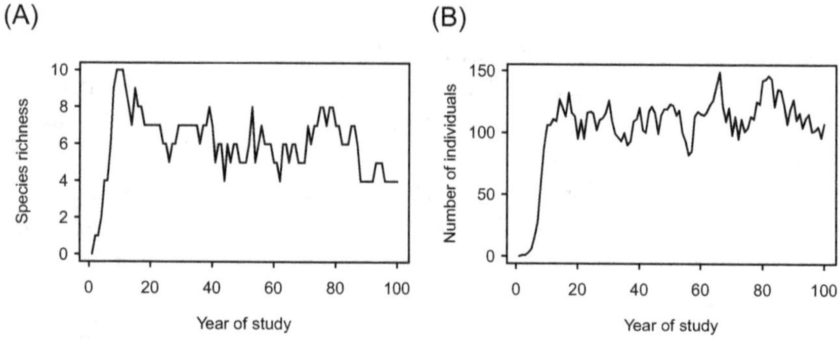

Figure 10.27 Variation in the number of species (panel A) and the number of individuals (panel B) observed by repeated surveys on a single study plot over the 100 year-long study period.

He also has data on species traits (**T**) and their phylogenetic relationships (shown in Figure 10.2), but no data on environmental covariates (**X**).

In the time-series data, both the species richness (Figure 10.27A) and the total count of individuals (Figure 10.27B) increase rapidly over the first decade of the study, and then stabilise to fluctuate around 4–8 species and around 80–140 individuals. Based on his knowledge about the study system, V. E. Time interprets Figure 10.27 to indicate that right after the species colonised the area, the populations were not limited by resources and thus they were able to grow exponentially. But soon they reached the carrying capacity of the environment, and thus population sizes became regulated by density-dependent processes related to resource competition (Hassell 1975).

10.3.2 Competitive Interactions Revealed by HMSC Analyses

As the data collected by V. E. Time does not involve variation in either habitat quality or climatic conditions, he cannot use the data to study environmental filtering, i.e. the abiotic axes of species niches. Instead, he aims to use the data to study density-dependent population processes (Hassell 1975). To evaluate density dependence, he plans to use the population sizes of the species in the previous years as predictors for population growth in the next years. But instead of using the population sizes of each species as separate predictors, he decides to use the counts summed over the species. He makes this choice to avoid the use of too many predictors and thus model overfitting. Furthermore, he considers this simplification ecologically justified, as the total number of individuals should be a relevant predictor of resource competition, assuming that all species compete an equal amount. However, he is aware that small- and large-bodied species may compete for different sets of resources, and thus he counts the individuals of these two sets of species separately.

He next prepares the explanatory variables of the summed counts of large- and small-bodied species, and constructs a dataframe XData that includes these two variables, which will be used as fixed effects in the model.

```
sum.small = rowSums(Y[1:99,TrData$B=="small"])
sum.large = rowSums(Y[1:99,TrData$B=="large"])
XData=data.frame(sum.large=sum.large, sum.small=sum.small)
```

Note that this script computes these predictors for the years 1–99, leaving out the final year 100. This is because the summed counts of the species from the year t are to be used as predictors for the year $t+1$. Thus, the counts from the years 1–99 are the predictors for the response variables from the years 2–100.

As the response variable, V. E. Time decides to use population growth rate rather than population size. He makes this choice because he expects the signature of density dependence to be imprinted in changes in population sizes rather than in absolute population sizes. His reasoning is that if a species has a high population size in a given year, it is likely to have a high population size also in the next year. This is because the population of the previous year acts as the parental generation for the population in the next year, and many parents are likely to produce many offspring. Furthermore, some of the individuals may have survived from the previous year, and thus the individuals from consecutive years may be the same. Thus, a high population size in the previous year is expected to lead to a high population size in the next year, which would not reveal density dependence. However, if competition is important, a high population size in the previous year can be expected to lead to a reduction in population growth.

V. E. Time decides to measure population-growth rate in proportional units, and thus computes the difference between log-transformed counts.

```
Yr = log(1+Y[2:100,]) - log(1+Y[1:99,])
```

When calculating the growth rate, V. E. Time has added one before taking the logarithm. This is done to prevent taking the logarithm from zero if the population was extinct.

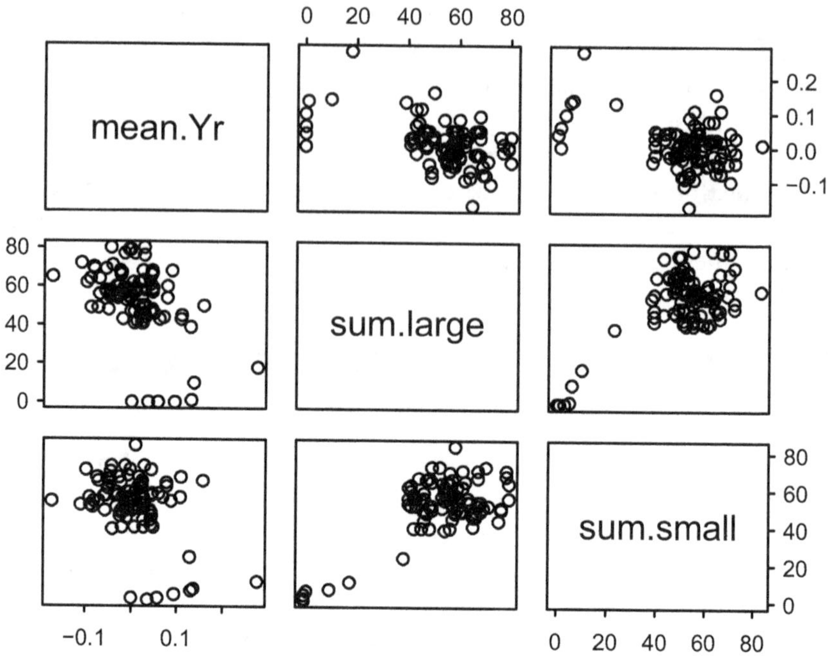

Figure 10.28 Predictors of density dependence (sum.large and sum.small) and the mean (over the species) population-growth rate (Yr) plotted with respect to each other.

To explore the data before moving to the HMSC analyses, V. E. Time plots the sum.large, sum.small and the population-growth rate (averaged over the species) with respect to each other (Figure 10.28). The figure shows that sum.large and sum.small had their lowest values in the same years when the population-growth rates were high. This is because during the initial transient phase, all population sizes were low and growth rates high. Other than that, Figure 10.28 does not show any other obvious correlations between these variables.

V. E. Time is now ready to set up the HMSC model. He includes sum.large and sum.small as fixed effects, and year as a temporally structured random effect. While he has access to many kinds of trait data, he includes only body size, as based on the literature he hypothesises that this trait influences density-dependence dynamics (Damuth 1981). Growth rate is not a count but a continuous-valued response variable that can be positive or negative, hence he fits a normally distributed model. As he is not interested in the absolute species-specific growth

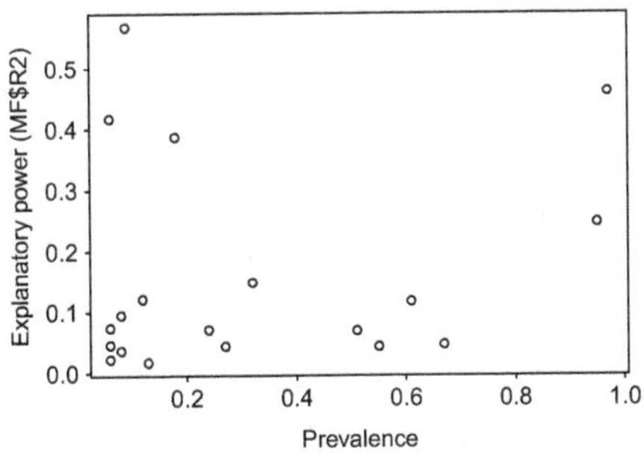

Figure 10.29 The explanatory power of the time-series model measured in terms of R^2.

rates but in their within-species variation, he selects to scale the growth rates to zero mean and unit variance (see the discussion on scaling the response matrix in Section 8.3.1). For modelling residual variation among the repeated surveys, he includes the sampling year as a random effect. To examine if the residual variation shows temporal autocorrelation, he uses a 'spatial' random effect on the one-dimensional space of time.

```
xy1 = subset(xy, year > 1)
row.names(XData) = row.names(xy1)
studyDesign = data.frame(sample = row.names(XData))
rL = HmscRandomLevel(sData = xy1)
XFormula = ~ sum.large + sum.small
TrFormula = ~ B
m = Hmsc(Y = Yr, YScale = TRUE, XData = XData,
  XFormula = XFormula, TrData = TrData, TrFormula = TrFormula,
  phyloTree = phy, studyDesign = studyDesign,
  ranLevels = list(sample = rL), distr = "normal")
```

After fitting the model and assessing its MCMC converge, V. E. Time evaluates the explanatory power of the model (Figure 10.29) and partitions the explained variance among fixed and random effects (Figure 10.30). The explanatory power is not very good for most species, suggesting that there is a high amount of stochasticity in the growth rates. The fixed effects (i.e. the effects related to density dependence) explain over half of the explained variation (Figure 10.30).

The estimates of the species-specific β parameters (Figure 10.31) show that the growth rates of some of the species respond negatively to the summed abundances of small-bodied species in the previous year,

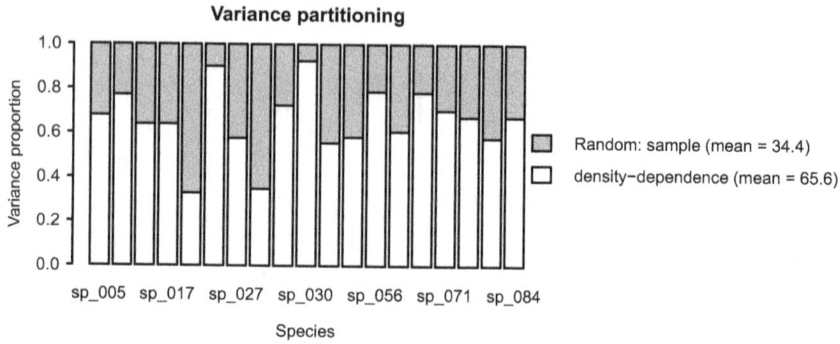

Figure 10.30 Variance partitioning among the fixed and random effects included in the model.

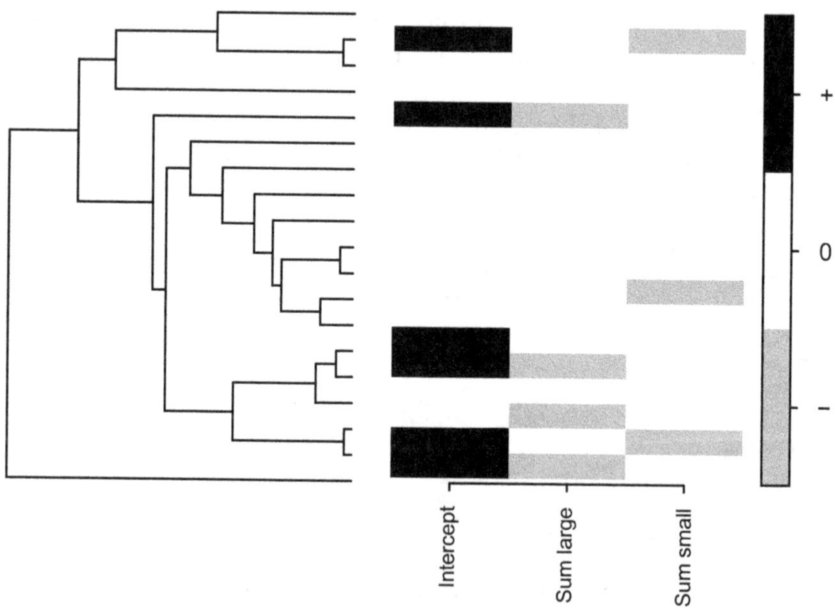

Figure 10.31 Heatmap of estimated β parameters, i.e. species niches. Black and grey colours show the parameters that are estimated to be positive and negative, respectively, with at least 0.95 posterior probability.

whereas the growth rates of other species respond negatively to the summed abundances of large-bodied species in the previous year. None of the species shows a positive response to either of these two predictors. In other words, there is evidence of density dependence restricting the growth rates of some of the species.

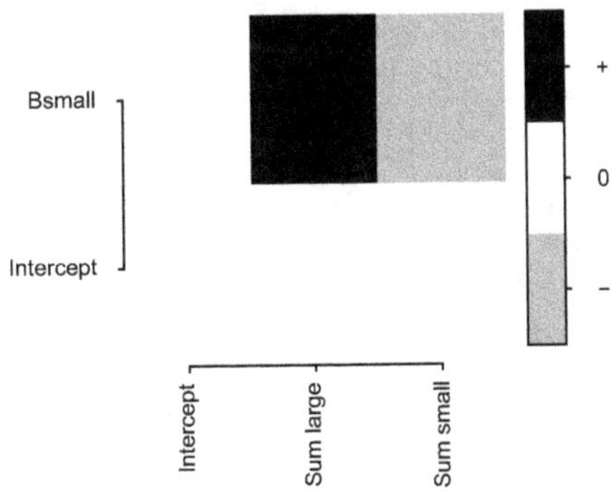

Figure 10.32 Heatmap of estimated γ parameters linking species traits to species niches. Black and grey colours show parameters that are estimated to be positive and negative, respectively, with at least 0.95 posterior probability.

The community-level γ parameters (Figure 10.32) show that the growth rate of small-bodied species is especially negative when their summed abundance is high in the previous year, while the growth rate of large-bodied species is especially negative when their summed abundance is high in the previous year. In other words, density dependence acts primarily within the size classes.

To further illustrate these results, V. E. Time constructs a gradient plot, where he explores variation in population-growth rates along a gradient of previous year's population densities. He selects sum.small as the focal variable, and uses the default options to set the value of sum.large to its conditional expectation based on the values of sum.small. He makes this choice to avoid extrapolation to situations outside the ranges in the data.

```
Gradient = constructGradient(m,focalVariable = "sum.small")
predY = predict(m,Gradient = Gradient, expected = TRUE)
sp1 = which(P==max(P))
q = c(0.25,0.5,0.75)
plotGradient(m, Gradient, pred=predY, measure="Y", index=sp1,
            showData = TRUE, q = q)
plotGradient(m, Gradient, pred=predY, measure="T", index=2,
            showData = TRUE, q = q)
plotGradient(m, Gradient, pred = predY, measure = "S",
            showData = TRUE, q = q)
```

Figure 10.33 shows the predicted community variation along the gradient of previous year's population densities of small-bodied species. As the focal species shown in Figure 10.33A, he has selected the species

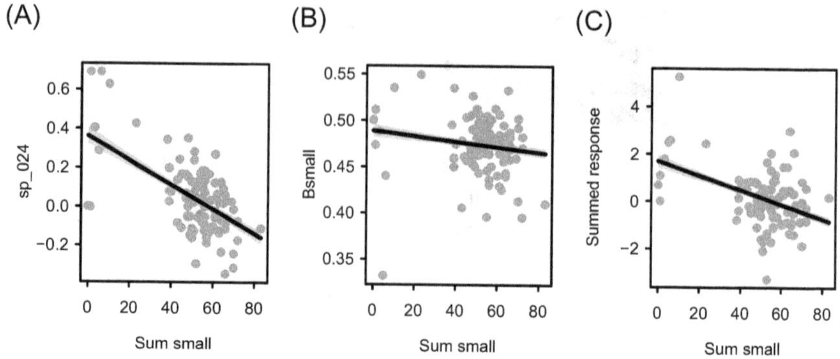

Figure 10.33 Predicted responses to variation in the previous year's population densities of small-bodied species. The panels show the expected growth rate of a single focal species (panel A), community-weighted mean value for the trait small body size (panel B), and expected total growth rate over all species (panel C). The dots show the species, the lines show the posterior median, and the shaded areas show the posterior interquartile range.

with the highest prevalence (see script), which happens to be a species with a small body size. The growth rate of this species shows a clear pattern of density dependence with respect to the summed abundance of all small-bodied species in the previous year (Figure 10.33A), and so does the total growth rate summed over all species (Figure 10.33C). We note that for capturing both of these, the data from the early years were especially valuable, as those data have low summed abundance in the previous years and high population-growth rates. The community-weighted mean trait of the proportion of small-bodied species (weighted by the exponentially transformed growth rates) shows that the relative growth rate of the small-bodied species expectedly decreases with increasing summed abundance of the small-bodied species (Figure 10.33B).

V. E. Time next examines the strength of the phylogenetic signal.

```
mpost = convertToCodaObject(m)
summary(mpost$Rho)[2]

## $quantiles
## 2.5%   25%   50%   75% 97.5%
## 0.64 0.85 0.91 0.96  1.00
```

The residual variation in species niches (here, density dependence) shows a phylogenetic signal, suggesting that it is also influenced by some other traits that are phylogenetically structured. Based on our knowledge of the

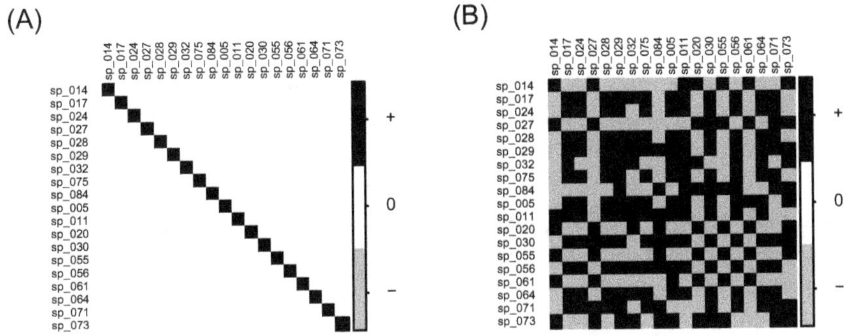

Figure 10.34 Residual species-to-species associations for those species pairs for which the support for either a positive or negative association is at least 0.95 (panel A), and for all species pairs (panel B). In both panels, the species are sorted according to their body size, so that small-bodied species are the first and large-bodied species the following.

agent-based model, we hypothesise that the missing trait is forest specialisation. In fact, forest specialisation controls for how efficiently the species use the resources and thus how much they are influenced by competition.

Finally, V. E. Time examines the structure of the variation attributed to the temporal random effect.

```
tmp = summary(mpost$Alpha[[1]])[2]
tmp$quantiles[1,]

## 2.5% 25% 50% 75% 97.5%
##    0   0   0   0     0
```

While V. E. Time assumed a temporally structured random effect, the leading site loading does not show any evidence of temporal structure. This suggests that the residual growth rates (after accounting for density dependence) are not auto-correlated over the time scale of these data. The residual association matrix shows no associations with high posterior support, and the associations do not seem to be strongly related to body size (Figure 10.34). As the random effect is largely independent over time and among species, it seems to capture mainly residual variation rather than structured variation.

10.4 What Did the Virtual Ecologists Learn from Their Data?

In this chapter, we have first used an agent-based model to simulate metacommunity processes over time and space (Section 10.1), and then applied HMSC analyses to spatial (Section 10.2) and temporal (Section 10.3) data virtually sampled from these simulations. In this section, we will summarise what we learned from these analyses, and more specifically, what kind of assembly processes the data and the HMSC analyses were and were not informative about.

In the underlying reality of the agent-based simulations, the simulated metacommunity had a high level of competitive interactions, as all of the 100 species competed for the four different resource types (small and large resource units in forest and open areas). In a neutral model, the species may have a hard time coexisting with such a limited number of resource types. Yet in the simulated model, their coexistence was facilitated by niche differences related not only to resource use, but also to tolerance to climatic conditions. The spatial snapshot data acquired by V. E. Space were highly informative about these niche differences, as her analyses essentially reconstructed the variation in the abiotic species niches assumed by the agent-based model. However, her analyses showed almost no signal for competitive interactions and thus about their biotic niches.

In contrast, the analyses of V. E. Time showed clear evidence of density dependence in population-growth rates, suggesting that the dynamics of the community are influenced by competitive interactions. The reason why V. E. Time succeeded while V. E. Space failed in extracting signals of competition is that V. E. Time had much more direct evidence for this, as he could assess not only population sizes but also their changes over time. In particular, his data included the initial period when he observed how the initial high rates decreased as population densities built up. In contrast, his analyses revealed nothing about the species responses related to variation in habitat quality or climatic conditions, simply because his data did not have variation along these axes.

As the results from the spatial snapshot data and the temporal time-series data revealed different aspects of assembly processes, the most comprehensive understanding of the system would have been obtained by combining the studies. This is generally the case in reality; to infer a specific assembly process, one may need to use a specific study design. Thus, it is often better to have smaller amounts of several types of data (temporal, spatial at small scale, spatial at large scale, etc.) rather than a very large amount of the same type of data. This is of course conditional on those smaller amounts of data being sufficiently large that each provides sufficiently high statistical power. We note, however, that having several types of data is not the same as having heterogeneous data that simultaneously include variation in everything, as such data are not generally ideal for finding out the relationships between the variables.

The structural assumptions of the fitted HMSC models were not likely to be perfectly in line with the properties of the data. The data were generated by a mechanistic agent-based model, whereas HMSC is a correlative GLMM. In spite of this, the results of the HMSC analyses could be successfully related back to the assembly processes. Thus, assuming that the researcher interpreting the results of the statistical analyses has expert knowledge about the ecology of the study system, there is potential for inferring the underlying assembly processes from the signatures that they leave in the data, even if doing so conclusively is not possible (Ovaskainen et al. 2017b; Ovaskainen et al. 2019). Of course, the real-world metacommunities are influenced by a much more complex set of drivers than our agent-based simulations, and thus the results presented here should be considered as a very simplified best-case scenario.

Let us also note that while above we restricted the analyses to a single simulation of the underlying reality, a much more comprehensive understanding can be obtained by simulating a variety of scenarios that differ in their underlying assumptions. Some steps in this direction were taken by Ovaskainen et al. (2019), who simulated a competitive metacommunity model that was similar to the one assumed in this chapter. However, unlike the model considered here, Ovaskainen et al. (2019) included a large number of parameterisations of the agent-based model, which included the four classical metacommunity paradigms (Section 1.5.1) as special cases, as well as the continuum of community types that fall in between the classical paradigms. Furthermore, while here we applied HMSC as the sole statistical approach, Ovaskainen et al. (2019) compared HMSC outputs to those of other statistical methods, such as variation partitioning and beta-diversity indices. In line with earlier research (e.g. Chave et al. 2002), Ovaskainen et al. (2019) showed that it is difficult to conclusively infer assembly processes from the signatures that they leave on snapshot data. In particular, none of the statistical methods applied were able to distinguish whether all species follow the same dispersal strategy, or if there is variation in the dispersal strategies among the species. However, many underlying mechanisms, such as whether the species were generalistic or specialised in their resource use, left clear signatures on the outputs of HMSC and some of the other statistical analyses.

11 · *Illustration of HMSC Analyses*

Case Study of Finnish Birds

In this chapter, we apply HMSC to a real dataset on Finnish birds, with the aim of using the case study to simultaneously demonstrate the many uses of HMSC: to illustrate the full workflow of a typical HMSC analysis, to show how the researcher can access the full posterior distribution to go beyond the default outputs of HMSC analyses, to show how predictions of HMSC can be used as a starting point for further analyses, as well as to compare HMSC outputs to results obtained by other statistical methods in community ecology.

We start this chapter by applying the five steps of the HMSC workflow as presented in Section 4.4. In Section 11.2 we show how the researcher can access the entire posterior distribution of model parameters or predictions, e.g. for examining the level of statistical support related to either of these. We then show how one may use HMSC predictions as a starting point for applied research, such as spatial conservation prioritisation (Section 11.3) or bioregionalisation (Section 11.4). Finally, in Section 11.5 we apply other widely used methods in statistical community ecology (as introduced in Chapter 3) to the same data, with the aim of comparing how their results relate to those obtained by HMSC.

11.1 Steps 1–5 of the HMSC Workflow

The data that we will analyse here is an extended version of the data used in Section 5.7 when illustrating single-species modelling with the bird *Corvus monedula*. We will now model a community consisting of the fifty most common bird species in Finland. These data originate from Lindström et al. (2015), and they were utilised as a case study to illustrate the functionality of the Hmsc software (Tikhonov et al. 2020b).

To simplify the data for illustrative purposes, we have made a number of choices while preprocessing the data. First, we have selected the data from the year 2014 only, even if the original data includes repeated surveys. Second, we have selected habitat type and spring temperature as candidate environmental covariates, even if many other environmental variables might also influence the community. Third, we have truncated the species data to presence–absence, and thus decided to ignore the variation in abundance.

The data that we have imported to R consist of the community data matrix Y, the environmental data frame XData, the coordinates of the sampling locations xy, the species trait data frame TrData, and the phylogenetic tree object phyloTree.

Let us first look at the community data.

```
dim(Y)

## [1] 137  50

head(Y[ ,c(1,25,50)] )

##    Phylloscopus_trochilus Parus_montanus Corvus_monedula
## 5                       1              0               1
## 15                      1              1               0
## 23                      1              0               1
## 27                      1              0               1
## 33                      1              0               1
## 37                      1              1               0

S = rowSums(Y)
P = colMeans(Y)
```

We observe that there are data on 50 species in 137 sampling units. As we have truncated the data to presence–absence, the data matrix Y consists of zeros and ones. With the above script, we computed the distributions of species richness and species prevalence illustrated in Figure 11.1. The prevalence of the species ranges from 27 per cent to 99 per cent, reflecting the fact that we have selected common species. Species richness varies greatly among the sampling units, ranging from zero to forty-seven.

Let us then look at the environmental and spatial data.

```
head(XData)

##   Route hab    clim
## 1    13  Op 7.57005
## 2    14  Co 7.74435
```

```
## 3     15 Urb 7.70685
## 4     16  Co 7.63350
## 5     20 Urb 6.89660
## 6     24  Co 7.11275

head(xy)

##    x-coordinate y-coordinate
## 13       5028.6       4167.9
## 14       5052.3       4171.7
## 15       5077.6       4178.0
## 16       5101.1       4183.9
## 20       4876.6       4153.3
## 24       4975.9       4176.3
```

In the environmental data, the sampling units are labelled with an identity number corresponding to the survey route. The categorical environmental covariate hab classifies habitat type into the five levels of broadleaved forests (including mixed forest), coniferous forests, open habitats (mountains and scrubland), urban habitats (human settlements and farmland), and wetlands (inland water ecosystems and peatlands). The habitat data was based on CORINE land cover data from the year 2012 and defined as the most common habitat type within a 300 m buffer about the census sites. The continuous environmental covariate clim (called henceforth climatic conditions) is the mean temperature in April and May in the year 2014, obtained from the European Agency of

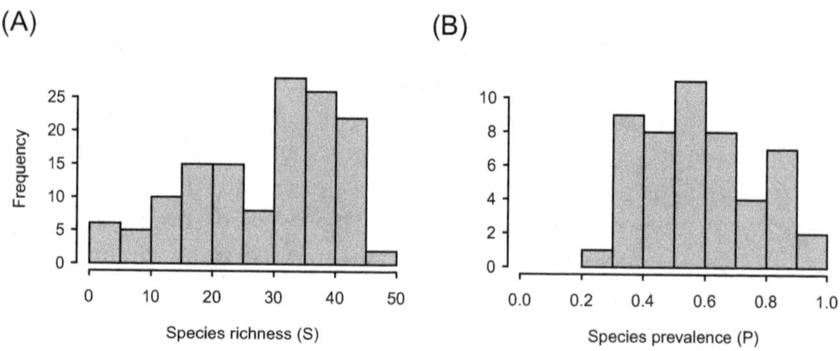

Figure 11.1 Basic summaries of the bird survey data. For species richness (panel A), the y-axis (Frequency) corresponds to the number of sampling units, and the x-axis to the number of species in each sampling unit. For species prevalence (panel B), the y-axis (Frequency) corresponds to the number of species, and the x-axis to the fraction of sampling units in which the species is present.

Climate (Haylock et al. 2008). The matrix xy contains the spatial coordinates of the survey routes.

Let us then inspect the data on species traits and phylogenetic relationships.

```
head(TrData)

##                    Species Migration  LogMass
## 1 Phylloscopus_trochilus           L 2.197225
## 2         Turdus_iliacus           S 4.094345
## 3      Fringilla_coelebs           S 3.091042
## 4     Turdus_philomelos            S 4.234107
## 5       Carduelis_spinus           S 2.564949
## 6            Parus_major           R 2.944439

phyloTree

##
## Phylogenetic tree with 50 tips and 49 internal nodes.
##
## Tip labels:
##  Pica_pica, Corvus_corone, Corvus_corax, Corvus_monedula,
  Garrulus_glandarius, Alauda_arvensis,...
##
## Rooted; includes branch lengths.
```

For each of the fifty species, the data frame TrData includes the log-transformed body mass (Cramp et al. 1977–1994) and migratory strategy classified as long-distance migrant (L), short-distance migrant (S) or resident species (R) (Saurola et al. 2013; Valkama et al. 2014). The object phyloTree contains a phylogenetic tree of the fifty species, acquired from birdtree.org (Jetz et al. 2012).

11.1.1 Step 1. Setting Model Structure and Fitting the Model

We are now ready for Step 1 of the HMSC workflow (Section 4.4): define and fit the HMSC models. Mirroring the single-species analysis of *C. monedula*, we will formulate three models. The Model FULL (m. FULL) includes both the environmental covariates and the spatial random effect of the route. As environmental covariates, we include the habitat type as a categorical variable and climatic condition as a continuous variable. We include the second-order term of the climatic covariate to allow species to have their climatic optimum at an intermediate temperature. The Model ENV (m.ENV) includes environmental covariates but no spatial random effect, whereas the Model SPACE (m.SPACE) includes the spatial random effect but no environmental covariates. All three models include the effects of species traits and phylogenetic relationships.

```
studyDesign = data.frame(route = XData$Route)
rL = HmscRandomLevel(sData = xy)
XFormula = ~ hab + poly(clim, degree = 2, raw = TRUE)
TrFormula = ~ Migration + LogMass
m.FULL = Hmsc(Y = Y, XData = XData, XFormula = XFormula,
            phyloTree = phyloTree, TrData = TrData,
            TrFormula = TrFormula, distr = "probit",
            studyDesign = studyDesign,
            ranLevels = list(route=rL))
m.ENV = Hmsc(Y = Y, XData = XData, XFormula = XFormula,
            phyloTree = phyloTree, TrData = TrData,
            TrFormula = TrFormula, distr = "probit")
m.SPACE = Hmsc(Y = Y, XData = XData, XFormula = ~1,
            phyloTree = phyloTree, TrData = TrData,
            TrFormula = TrFormula, distr = "probit",
            studyDesign = studyDesign,
            ranLevels = list(route = rL))
```

To complete Step 1, we combine all three models into the object models, and fit them to the data.

```
models = list(m.FULL, m.ENV, m.SPACE)
for (i in 1:3){
   models[[i]] = sampleMcmc(models[[i]], thin = thin,
            samples = samples, transient = transient,
            nChains = nChains,
             verbose = verbose, initPar = "fixed effects")}
```

11.1.2 Step 2. Examining MCMC Convergence

Step 2 of the HMSC workflow (Section 4.4) consists of examining MCMC convergence. To keep the script notation short and avoid repetition, we will show the MCMC convergence for Model FULL only. As this model is the most complex, it is actually likely to show the worst MCMC convergence.

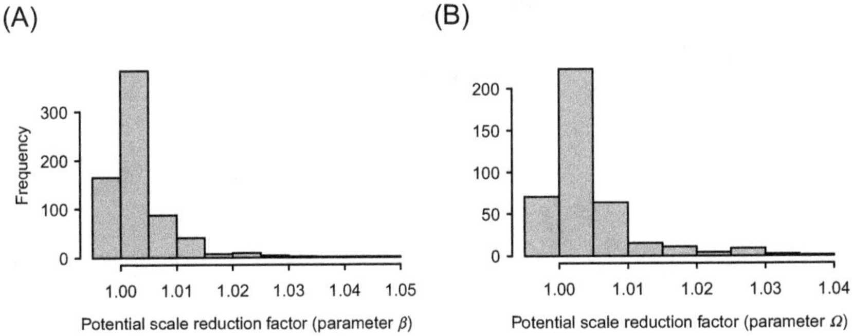

Figure 11.2 MCMC convergence diagnostics for Model FULL, which contains environmental covariates and spatial random effects. The panels show the distribution of the potential scale reduction factor for the β (panel A) and Ω (panel B) parameters.

As usual, we first convert the HMSC object to a coda-object. Here, we restrict the study of MCMC convergence to examining the potential scale reduction factor of the β and Ω parameters. For the latter, we take a subsample of 200 randomly selected species pairs to avoid excessive computation.

```
mpost=convertToCodaObject(models[[1]], spNamesNumbers=c(T,F),
                                      covNamesNumbers = c(T,F))
psrf.beta = gelman.diag(mpost$Beta, multivariate = FALSE)$psrf
tmp = mpost$Omega[[1]]
z = dim(tmp[[1]])[2]
sel = sample(1:z, size = 200)
for(i in 1:length(tmp)){
   tmp[[i]] = tmp[[i]][,sel]
}
psrf.omega = gelman.diag(tmp, multivariate = FALSE)$psrf
```

The MCMC convergence diagnostics can be considered satisfactory, as for most parameters the potential scale reduction factor is close to the ideal value of one (Figure 11.2).

11.1.3 Step 3. Evaluating Model Fit and Comparing Models

Step 3 of the HMSC workflow (Section 4.4) consists of evaluating model fit and comparing models. We first evaluate model fit in terms of explanatory power and predictive power, and then compare the models based on WAIC.

In the script below, we derive predictive power based on two-fold cross-validation.

```
partition = createPartition(models[[1]], nfolds = 2,
  column = "route")
MF = list()
MFCV = list()
for (i in 1:3){
   preds = computePredictedValues(models[[i]])
   MF[[i]] = evaluateModelFit(hM = models[[i]], predY = preds)
   preds = computePredictedValues(models[[i]],
     partition = partition)
   MFCV[[i]] = evaluateModelFit(hM = models[[i]], predY = preds)
}

##                    AUC  AUC (CV)     TjurR2 TjurR2 (CV)
## Model FULL  0.8805655 0.7828792 0.3996951   0.2782083
## Model ENV   0.8349936 0.7843636 0.3559868   0.2845966
## Model SPACE 0.8735123 0.7554882 0.3875864   0.2444525
```

In cross-validation, the Model SPACE does somewhat worse than the two other models, and the Model ENV performs marginally better than the FULL model. Overall, all three models perform very similarly, whether we consider explanatory power or predictive power, and whether we evaluate model fit in terms of AUC or Tjur R^2.

As a complementary measure, we next compare the three models in terms of their WAIC.

```
WAIC = unlist(lapply(models, FUN = computeWAIC))

##  Model FULL Model ENV  Model SPACE
##    20.80341  21.59316     21.85181
```

Model comparison based on WAIC also suggests that the differences among the three models are small. However, unlike the cross-validation approach performed above, WAIC ranks Model FULL as the best. We recall that WAIC approximates leave-one-out cross-validation, whereas the 'brute-force' cross-validation that we conducted above was based on splitting the data into two folds only. Thus, part of the difference might be because an appropriate estimation of the spatial random effect requires more data than were available for two-fold cross-validation. Furthermore, we recall that model comparison in our two-fold cross-validation was based on the discrimination power measured by AUC and Tjur R^2, whereas WAIC measures the goodness-of-fit in terms of the log posterior predictive score.

The results from model comparison suggest that the environmental and spatial predictors convey overlapping information, and thus the choice between Models FULL and ENV should be based on the purpose for which the model will be used, rather than the differences in predictive power.

11.1.4 Step 4. Exploring Parameter Estimates

Step 4 of the HMSC workflow (Section 4.4) consists of exploring the parameter estimates. For this, we first perform a variance partitioning of the three models. As Model SPACE contains only the spatial random effect, all the explained variation will be attributed to the random effect. The comparison between Models FULL and ENV will thus be more informative about the importance of the environmental covariates and spatial effects in explaining the bird community.

To be able to group the environmental variables, we look at the design matrix X that Hmsc has constructed by applying the XFormula to the XData.

```
head(models[[1]]$X)

## (Intercept) habCo habOp habUrb habWe poly(clim)1 poly(clim)2
## 1          1     0     1      0     0      7.57005     57.30566
## 2          1     1     0      0     0      7.74435     59.97496
## 3          1     0     0      1     0      7.70685     59.39554
## 4          1     1     0      0     0      7.63350     58.27032
## 5          1     0     0      1     0      6.89660     47.56309
## 6          1     1     0      0     0      7.11275     50.59121
```

We observe that columns 2–5 relate to habitat variation and columns 6–7 to climatic variation. Arbitrarily, we include the intercept in the habitat variables, and thus perform the variance partitioning with the following grouping of the columns of the X matrix.

```
groupnames = c("habitat", "climate")
group = c(1,1,1,1,1,2,2)
VP = list()
for (i in 1:2){
```

```
    VP[[i]] = computeVariancePartitioning(models[[i]],
              group = group, groupnames = groupnames)
}

##                 habitat   climate    random
## Model FULL  0.1578784 0.7416933 0.1004283
## Model ENV   0.1899937 0.8100063 0.0000000
## Model SPACE 0.0000000 0.0000000 1.0000000
```

The results of the variance partitioning show that Model FULL relies only little on the spatial random effect, which also explains why it performed so similarly to the Model ENV. Climatic variation explains almost twice the variance than does habitat variation. We note that these results show average variance proportions over the species, and thus for some individual species the habitat may matter more than climatic conditions.

To examine variation among the species, we visualise their niches in Figure 11.3 by generating a plot for Model FULL with the function plotBeta. In Figure 11.3, the habitat preferences of the species are compared to the reference level of broadleaved forests (Br). We can observe that there are only few responses to coniferous forests, which means that most species do not strongly prefer coniferous forests over broadleaved forests, and vice versa. Concerning the other habitat types of wetland, open habitats and urban habitats, we observe statistically supported responses for many species, reflecting the fact that these habitats are very different from forests. For most species, the linear response to spring temperature is positive whereas the quadratic effect is primarily negative. As we will later see in the prediction plots, this means that the occurrence probability of most species peaks at warm climatic conditions, and thus so too does the species richness.

We next examine whether the species niches are linked to their traits and phylogenetic relationships. Figure 11.4, which is generated for Model FULL with the function plotGamma, does not suggest that the responses of the species are tightly linked to the trait data (panel A), but it confirms that on average, the species respond positively to temperature. When relaxing the level of statistical support, we observe that resident migratory behaviour appears to be positively correlated with coniferous habitat use, and that short migratory behaviour is negatively correlated with wetland use (panel B).

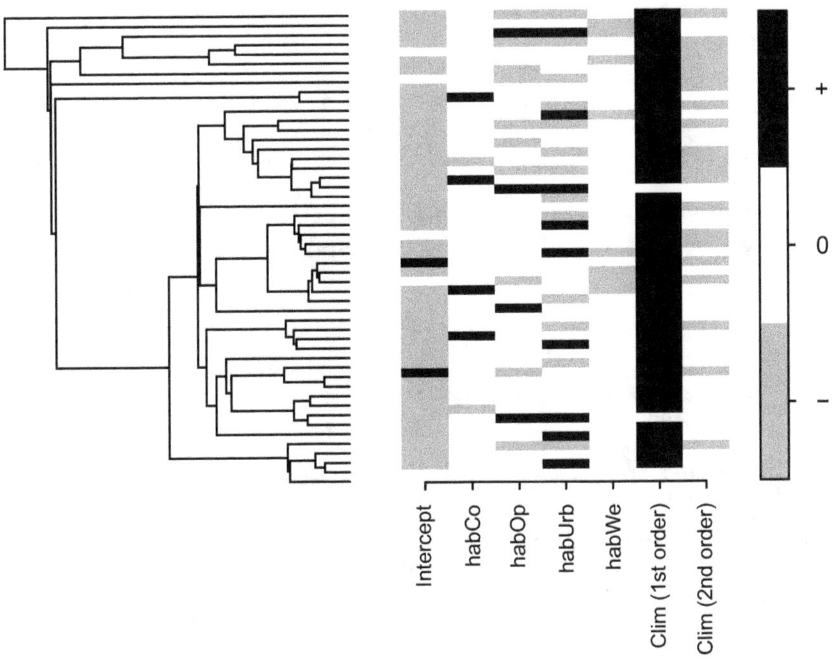

Figure 11.3 Heatmap of estimated parameters β, i.e. species niches. Black and grey colours show parameters that are estimated to be positive and negative, respectively, with at least 0.95 posterior probability in Model FULL.

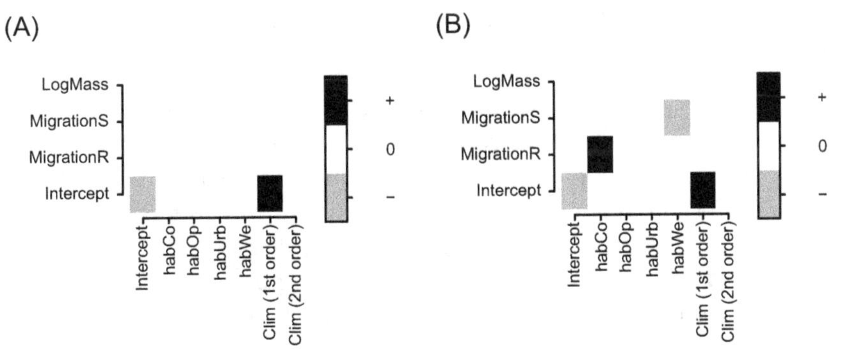

Figure 11.4 Heatmap of estimated γ parameters linking species traits to species niches. Black and grey colours show parameters that are estimated to be positive and negative, respectively, with at least 0.95 (panel A) or 0.85 (panel B) posterior probability in the FULL model.

```
postGamma = getPostEstimate(models[[1]], parName = "Gamma")
par(mfrow = c(1,2))
plotGamma(models[[1]], post = postGamma, param = "Support",
  supportLevel = 0.95)
plotGamma(models[[1]], post = postGamma, param = "Support",
  supportLevel = 0.85)
```

Another way of examining the influence of traits is to assess how much of the variation they explain among the responses of the species to their covariates.

```
VP[[1]]$R2T$Beta

## (Intercept) habCo habOp habUrb habWe poly(clim)1 poly(clim)2
##       0.093 0.115 0.095  0.033 0.288       0.092      0.0939
```

These results are consistent with Figure 11.4: the traits explain only a minor part of the variation. The same pattern is observed when quantifying the proportion of variation in species occurrence that the traits explain:

```
VP[[1]]$R2T$Y

## [1] 0.09822434
```

Furthermore, there is no evidence of phylogenetic signal in species niches.

```
mpost = convertToCodaObject(models[[1]])
round(summary(mpost$Rho, quantiles = c(0.025, 0.5, 0.975))
    [[2]],2)

##  2.5%   50% 97.5%
##  0.00  0.00  0.33
```

The results reported above indicate that the fifty species included in our analyses respond mostly individualistically to environmental variation, without dependences on the included traits and phylogenetic relationships.

Figure 11.5 illustrates the species associations revealed by the random effects. We observe a much stronger pattern of associations in Model SPACE (Figure 11.5A) than in Model FULL (Figure 11.5B). This is to be expected, as the associations revealed by Model SPACE are raw associations, whereas the associations revealed by Model FULL are residual associations that account for the responses of the species to the habitat and climatic conditions. By comparing the associations inferred from Model SPACE versus Model FULL, we can see that many of the positive associations revealed by Model SPACE can be explained by the shared responses of the species to the environmental covariates. The bird data

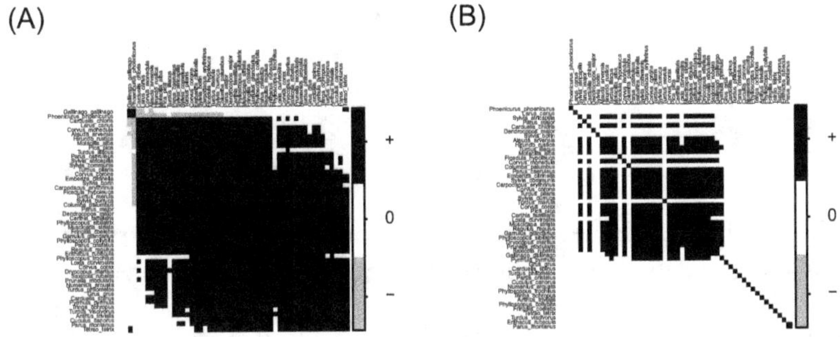

Figure 11.5 Residual species-to-species associations shown for Model SPACE (A), and Model FULL (B). The black and grey colours indicate those species pairs with at least 0.95 support for either a positive or negative association, respectively. In the version of the figure shown in the Colour Plate, the intensity of the colour represents the strength of the association, measured in units of correlation.

are assemblage data where the sampling-unit scale is not necessarily linked to biotic assembly processes, and so the positive associations in Model FULL may well relate to the species responses to missing environmental variables, such as more refined measures of habitat quality than the broad classification used in our example. Another plausible explanation for some of those residual associations is unaccounted variation in the observation probability, such as the differences in bird activity that relate to the specific weather conditions on the day of the survey. Finally, some of the positive associations may reflect true interactions among bird species, such as heterospecific attraction (Mönkkönen & Forsman 2002). As discussed extensively in Chapter 7, disentangling these possibilities is unfortunately not possible in the absence of additional data.

Let us next examine the spatial scale at which the variation is captured by the random effect. To do so, we examine the spatial scale paramerer α_h for the two first factors ($h = 1, 2$) included in the model, out of which the first factor ($h = 1$) is more important in the sense of being the leading factor that explains most of the variation.

```
mpost = convertToCodaObject(models[[1]])
round(summary(mpost$Alpha[[1]], quantiles
    = c(0.025,0.5,0.975))[[2]][1:2,],2)

##                    2.5%    50%   97.5%
## Alpha1[factor1]     0    0.00    0.00
## Alpha1[factor2]     0  265.43 1175.46
```

```
mpost = convertToCodaObject(models[[3]])
round(summary(mpost$Alpha[[1]], quantiles
    = c(0.025,0.5,0.975))[[2]][1:2,],2)
##                      2.5%    50%  97.5%
## Alpha1[factor1]  101.11 151.67 252.79
## Alpha1[factor2]   25.28  37.92 164.31
```

In Model FULL, the leading factor is non-spatial, and thus the variation is independent among the survey routes. In Model SPACE, the variation in the leading factor occurs at the scale of ca. 150 km, reflecting the scale at which the relevant environmental conditions vary.

11.1.5 Step 5. Making Predictions

Step 5 of the HMSC workflow (Section 4.4) consists of using the fitted model to make predictions. We start by creating gradient plots that illustrate how the bird communities are predicted to vary among the environmental variables. Both Figures 11.6 and 11.7 are generated by applying the constructGradient and the plotGradient functions to Model FULL.

Figure 11.6 shows examples of the many kinds of predictions that can be performed with HMSC. We show the predictions for occurrence probability of the species *C. monedula*, overall species richness, proportion of resident species and community-weighted mean of log-transformed body weight along the climatic gradient present in the data. We observe that the occurrence probability of the species *C. monedula* increases with the climatic variable (Figure 11.6A), consistent with the single-species analysis performed in Section 5.7. The species richness (Figure 11.6B) increases with increasing temperature, reflecting the fact that most species respond positively to temperature (Figure 11.4). Concerning the traits, we observe that the mean proportion of resident species increases with temperature (Figure 11.6C), whereas the average body mass slightly decreases with it (Figure 11.6D).

Figure 11.7 shows predictions for the same community features illustrated in Figure 11.6, but in this case for the habitat gradient. Consistent with the single-species analysis of Section 5.7, the occurrence probability of the species *C. monedula* is highest in urban habitats (Figure 11.7A). Species richness is lowest in open habitats and wetlands (Figure 11.7B), and these habitats also have the lowest proportions of resident species (Figure 11.7C). These results partially reflect the responses of the species

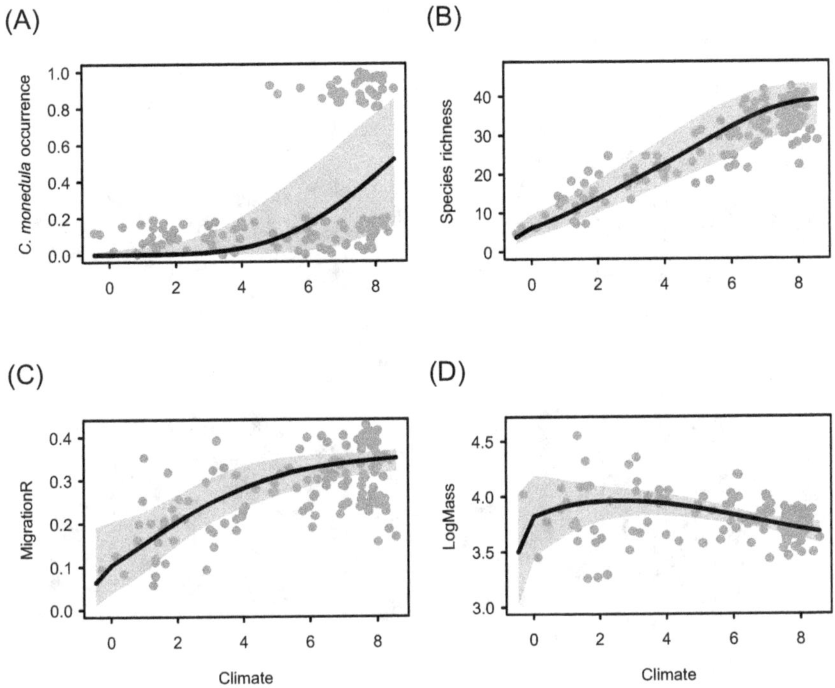

Figure 11.6 Predictions over the climatic gradient present in the data. The predictions have been made by varying the climatic conditions and setting habitat type to the most common habitat type conditional on the climatic variable. The panels show the predicted occurrence probability of *C. monedula* (A), species richness (B), proportion of resident species (C) and community-weighted mean of log-transformed body weight (D).

to climatic conditions, as these two habitats are the most common in northern Finland where the climate is the coldest. In contrast, the average body size is essentially independent of habitat type (Figure 11.7D), even if it shows some dependency on the climate (Figure 11.6D).

In addition to the predictions over the environmental gradients, we will perform spatial predictions with Model FULL. To do so, we proceed as in the single-species case study in Section 5.7. We first import a grid of spatial coordinates, habitat types and climatic conditions for 10,000 locations where we wish to predict the communities. We then apply the prepareGradient function to these data to prepare a spatial gradient.

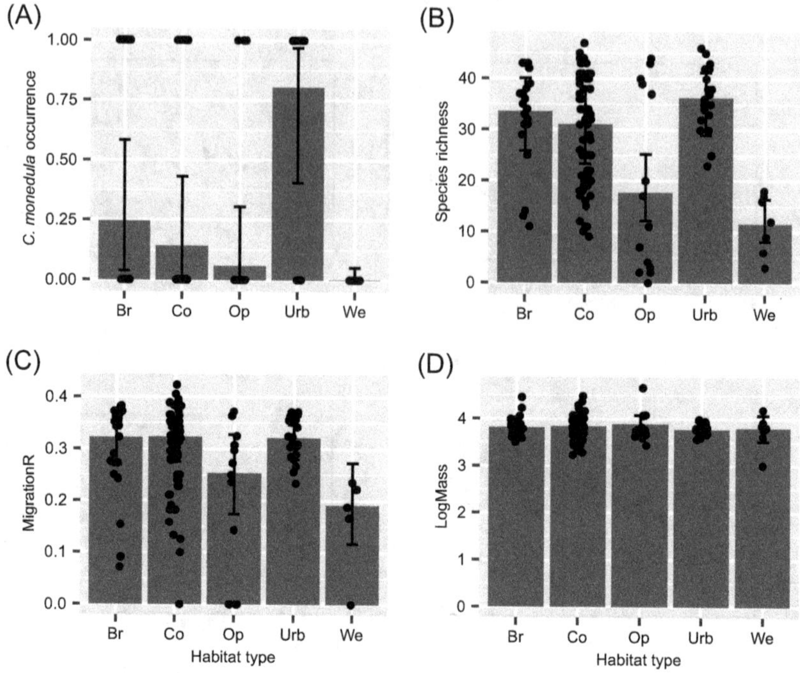

Figure 11.7 Predictions over the habitat gradient present in the data. The predictions have been made by varying the habitat type and setting climatic conditions to their expectation conditional on the habitat type. The panels show the occurrence probability of *C. monedula* (A), species richness (B), proportion of resident species (C), and community-weighted mean of log-transformed body weight (D).

```
grid = read.csv(file.path(data.directory,
  "bird data\\grid_10000.csv"))
grid = droplevels(subset(grid,!(Habitat=="Ma")))
xy.grid = as.matrix(cbind(grid$x,grid$y))
XData.grid = data.frame(hab = grid$Habitat,
  clim = grid$AprMay)
m = models[[1]]
Gradient = prepareGradient(m, XDataNew = XData.grid,
                           sDataNew = list(route = xy.grid))
```

We next compute the posterior predictive distribution, which we compress to the posterior mean prediction.

```
predY = predict(m, Gradient = Gradient, predictEtaMean = TRUE)
EpredY = Reduce("+", predY)/length(predY)
```

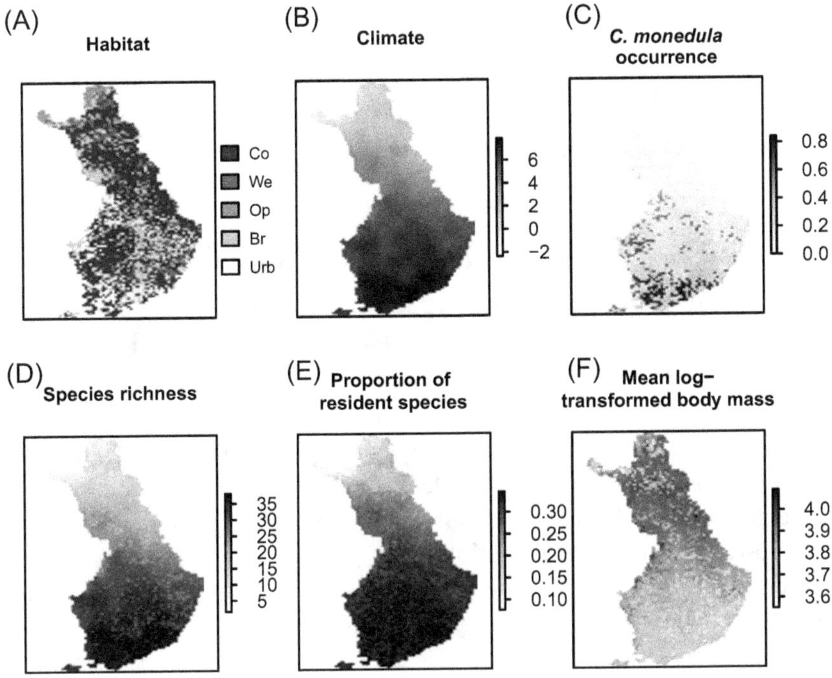

Figure 11.8 Environmental data and predicted community features across Finland. The panels show the habitat (A) and climatic (B) variables for 10,000 prediction locations, the predicted occurrence of *C. monedula* (C), species richness (D), proportion of resident species (E), and the community-weighted mean body size (F). For a colour version of the figure, see the Colour Plate.

In the next step, we postprocess the predictions for those community features that we wish to illustrate over the prediction space. With the script below, we derive from the predictions the occurrence probability of *C. monedula* (species number 50), species richness and community-weighted mean traits.

```
Cm = EpredY[ ,50]
S = rowSums(EpredY)
CWM = (EpredY %*% m$Tr)/matrix(rep(S, m$nt), ncol = m$nt)
xy = grid[,1:2]
```

Figure 11.8A and B shows the variation in the habitat and climatic conditions on which the predictions are based. Figure 11.8C shows the predicted distribution of *C. monedula* across Finland, which is reassuringly very similar to that based on the single-species model in Section 5.7.

The lower row of panels in Figure 11.8 illustrates community-level predictions. Reflecting the results over the climatic gradient, species richness (Figure 11.8D), and proportion of resident species (Figure 11.8E) are highest in southern Finland, and community-weighted body size (Figure 11.8F) is smallest.

11.2 Measuring the Level of Statistical Support and Propagating Uncertainty into Predictions

With maximum likelihood inference, the best estimate for a parameter is given by the maximum likelihood estimate, the level of statistical evidence is measured by statistical significance given by a *p*-value and uncertainty in parameter values is quantified by confidence intervals. These concepts do not apply to HMSC, since it is implemented in the Bayesian context. Instead, the best estimate for a parameter is given by the posterior mean (or sometimes the posterior median or posterior mode), the level of statistical support is measured by posterior probability and the uncertainty in a parameter value is quantified by credible intervals. Thus, in order to be able to correctly interpret the HMSC outputs, it is important to understand the meanings of these concepts. While the formal theory of Bayesian inference is extensively covered e.g. in Gelman et al. (2013), here we will illustrate these concepts in terms of how they apply specifically to HMSC. We will also show how the user can access the full posterior distribution in Hmsc, and postprocess it in a more flexible way than may be possible when applying the ready-made functionality. For this, we will use the Model FULL as the example.

HMSC is implemented in the Bayesian context and thus the parameters are estimated by the posterior distribution. As illustrated with the script below, the posterior distribution is stored as the variable postList in the HMSC model object. The variable postList is a list, with each element corresponding to one MCMC chain. Each MCMC chain further forms its own list, with each element corresponding to one posterior sample. As we fitted the model with two chains and obtained 1,000 (thinned) samples for each, there are 2,000 samples in total. In the script below, we use the function poolMcmcChains to merge the two chains into one list of 2,000 posterior samples. Each of these samples is a named list, which contains the parameters in the notation used in this book.

```
m = models[[1]]
length(m$postList)

## [1] 2

length(m$postList[[1]])

## [1] 1000

post = poolMcmcChains(m$postList)
length(post)

## [1] 2000
names(post[[1]])

## [1] "Beta"   "Gamma" "V"     "rho"  "sigma" "Eta"  "Lambda"
## [8] "Alpha" "Psi"   "Delta"
```

As one example, in Figure 11.3 we applied the plotBeta function to illustrate how the species respond to environmental variation. A visual examination of that plot shows that the species number fifty (*C. monedula*, which appears as the lowest species in the figure) responds more positively to the urban habitat type than to the reference habitat type of broadleaved forests. We will illustrate how one can extract this result from the posterior distribution. The species responses to environmental covariates are found from the Beta (β) parameters. Let us assess the dimension of these parameters from the first posterior sample.

```
dim(post[[1]]$Beta)
## [1] 7 50
```

The seven rows of the Beta (**B**) matrix correspond to the environmental covariates that define the columns of the **X** matrix.

```
colnames(m$X)

## [1] "(Intercept)"
## [2] "habCo"
## [3] "habOp"
## [4] "habUrb"
## [5] "habWe"
## [6] "poly(clim, degree = 2, raw = TRUE)1"
## [7] "poly(clim, degree = 2, raw = TRUE)2"
```

The fifty columns of the Beta matrix correspond to the species, out of which species number fifty is our target.

```
m$spNames[50]

## [1] "Corvus_monedula"
```

Thus, the element [4,50] of the Beta matrix ($\beta_{4,50}$) measures the influence of the urban habitat type (compared to the habitat type of broadleaved forests) on *C. monedula*. We next form a vector that contains the 2,000 posterior samples specifically for this parameter. This can be done in many ways using the basic R functionality. In the script below, we define the function getvalue that takes one posterior sample of all parameters as the input, and returns the value of $\beta_{4,50}$. We then apply this function to all elements of the list post of posterior samples, and finally apply the function unlist to convert the list into a vector called beta_4_50. The vector beta_4_50 includes 2,000 values that define the marginal posterior distribution of the parameter $\beta_{4,50}$.

```
getvalue = function(p){ return(p$Beta[4,50] )}
beta_4_50 = unlist(lapply(X = post, FUN = getvalue))
```

Now it is easy to compute the posterior mean estimate.

```
mean(beta_4_50)

## [1] 1.098784
```

This can be written mathematically as $E[\beta_{4,50}] = 1.099$, where $E[\cdot]$ denotes the expectation over the posterior distribution. We note that to emphasise that the expectation is over the posterior rather than the prior distribution, we could write in a fuller notation $E[\beta_{4,50} | y]$ instead of as $E[\beta_{4,50}]$, where y denotes the data, which in the case of Hmsc is the collection of all data matrices (Section 8.1.4).

It is equally easy to compute posterior median, which equals 50 per cent quantile, as well as any other posterior quantiles that the user defines.

```
median(beta_4_50)

## [1] 1.089477

quantile(beta_4_50,probs = c(0.025,0.5,0.975))

##      2.5%      50%     97.5%
## 0.364932 1.089477 1.879125
```

These can be written mathematically as $\Pr[\beta_{4,50} < 0.36] = 0.025$, $\Pr[\beta_{4,50} < 1.09] = 0.5$, and $\Pr[\beta_{4,50} < 1.88] = 0.975$, where $\Pr[\cdot]$ denotes posterior probability. We may further characterise the 95 per cent credible interval with the help of the posterior probability $\Pr[0.36 < \beta_{4,50} < 1.88] = 0.95$.

As a further example, let us compute the level of statistical support by which $\beta_{4,50}$ is greater than zero, as well as the level of statistical support by which it is greater than two.

```
mean(beta_4_50 > 0)

## [1] 0.999

mean(beta_4_50 > 2)

## [1] 0.013
```

These results can be written as $\Pr[\beta_{4,50} > 0] = 0.999$ and $\Pr[\beta_{4,50} > 2] = 0.013$. The first part means that there is a very high level of evidence that the parameter is positive, i.e. that the species prefers urban habitats over broadleaved forests. This is why this parameter was coloured black in Figure 11.3. The second part reveals that it is very unlikely that the preference for urban areas is so strong that the parameter value would be greater than two, as it implies that $\Pr[\beta_{4,50} \leq 2] = 1 - 0.013 = 0.987$. In general, posterior probabilities close to zero and close to one indicate a high level of statistical support.

We next illustrate how parameter uncertainty can be propagated into model predictions. To do so, let us return to the spatial predictions on Finnish bird communities presented in Figure 11.8. These predictions were generated with the following lines of code:

```
predY = predict(m, Gradient = Gradient, predictEtaMean = TRUE)
EpredY = Reduce("+", predY)/length(predY)
```

Here predY is the full posterior distribution of predictions. Thus, it is a list of length 2,000, where each element of the list is a matrix of the predicted occurrence probabilities of the fifty species in the prediction locations. In Figure 11.8, we showed predictions based on the posterior mean, denoted by EpredY in the above code. Thus, Figure 11.8 shows the 'best' predictions (in terms of the posterior mean) but ignores their uncertainty.

We will illustrate prediction uncertainty using species richness (Figure 11.8D) as an example. In the script below, we first store the posterior distribution of species richness in the array aS.

```
getS = function(p){ return(rowSums(p))}
aS = simplify2array(lapply(X = predY, FUN = getS))
dim(aS)

## [1]  9454 2000
```

The rows of this array are the 9,454 prediction locations (note that we could not make predictions for the full set of the 10,000 locations as some of them presented habitat type not included in the training data), and the columns are the posterior samples. In the next script we derive three summaries from the posterior distribution of species richness: posterior mean (ES), posterior standard deviation (sdS) and the posterior probability of the prediction location having more than twenty-five species (S25).

```
ES = apply(aS, 1, mean)
sdS = sqrt(apply(aS, 1, var))
S25 = apply(aS > 25, 1, mean)
```

Figure 11.9 illustrates these three vectors as maps. The map of the posterior mean (Figure 11.9A) is identical to that shown in Figure 11.8. The map of the posterior standard deviation (Figure 11.9B) illustrates the amount of uncertainty around the posterior mean. Informally, one may think of the prediction as the posterior mean plus/minus the posterior standard deviation. The level of uncertainty is typically in the order of two to three species, being lowest in northern Finland. Figure 11.9C is a map of the posterior probability of the prediction location having more

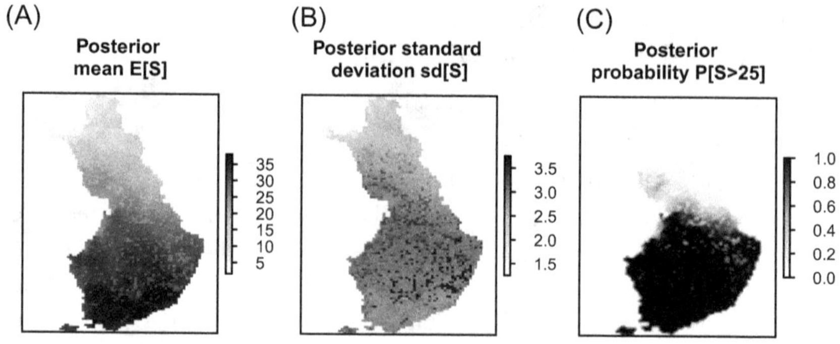

Figure 11.9 Illustration of uncertainty in model predictions. The panels show the posterior mean estimate of species richness (A), the posterior standard deviation of species richness (B), and the posterior probability by which there are more than twenty-five species (C). For a colour version of the figure, see the Colour Plate.

than tweny-five species. It shows that in southern Finland one observes more than twenty-five species almost surely, whereas in northern Finland one observes less than twenty-five species almost surely. Only in central Finland is it difficult to predict whether the number of species will be more or less than twenty-five, as in this area the posterior probability is not close to either zero or one.

When assessing uncertainty, it is crucial to specify the type of uncertainty that is being assessed. For example, the uncertainty measured by the standard deviation of the posterior distribution in Figure 11.9B solely measures parameter uncertainty. If we would have a very large amount of data, parameter uncertainty would vanish and thus the posterior standard deviation would be close to zero. However, this would not mean that bird communities would be deterministic. When surveying a community where the predicted species richness is twenty-five, the actual number of species found may be twenty or thirty species. This is because we have made our prediction not for the realised number of species, but for the expected number of species. This choice was hidden when applying the predict function, as the setting expected = TRUE is the default, thus was not made explicit in the above scripts. Other aspects of uncertainty that we have ignored in our assessment above, and in which the researcher might be interested, is uncertainty related to model structure and uncertainties related to possible errors in the data.

11.3 Using HMSC for Conservation Prioritisation

The ongoing global change, including habitat loss and climate change, has resulted in the so-called sixth mass extinction. The rate of biodiversity loss is more alarming than ever; the average rate of vertebrate species loss over the last century is up to 100 times higher than the background rate (Ceballos et al. 2015). To find solutions to this problem, one of the overarching aims in conservation biology is to design effective methods of protecting species, habitats and ecosystems in an integrative way, namely systematic conservation planning (Margules & Pressey 2000; Margules & Sarkar 2007). One part of systematic conservation planning is conservation prioritisation, which is aimed at providing decision support about when, where and how can we most efficiently achieve conservation goals (Wilson et al. 2007). In particular, spatial conservation prioritisation aims to answer the 'where' part of the question, by identifying where the most important areas for biodiversity are (Kukkala & Moilanen 2013). In order to answer this question, one first needs to know how biodiversity is structured across space. This is where HMSC comes into play, as it can be used to make spatial predictions that can then be fed into a spatial prioritisation algorithm.

We will exemplify spatial conservation prioritisation with the Zonation software (see Lehtomäki & Moilanen 2013 for references). As input for the Zonation analysis, we use the predicted occurrence probabilities of the 50 bird species over the ca. 10,000 prediction locations (Section 11.1.5). When making a conservation prioritisation, one needs to make a number of choices related to the specific objectives of the prioritisation. In the context of Zonation, such choices influence how much each biodiversity feature is weighted in the priority ranking. In our example, we apply the two alternative methods of the additive benefit function formulation (ABF) and the core-area Zonation (CAZ). The difference between these two is that ABF implicitly gives higher preference to areas that have many species, whereas CAZ attempts to represent all species in the prioritised areas. Thus, use of ABF will translate into elevated priorities for areas with high species richness, while CAZ may prioritise species-poor areas as well if they are important for some specific species (Moilanen 2007).

The difference between these two prioritisation algorithms is reflected in the prioritisation maps that the Zonation software displayed (Figure 11.10). As conservation priority areas, the ABF method highlighted southern Finland, where the species richness is the highest. On the other hand, the CAZ method also ranked some parts from northern Finland, where more unique species occur. Clearly, the conservation goal will influence which ranking method is to be used. ABF may be a more appropriate choice when the number of species indicates a larger species pool, whereas CAZ might be more appropriate when the coverage of every species included in the analyses is of interest.

While the maps of Figure 11.10A and B describe where the most important areas are, another question that can be addressed with a Zonation analysis is how much of the biodiversity features are lost when only the top ranked areas are conserved. This question is addressed by the so-called performance curves, which are exemplified in Figure 11.10C. The x-axis of Figure 11.10C shows the proportion of the landscape that is lost (0 indicating that the entire landscape is conserved and 1 that the entire landscape is lost), and the y-axis shows the proportion of the species distributions that are lost (0 indicating the entire distributions are conserved and 1 that the entire distributions are lost). The performance curves can thus be used to ask e.g. what proportion of the original

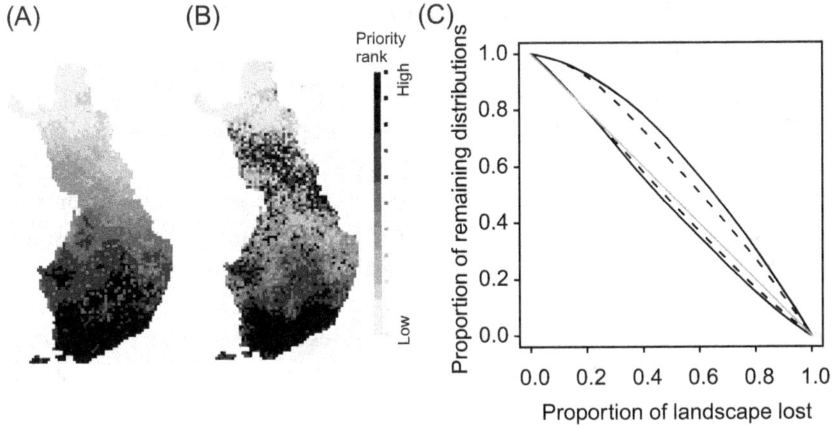

Figure 11.10 Conservation prioritisations produced by applying Zonation to HMSC-predicted distribution maps on the fifty bird species. The maps show the result based on the ABF method (A) and the CAZ method (B). In the performance curves of panel C, the continuous curves correspond to the ABF method and the dashed curves to the CAZ method. The grey line shows the null expectation that the proportion of remaining distribution is proportional to the proportion of remaining landscape. Curves above and below the grey line refer to average occurrence of all species, and to the species that is the least represented in the prioritisation, respectively. For a colour version of the figure, see the Colour Plate.

distributions can be conserved, if the conserved 10 per cent is selected according to the top 10 per cent suggested by the prioritisation map. In particular, we may ask how much better is the prioritisation than the null expectation that conserving 10 per cent of the area would conserve 10 per cent of the species distributions. If considering the average over all species, the mean performance curves in Figure 11.10C are clearly above the null expectation, meaning that both the ABF and the CAZ methods have successfully chosen areas that represent biodiversity as priority areas better than would be expected by choosing the areas randomly. This is not surprising, as it is exactly what Zonation has been designed to do. However, both the ABF and CAZ curves are slightly worse than expected by chance for the species that is least represented in the prioritisations (the lower curves in Figure 11.10C), illustrating that it is not possible to simultaneously select the most important part of the landscape for all fifty species. In line with the discussion above, the performance curves suggest that the ABF method does better in terms of the average species, whereas the CAZ method does slightly better in terms of the least represented species.

11.4 Using HMSC for Bioregionalisation: Regions of Common Profile

The concept of bioregionalisation derives back to the work of Wallace (1876), who classified the world to zoogeographical regions by combining the biological collections with expert knowledge, and to Köppen (1884), who classified the world in terms of climatic regions, based on the types of vegetation that grow under different climatic conditions. In bioregionalisation, the complexity of real-world species communities is simplified by classifying them into distinct types that occupy distinct spatial units. The classification of communities into biogeographical units (bioregions) provides an essential tool for management of natural resources, as well as for conservation planning (Hill et al. 2017; Ladle & Whittaker 2011; Lomolino et al. 2010).

While bioregionalisation would ideally be based on complete data on ecological communities, such data are not typically available. Because of this, data on environmental covariates (matrix **X**, Figure 2.1) is often used as a proxy for community data (matrix **Y**, Figure 2.1), based on the assumption that environmental filtering shapes variation in species communities. In this way, access to high-resolution environmental data enables mapping the bioregions also at sites where community data are not available.

Typical questions that are addressed in bioregionalisation include how many distinct types of bioregions the data support, what is their spatial distribution over the study area and what species and environmental characteristics are associated with each bioregion. Two kinds of model-based approaches have been developed specifically for addressing these questions. The approach that is perhaps the most directly aimed for bioregionalisation is the Regions of Common Profile (RCP, Foster et al. 2013). RCP is a mixture model that groups the sites based on the community data, and then models the groups as a function of environmental covariates. Another approach is the Species Archetype Model (SAM, Dunstan et al. 2011). SAM is also a mixture model, but instead of grouping the sites, it groups the species based on their responses to the ecological covariates.

Here we illustrate how HMSC can be applied to bioregionalisation. We note that this is somewhat similar to applying SAM, as both SAM and HMSC are JSDMs. However, while SAM clusters the species responses to environmental variation into distinct groups, HMSC assumes that there is continuous variation among the species in their responses. Thus, in HMSC-based bioregionalisation, the clustering of species and sites is conducted as a postprocessing step on the predicted communities.

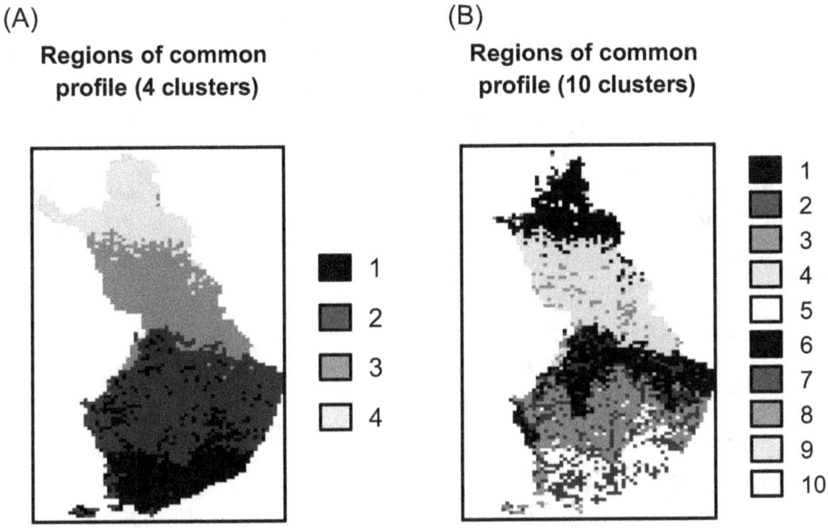

Figure 11.11 RCPs derived from the Finnish bird data. Each colour corresponds to a distinct species community type. The figure has been generated by applying clustering to the species distributions predicted by HMSC. In panel A, the number of clusters has been set to 4, whereas in panel B it has been set to 10. For a colour version of the figure, see the Colour Plate.

Among the many methods available for clustering, we apply the k-means clustering, as implemented by the kmeans function of the R-package stats. As input arguments for the clustering, we use the predicted species communities (EpredY), as well as the number of clusters we wish to generate. To illustrate this, we generate one clustering with four groups and another with ten groups. In the script below, we call each cluster RCP, referring to the 'Region of Common Profile' terminology.

```
RCP4 = kmeans(EpredY, 4)
RCP10 = kmeans(EpredY, 10)
```

The RCPs are largely aligned along the South–North gradient (Figure 11.11) and thus reflect the fact that bird communities respond strongly to climatic variation (Figure 11.6). The main patterns are very similar, whether the data are clustered into four clusters (Figure 11.11A) or ten clusters (Figure 11.11B). Naturally, adding more clusters results in a more fine-scaled partitioning of the study area.

How many clusters should then be included? Sometimes the optimal number of clusters is set by the purpose of the bioregionalisation, for example when the number of management units is preset. But often the optimal number of clusters is part of the research question: how many clusters do the data best support? Among the many methods that can be used to address this question, in the script below we apply the Elbow method and the Silhouette method, as implemented in the fviz_nbclust function of the factoextra and Nbclust R-packages (Charrad et al. 2014, Kassambara & Mundt, 2019).

```
# Elbow method
p1 = fviz_nbclust(EpredY, kmeans, method = "wss") +
     geom_vline(xintercept = 4, linetype = 2) + labs(sub
     title = "Elbow method")

# Silhouette method
p2 = fviz_nbclust(EpredY, kmeans, method = "silhouette") +
     labs(subtitle = "Silhouette method")
```

While the silhouette method suggests that the optimal number of clusters is two, the Elbow method suggests that the optimal number is four (Figure 11.12). This demonstrates that the optimal number of clusters depends on the criteria used, and thus there is no single correct answer. Since the aim of this section is only to show how bioregionalisation is performed with HMSC, we will not further discuss the choice of optimal number of clusters; rather, we continue with four clusters in our example.

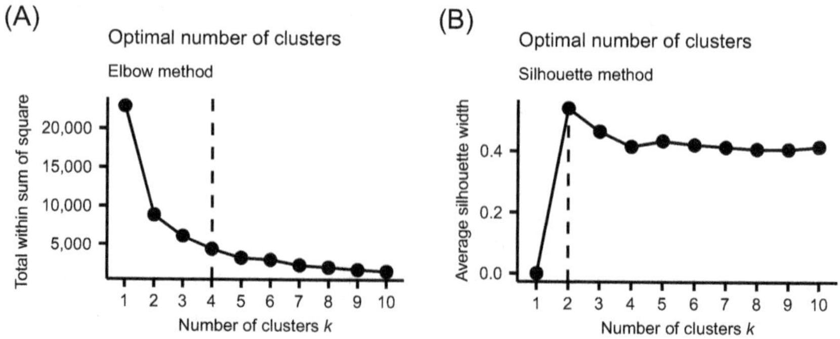

Figure 11.12 Optimal number of clusters (RCP) determined by the Elbow (A) and Silhouette (B) methods.

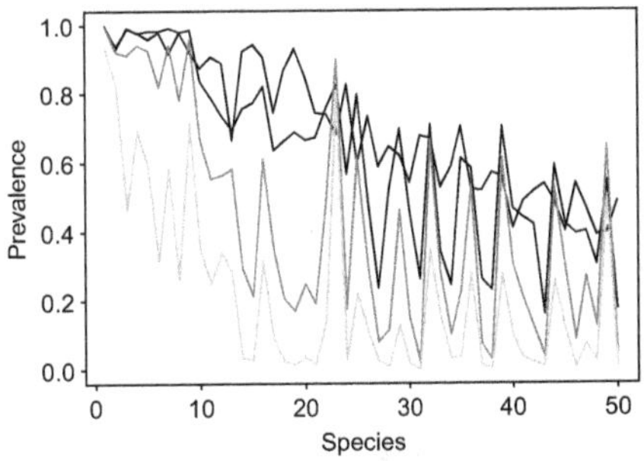

Figure 11.13 Species profiles in the four bioregions shown in Figure 11.11A. The colours correspond to the four bioregions, and the species are ordered according to their overall prevalence in the data.

Next, we will ask which species characterise each bioregion. For this, we may simply look at the centres of the clusters, which represent the average occurrence probabilities of each species in each cluster. We illustrate this for two species: *Phylloscopus trochilus* (species 1) and *Alauda arvensis* (species 43).

```
RCP4$centers[ , c(1,43)]
##    Phylloscopus_trochilus Alauda_arvensis
## 1               0.9932634     0.537671826
## 2               0.9981295     0.160264228
## 3               0.9948072     0.037686714
## 4               0.9345992     0.008665761
```

We observe that *P. trochilus* is present in all clusters with very high prevalence, and thus this species does not separate bioregions. In contrast, *A. arvensis* is common in the most southern bioregion (Bioregion 1), relatively common in the second-most southerly bioregion (Bioregion 2) and very rare in the two northernmost bioregions (Bioregions 3 and 4), and thus it contributes to separating the bioregions.

We can also plot the species profile of each bioregion, as is done in Figure 11.13. As expected, the darkest and lightest lines are most separate from each other, as they represent the southernmost and northernmost bioregions.

For another view of the species communities, we may simply list the most common species in each region, as is done with the script below for the five most common species in each bioregion.

```
MCS = matrix(NA,m$ns,4)
for (i in 1:4){
    MCS[,i] = names(rev(sort(RCP4$centers[ i,] )))
}
colnames(MCS) = c("Region 1","Region 2","Region 3","Region 4")
MCS[1:5,]
##       Region 1                  Region 2
## [1,] "Phylloscopus_trochilus" "Phylloscopus_trochilus"
## [2,] "Fringilla_coelebs"      "Fringilla_coelebs"
## [3,] "Parus_major"            "Anthus_trivialis"
## [4,] "Turdus_philomelos"      "Cuculus_canorus"
## [5,] "Erithacus_rubecula"     "Carduelis_spinus"

##       Region 3                  Region 4
## [1,] "Phylloscopus_trochilus" "Phylloscopus_trochilus"
## [2,] "Cuculus_canorus"        "Turdus_iliacus"
## [3,] "Anthus_trivialis"       "Phoenicurus_phoenicurus"
## [4,] "Turdus_philomelos"      "Cuculus_canorus"
## [5,] "Carduelis_spinus"       "Turdus_philomelos"
```

We observe that *P. trochilus* is the most common species in all regions, but the commonness ranking of the other species varies among bioregions. For example, while *Turdus iliacus* is the second-most common species in the northernmost region (Bioregion 4), it is not ranked in the top five in the other regions.

To obtain a more functional view on how the species communities differ among the bioregions, we may look at the distribution of community-weighted mean traits, as is done in Figure 11.14. In line with our earlier analyses, we observe that the proportion of resident

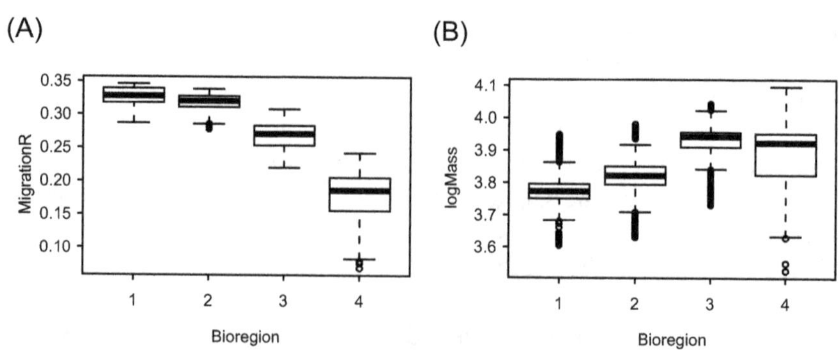

Figure 11.14 Species traits characterising each bioregion. The panels show the distributions of proportion of resident species (A) and the community-weighted mean of log-transformed body weight (B) over the locations from each bioregion.

species is the smallest in the northernmost Bioregion 4, and that the average body size is smallest in the southernmost Bioregion 1.

11.5 Comparing HMSC to Other Statistical Methods in Community Ecology

In this section, we will provide a brief comparison between HMSC and some other commonly applied statistical methods in community ecology (see Chapter 3). To do so, we will apply alternative methods to the bird data that are used throughout this chapter. Following the order in which the different methods were introduced in Chapter 3, we will first apply ordination analyses (including RDA, RDA-based variance partitioning, and fourth-corner methods), followed by co-occurrence analyses and finally univariate analyses of species richness. As we wish to keep our treatment brief, we will apply only methods that are qualitatively different from HMSC. Thus, we will not contrast HMSC to other SDMs or JSDMs (see Norberg et al. (2019) for such comparisons).

As a starting point, we extract the species data, environmental covariate data, spatial data and trait data from the HMSC object of the full model.

```
m = models[[1]]
Y = m$Y
XData = m$XData
TrData = m$TrData
xy = m$rL$route$s
```

11.5.1 Redundancy Analysis and Variance Partitioning

We start with redundancy analysis (RDA, Legendre & Anderson 1999), which is a very widely applied method of constrained ordination (Section 3.1). To perform this, we apply the rda function from the vegan R-package (Oksanen et al. 2019), setting the community data as the response and the habitat and climatic variables as the constraining explanatory variables.

```
library(vegan)
myrda = rda(Y ~ hab + clim, data = XData)
plot(myrda, display = c('bp','site','sp'))
```

The RDA analysis identifies climate to be related as the main axis of variation (RDA1), and places most species to the positive side of

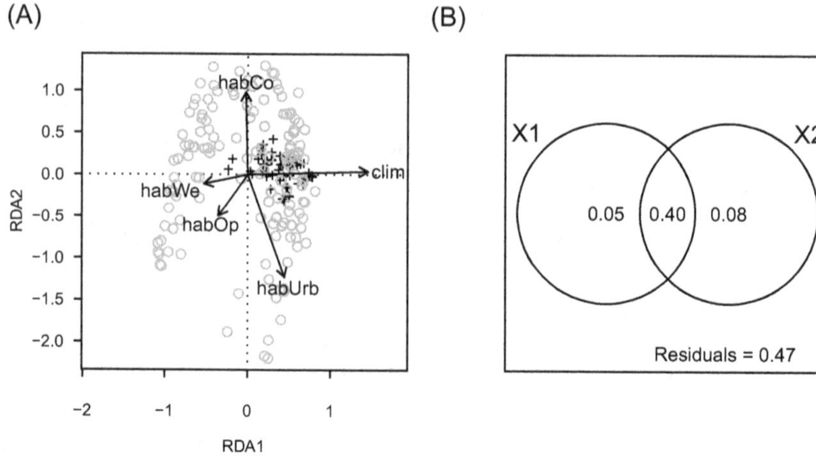

Figure 11.15 Results of RDA analyses. Panel A shows an RDA triplot for the bird community data (circles represent the study sites, vectors represent the environmental covariates and crosses represent the species). Panel B shows the results of RDA-based variance partitioning, indicating the amount of variance explained by the environmental (X1) and spatial (X2) predictors. For a colour version of the figure, see the Colour Plate.

the climatic gradient (Figure 11.15A). The habitat-urban gradient (RDA2) is almost orthogonal to the climatic gradient, whereas the wetlands and open areas are associated with the cold climate. These results are well in line with the HMSC results reported in Section 11.1.

We next apply distance-based RDA to determine the proportion of the variation that is explained by environmental or spatial predictors, following McArdle and Anderson (2001). In the script below, we use the varpart function of the vegan R-package, where we use a distance matrix derived from the community data as a response variable, and environmental and spatial predictors as explanatory variables. We have generated the spatial predictors with the tools developed by Dray et al. (2006) and Bauman et al. (2018), as implemented in the adespatial R-package (Dray et al. 2019). We first use the function listw.candidates to generate a set of candidate predictors, and then select the final set of predictors with the function listw.select. The results of this variance partitioning (Figure 11.15B) are also fully in line with the results obtained in the HMSC analyses, as they suggest that both the environment and space explain a substantial part of the community variation, with most of the explanatory power being shared between the two types of predictors.

```
library(ade4)
library(adespatial)
candidates = listw.candidates(xy, nb = c("gab"),
  weights = c("binary","flin"))
modsel.Y = listw.select(Y, candidates, method = "FWD",
                MEM.autocor = "positive", p.adjust = TRUE)
MEM.spe = modsel.Y$best$MEM.select
vY = vegdist(Y, method = "bray")
VP = varpart(vY, XData, MEM.spe)
plot(VP)
```

11.5.2 Fourth-Corner Analysis

In order to bring traits – and thus a more functional perspective – to the analyses, we next perform a fourth-corner analysis (Legendre et al. 1997). The input for the fourth-corner method includes the environmental data (called **X** in the HMSC context and **R** in the fourth-corner method context), the community data (called **Y** in the HMSC context and **L** in the fourth-corner context) and the species trait data (called **T** in the HMSC context and **Q** in the fourth-corner context). We perform the analyses with the function fourthcorner from the ade4 R-package (Dray et al. 2018). We follow the procedure proposed by Dray and Legendre (2008), which includes fitting two kinds of models.

The first model that we apply is the permutation model 2 proposed by Dray and Legendre (2008), for which the null model is obtained by permuting the sites (i.e. all rows of the community data matrix **L**). This null model preserves the link between community data and

```
library(ade4)
four2 = fourthcorner(XData, as.data.frame(Y), TrData,
                nrepet = 99, model type = 2)
summary(four2)
## Fourth-corner Statistics
## ------------------------------
## Permutation method 2 ( 99  permutations)
##
## Adjustment method for multiple comparisons:   holm
##              Test Stat    Obs Std.Obs   Alter Pvalue Pvalue.adj
## 1 hab / Migration Chi2  7.530   3.024   greater   0.03      0.06
## 2 clim / Migration  F 17.901  34.113   greater   0.01      0.04
## 3    hab / LogMass  F  0.027  -1.379   greater   1.00      1.00
## 4   clim / LogMass  r -0.047  -4.563 two-sided   0.01      0.04
```

trait data, but not between the community and environmental data. Thus, it tests the null hypothesis that there is no link between the species trait data and environmental data (Dray & Legendre 2008).

This part of the fourth-corner analysis suggests that the underlying null hypothesis can be rejected, thus the environmental data and the trait data are linked. More specifically, the results show that climate is negatively related to the log-transformed body size of the species, which is in accordance with the HMSC result suggesting that body size increases towards the colder climates in the North (Figure 11.6D). The results further show that climate and migratory strategies are related, which is also in line with the HMSC result showing that the resident migratory strategy is more common in warmer climates (Figure 11.6C). Concerning the effect of habitat, the fourth-corner analysis suggests a relationship between habitat type and migratory strategy, but not with body mass, as the HMSC results also suggested (Figure 11.7C and D).

The second model that we apply is the permutation model 4 proposed by Dray and Legendre (2008), for which the null model is obtained by permuting the species (i.e. all columns of the community data matrix **L**). This null model preserves the link between the community and environmental data, but not between the community and trait data. Thus, it tests the null hypothesis that there is no link between the community data and species trait data (Dray & Legendre 2008).

```
four4 = fourthcorner(XData, as.data.frame(Y), TrData, nrepet = 99,
                     modeltype = 4)
summary(four4)
## Fourth-corner Statistics
## ---------------------------
## Permutation method 4 ( 99  permutations)
##
## Adjustment method for multiple comparisons:   holm
##            Test Stat   Obs Std.Obs      Alter Pvalue  Pvalue.adj
## 1  hab / Migration Chi2 7.530  -0.025    greater   0.46        1.00
## 2 clim / Migration    F 17.901   1.274    greater   0.12        0.48
## 3    hab / LogMass    F  0.027  -1.321    greater   1.00        1.00
## 4   clim / LogMass    r -0.047  -0.982 two-sided   0.34        1.00
```

The results of this analysis do not reveal any associations between the community and trait data. Thus, we conclude that there is no evidence that the community data and species trait are linked beyond the link generated by the environmental data. We note that when both of the above analyses reveal significant associations between the data matrices,

Dray and Legendre (2008) propose performing a third step that combines the two analyses.

11.5.3 Co-occurrence Analysis

We next move to co-occurrence analyses, which are used to ask whether the species show aggregated or segregated distributional patterns with respect to each other (Section 3.2). To do so, we apply the methods of Gotelli (2000), implemented in the function cooc_null_model of the EcoSimR R-package (Gotelli et al. 2015). In the script below, we first run the analysis using the default settings.

```
library(EcoSimR)

summary(cooc_null_model(speciesData = t(Y)))

## Metric:  c_score
## Algorithm:  sim9
## Observed Index:  423.61
## Mean Of Simulated Index:  393.11
## Variance Of Simulated Index:  0.5492
## Lower 95% (1-tail):  391.93
## Upper 95% (1-tail):  394.69
## Lower 95% (2-tail):  391.79
## Upper 95% (2-tail):  394.87
## Lower-tail P >  0.999
## Upper-tail P <  0.001
## Observed metric > 1000 simulated metrics
## Observed metric < 0 simulated metrics
## Observed metric = 0 simulated metrics
## Standardised Effect Size (SES):  41.154
```

Note that we have used here the transposed community data as the input argument, because cooc_null_model assumes that the species are placed as rows and the sites as columns of the matrix. With the default settings, the output metric is the checkerboard score (C-score). The observed C-score is compared to the expected distribution under the null model, which permutes the data matrix in a way that keeps both the row sums and the column sums constant. Consequently, this analysis identifies co-occurrence patterns beyond those determined by how species richness varies over the sites, and how species prevalences vary among the species. The observed C-score is higher than the simulated C-score, indicating that the species co-occur less than expected by chance, which is expected in competitively structured communities (Gotelli 2000).

The above result is in apparent contradiction with the HMSC analyses, which identified mainly positive raw associations among the species. However, this is because the raw species associations in HMSC are based on fitting an intercept-only model. The intercept-only model estimates species incidences, and thus it constrains species prevalences, but it does not constrain how species richness varies over the sites. Thus, a more direct comparison between raw associations of HMSC and Gotelli's co-occurrence analysis is obtained by using the algorithm 'sim2' instead of the default choice of 'sim9'. The 'sim2' algorithm preserves the differences among species in incidence, but assumes that all sites are equiprobable (Gotelli 2000).

```
summary(cooc_null_model(speciesData = t(Y), algo = "sim2"))

## Metric:  c_score
## Algorithm:  sim2
## Observed Index:  423.61
## Mean Of Simulated Index:  824.5
## Variance Of Simulated Index:  25.869
## Lower 95% (1-tail):  815.41
## Upper 95% (1-tail):  831.91
## Lower 95% (2-tail):  814.04
## Upper 95% (2-tail):  833.34
## Lower-tail P >  0.001
## Upper-tail P <  0.999
## Observed metric > 0 simulated metrics
## Observed metric < 1000 simulated metrics
## Observed metric = 0 simulated metrics
## Standardised Effect Size (SES):  -78.82
```

In this analysis, the observed C-score is lower than the simulated C-score. Thus, in line with the raw associations estimated by HMSC, these results indicate that the species co-occur more often than expected by chance. An interesting question is whether the segregation among species, as identified by the 'sim9' analysis, could be replicated by a HMSC analysis. As 'sim9' constrains both row and column sums, a comparable HMSC analysis might be an intercept-only model with a non-structured random effect implemented through a single latent factor. In this model, the species-specific intercept would control for species incidences, whereas the single latent factor would capture variation in species richness. However, we will not pursue this topic further; a more profound comparison of HMSC analyses and co-occurrence analyses is left for future work.

11.5.4 Analysis of Species Richness

As a final method, we apply univariate generalised linear modelling of diversity metrics (Section 3.3). While other diversity metrics could also be considered, here we use species richness as the response variable – despite being the simplest possible diversity metric, it is the most commonly applied (Magurran 2004). We model species richness by Poisson regression, where the candidate explanatory variables were the same variables that we used in the HMSC analyses: habitat type as categorical variable, and climate as continuous variable. In the scripts below, we select the most parsimonious model based on the AIC criterion (Burnham & Anderson 2002).

We first test whether there is support for the second-order effect of climate. To do so, we fit two models, one of which includes the second-order response to climate, and the other that includes only the linear response to it. Both models control for the effect of habitat.

```
S = rowSums(Y)
m1 = glm(S ~ poly(clim, degree = 2) + hab, data = XData,
    family = "poisson")
m2 = glm(S ~ clim + hab, data = XData, family = "poisson")
AIC(m1, m2)

##      df      AIC
## m1   7 863.2783
## m2   6 897.6404
```

The AIC-based comparison between the two models supports the inclusion of the second-order effect. We next compare models with and with without the effect of habitat, both of which include the second-order effect of climate.

```
m3 = glm(S ~ poly(clim, degree = 2), data = XData,
    family = "poisson")
AIC(m1,m3)

##      df      AIC
## m1   7 863.2783
## m3   3 859.1176
```

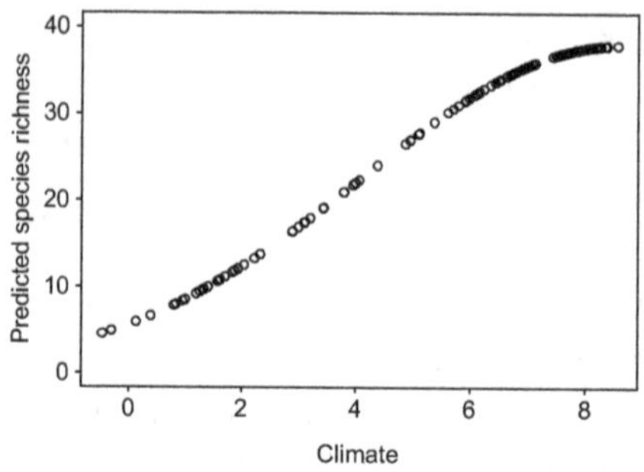

Figure 11.16 Predicted species richness as a function of climate (mean temperature in April and May).

These results suggest that habitat does not influence species richness when the effect of climate is accounted for. Thus, the most parsimonious model is the one which includes the first and second-order effects of climate but excludes the effect of habitat type.

To illustrate the results, we plot predicted species richness as a function of the climatic variable from the selected model.

```
plot(XData$clim, exp(predict(m3)))
```

Note that in the script above, we have taken the exponential of model predictions to convert the results from the scale of the linear predictor to the scale of the data, and thus to counteract the effect of the log link function that is included in Poisson regression. The results are fully in line with the HMSC analyses: species richness peaks at the warmest climate (Figure 11.16). Note that the second-order effect is visible by the steeper increase in species richness in colder climates than in warmer climates.

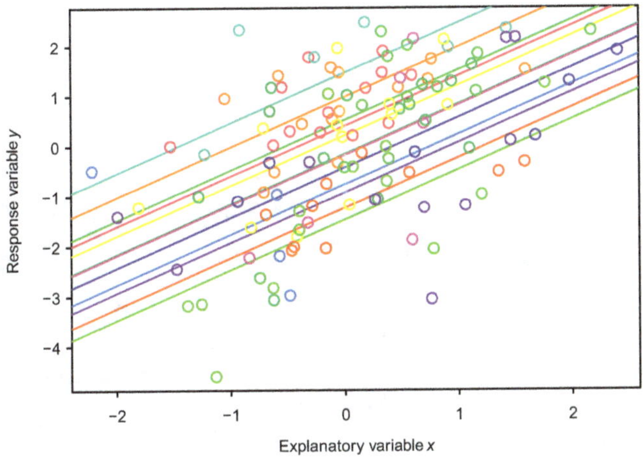

Figure 5.6 Scatterplot of the simulated data originating from twelve plots. The lines show the plot-specific models that were used to generate the data. Each plot is indicated by one colour.

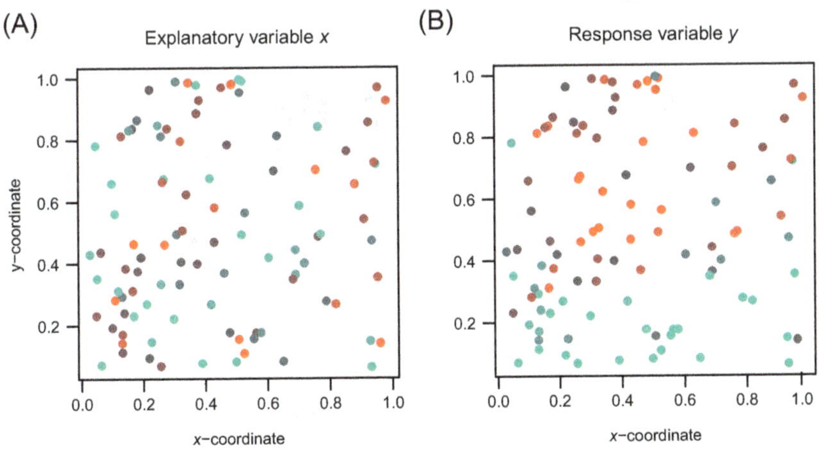

Figure 5.7 Plots of simulated spatially structured data. The panels show the spatial variation in the explanatory variable x (A) and in the response variable y (B). Lowest values are shown in blue, highest in red.

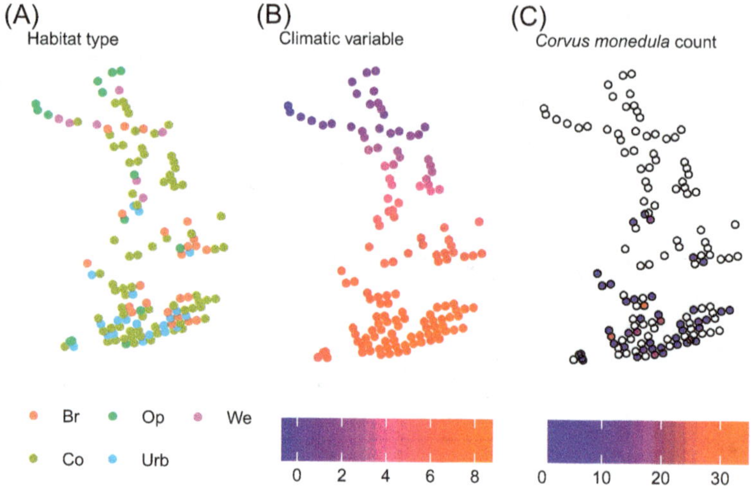

Figure 5.9 An illustration of environmental and species data used in this example. The panels show spatial variation in habitat type (A), climatic conditions (B), and the counts of the target species across Finland (C).

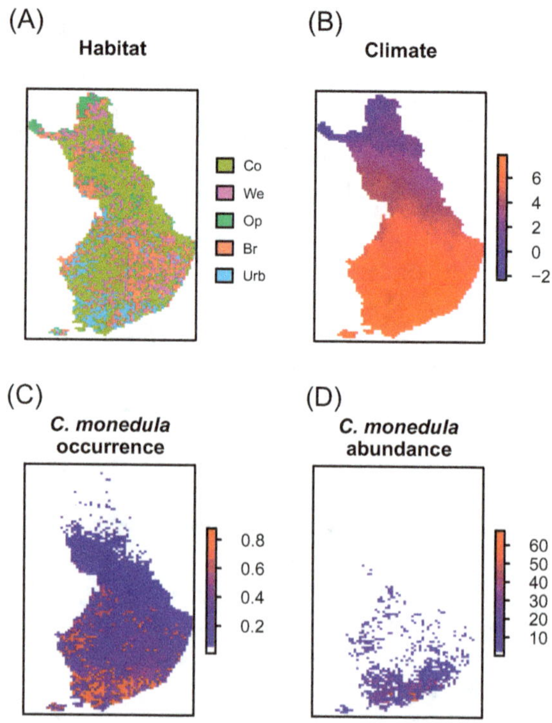

Figure 5.13 Environmental data and predicted species distribution over all of Finland. The panels show habitat (A) and climatic (B) variables for 10,000 prediction locations, and the predicted occurrence (C) and abundance (D) of the target species.

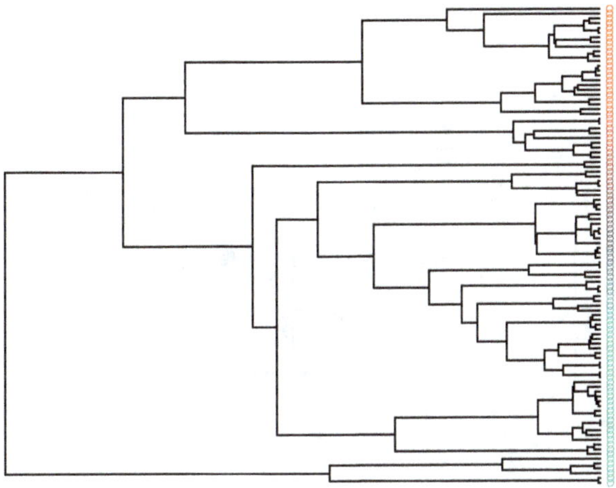

Figure 6.3 A simulated phylogeny of 100 species. The species are at the tips of the tree, and the branch lengths correspond to the duration of time that the species have belonged to the same ancestral state during their evolutionary history. The species have been coded with different colours for later reference.

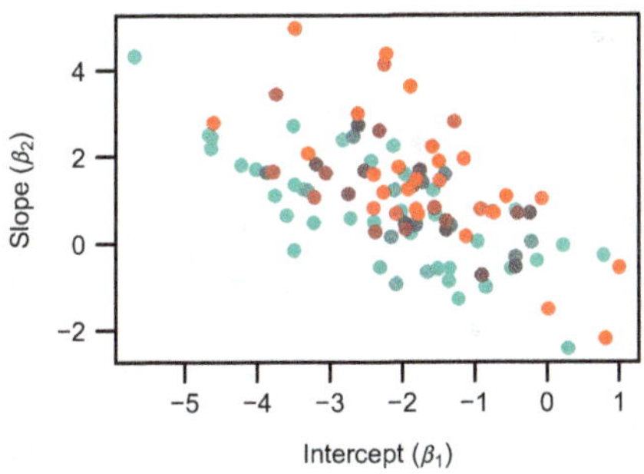

Figure 6.4 Illustration of variation in species niches structured by species phylogeny. The colours of the dots refer to the species location in the phylogeny shown in Figure 6.3.

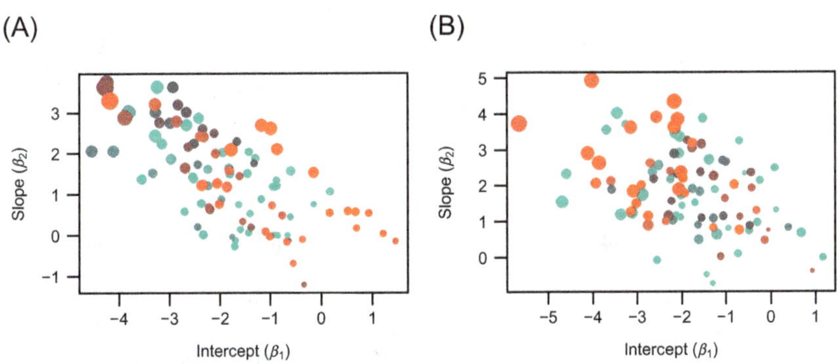

Figure 6.5 Illustration of variation in species niches structured by species phylogeny (panel A) and traits (panel B). In both panels, the colours of the dots refer to the species location in the phylogeny shown in Figure 6.3, and the size of the dot refers to a measured trait of the species.

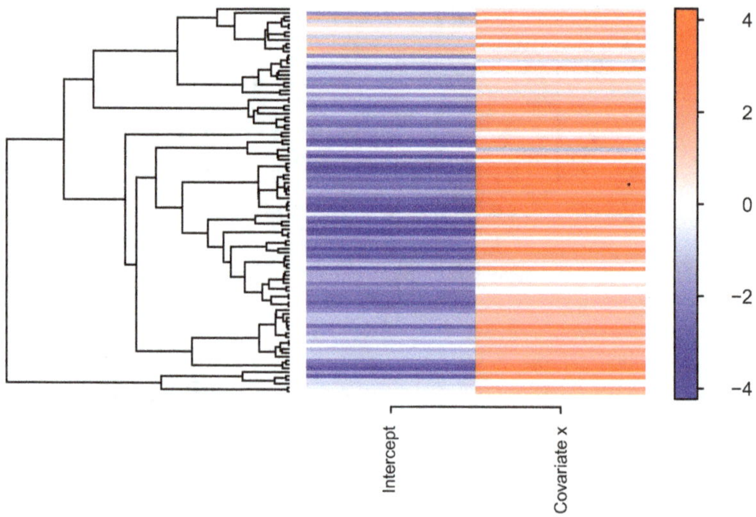

Figure 6.10 Heatmap of estimated species niches. Red and blue colour indicates parameters that are estimated to be positive and negative, respectively, with at least 0.95 posterior probability. The intensity of the colour represents the posterior mean estimate of the parameter, obtained by setting param = "Mean".

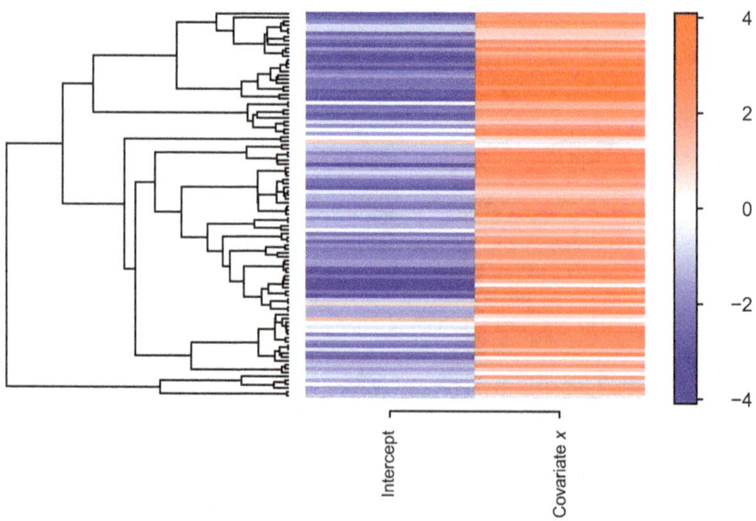

Figure 6.14 Heatmap of estimated species niches. Red and blue show parameters that are estimated to be positive and negative, respectively, with at least 0.95 posterior probability. The intensity of the colour represents the posterior mean estimate of the parameter, obtained by setting param = "Mean". The results shown here are for the Community B, while Figure 6.10 shows the same results for the Community A.

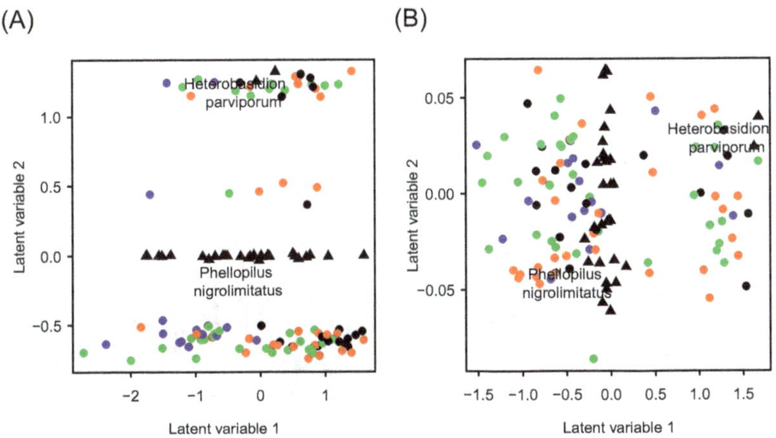

Figure 7.10 Biplots illustrating the latent variables modelling the random effect part of the HMSC model. In the biplots, the site scores (each corresponding to one sampling unit) are shown by the circles, the colour showing the decay class of the log. The triangles show to the species scores (each corresponding to one species), of which only *H. parviporum* and *P. nigrolimitatus* are labelled in the figure, to avoid excessive overlap of text. Panel A corresponds to the model that includes only sequencing depth as a covariate and thus shows raw associations. Panel B corresponds to the model that includes both sequencing depth and decay class as covariates and thus shows residual associations.

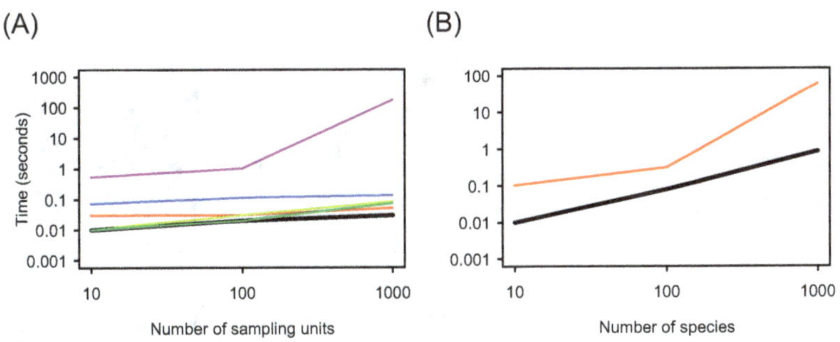

Figure 8.9 Time required for fitting an HMSC model for ten MCMC iterations for different numbers of sampling units (panel A) and species (panel B). In panel A, the baseline case (thick continuous black line) is a normal model fitted for ten species. The baseline includes intercept only ($n_c = 1$), and has no traits, phylogeny or random effects. The other cases deviate from the baseline by having environmental covariates ($n_c = 10$; red line), being based on a probit (green line) or Poisson (yellow line) model, including non-structured random effects (blue line), or spatial random effects (purple line). In panel B, the baseline case (thick continuous line) is the baseline case of panel A, fitted for ten sampling units. The red line deviates from the baseline by including a phylogeny. The models were fitted with an HP EliteBook 840 G3 laptop, with two 2.40 GHz cores and 16.0 GB RAM.

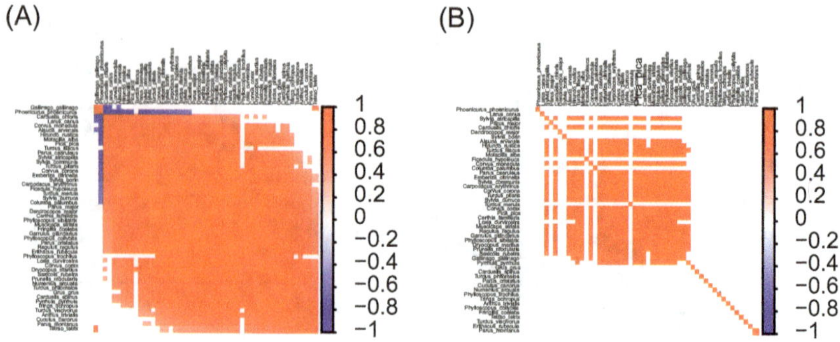

Figure 11.5 Residual species-to-species associations shown for Model SPACE (A), and Model FULL (B). The red and blue colours indicate those species pairs with at least 0.95 support for either a positive or negative association, respectively. The intensity of the colour represents the strength of the association, measured in units of correlation.

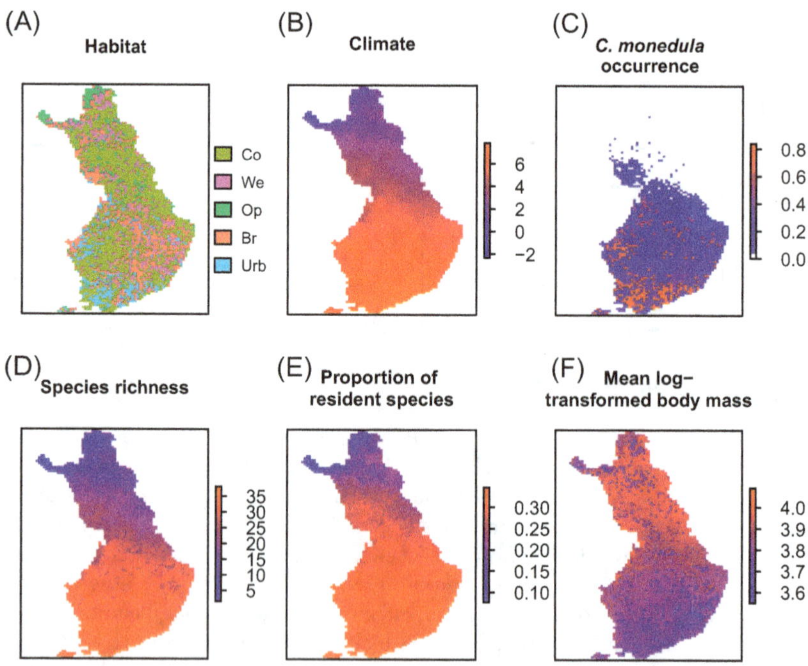

Figure 11.8 Environmental data and predicted community features across Finland. The panels show the habitat (A) and climatic (B) variables for 10,000 prediction locations, the predicted occurrence of *C. monedula* (C), species richness (D), proportion of resident species (E), and the community-weighted mean body size (F).

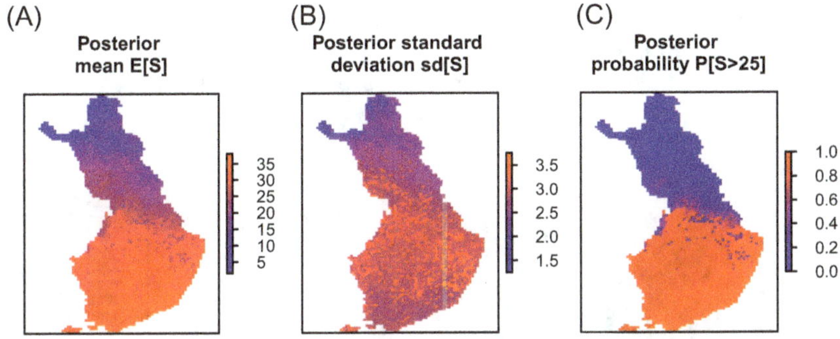

Figure 11.9 Illustration of uncertainty in model predictions. The panels show the posterior mean estimate of species richness (A), the posterior standard deviation of species richness (B), and the posterior probability by which there are more than twenty-five species (C).

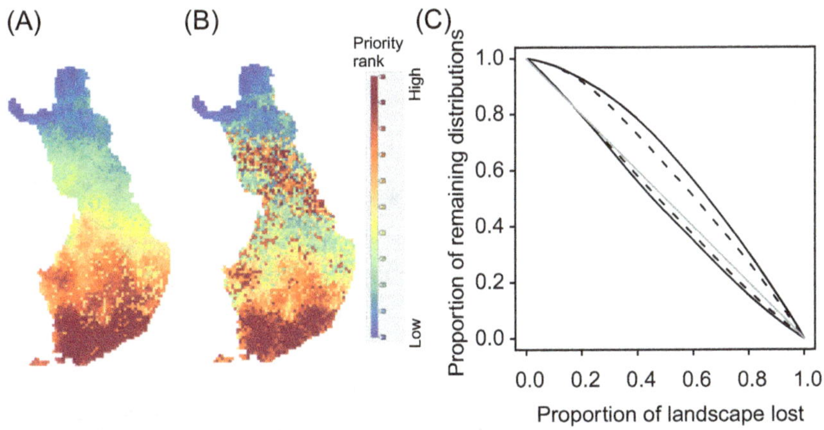

Figure 11.10 Conservation prioritisations produced by applying Zonation to HMSC-predicted distribution maps on the fifty bird species. The maps show the result based on the ABF method (A) and the CAZ method (B). In the performance curves of panel C, the continuous curves correspond to the ABF method and the dashed curves to the CAZ method. The grey line shows the null expectation that the proportion of remaining distribution is proportional to the proportion of remaining landscape. Curves above and below the grey line refer to average occurrence of all species, and to the species that is the least represented in the prioritisation, respectively.

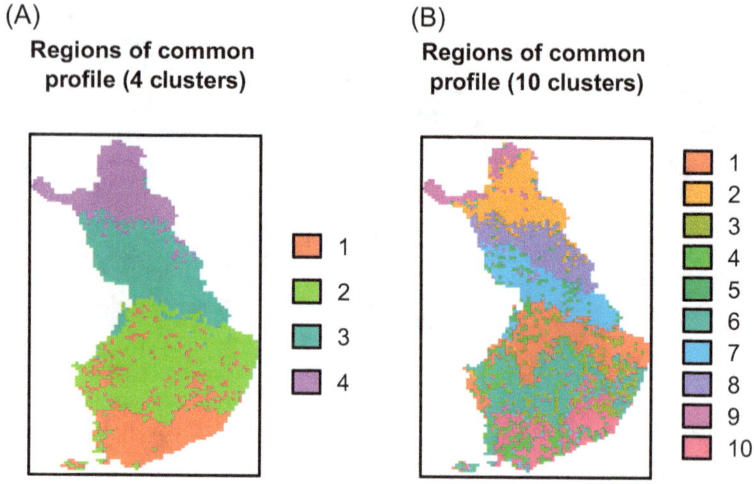

Figure 11.11 RCPs derived from the Finnish bird data. Each colour corresponds to a distinct species community type. The figure has been generated by applying clustering to the species distributions predicted by HMSC. In panel A, the number of clusters has been set to 4, whereas in panel B it has been set to 10.

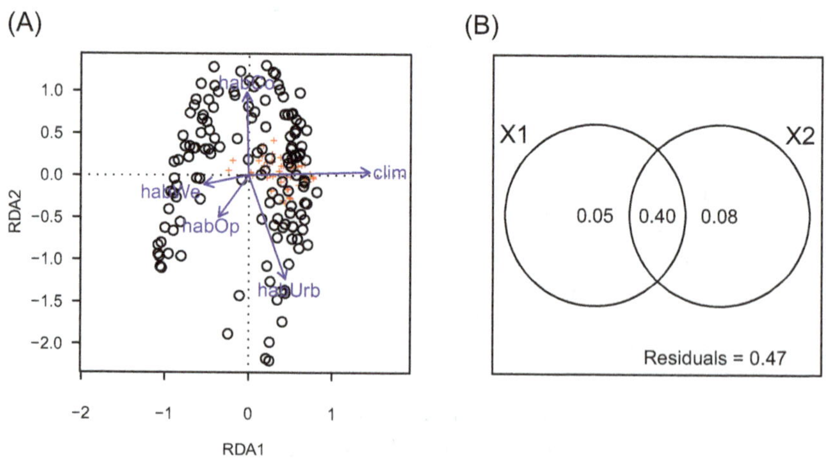

Figure 11.15 Results of RDA analyses. Panel A shows an RDA triplot for the bird community data (circles represent the study sites, vectors represent the environmental covariates and crosses represent the species). Panel B shows the results of RDA-based variance partitioning, indicating the amount of variance explained by the environmental (X1) and spatial (X2) predictors.

12 · *Conclusions and Future Directions*

In this book, we have introduced joint species distribution modelling with HMSC. In this concluding chapter, we will discuss the strengths and limitations of HMSC. We start in Section 12.1 by listing the ten strengths of HMSC originally introduced by Ovaskainen et al. (2017b), each of which we discuss in light of the results presented in this book. We then move to the limitations (Section 12.2), which we discuss in terms of the prospects for future development.

12.1 The Ten Key Strengths of HMSC

In our view, the main strength of HMSC is that it is a single encompassing modelling framework that allows for integrating many kinds of data and asking many kinds of questions in community ecology. Thus, instead of being a single test of whether the community is niche-based or neutral, whether the species co-occur more or less often than expected by chance or whether species traits or phylogenies are associated with their responses to environmental variation, it allows one to address all of these – and many other – questions simultaneously. Ovaskainen et al. (2017b) listed the ten strengths of HMSC, which we quote below with *italic font,* and relate them to the material presented in this book.

Strength 1. HMSC is a unifying framework that encompasses classic approaches such as single-species distribution models and model-based ordinations as special cases. We have illustrated this by applying HMSC as a single-species distribution model in Chapter 5, and as an ordination in Section 7.9.4. While there are comprehensive methods specifically for applying single-species distribution models (Franklin 2009; Guisan et al. 2017; Peterson et al. 2011) and ordination approaches (Borcard et al. 2011; Legendre & Legendre 2012), we hope that having both as special cases of a single modelling framework helps to bring these two largely separately evolved fields closer together.

Strength 2. HMSC provides simultaneous inferences at the species and community levels. We have illustrated this throughout the book, and in particular when building the modelling approach in Chapters 5–7. Concerning the fixed effects, in the single-species context in Chapter 5 we introduced species-level inference based on the assessment of the β parameters that model species niches. In Chapter 6, we showed how to derive community-level inference based on the γ and the ρ parameters that relate to how species niches depend on species traits and their phylogenetic relationships. Concerning the random effects, in the single-species context in Section 5.4 we introduced spatial and hierarchical models where the random effect may model processes that for example relate to dispersal limitation. In Chapter 7, we added a community-level perspective to the random effects, by showing how the random effects model residual associations among the species.

Strength 3. HMSC offers the general advantages of model-based approaches, such as tools for model validation and prediction. This has also been illustrated throughout the book's examples, particularly in Chapters 9 and 11. In Chapter 9, we introduced many kinds of tools for model validation and comparison, which allow one to critically test how well a HMSC model performs when making predictions for new sampling units, as well as to compare the predictive performance of different models. In Chapter 11, we illustrated not only how HMSC can be used to generate species (and trait) distributions, but also how such predictions can be post-processed for addressing more applied questions such as conservation prioritisation and bioregionalisation.

Strength 4. HMSC overcomes previous problems of modelling communities with sparse data. Modelling of rare species is generally difficult because there is usually a limited amount of data about them. For example, if a species has been recorded as present 5 times and absent 995 times, it is very hard to find out which environmental covariates influence the occurrence of this species. Applying a single-species model to such data is especially likely to fail, as the inference will be based solely on those 5 presences and 995 absences. HMSC will also have a hard time in drawing reliable inference from such data, and that is one reason why unless the research interest is in the very rarest species, we recommend leaving them out of the analyses. However, when rare species are to be modelled, HMSC is likely to be a good option for doing so. A comprehensive evaluation of a large number of single-species and JSDMs (Norberg et al. 2019) suggested that HMSC is generally the best-performing modelling approach in terms of its predictive power for case studies that involved a limited amount of data and communities with many rare species. The reason why HMSC can be expected to perform well in such cases is that it allows borrowing information from other species, as illustrated in Ovaskainen and Soininen (2011) and discussed in Chapter 6 of this book.

Strength 5. HMSC overcomes the long-standing challenge in species distribution modelling of how to account for species interactions in explaining and predicting species occurrences. One hot topic in the species distribution literature is whether and how species interactions can be used to explain and predict species occurrences (e.g. Araújo & Luoto 2007; Dormann et al. 2018; Kissling et al. 2012; Zurell et al. 2018). In Chapter 7, we introduced residual species association matrices as a way of accounting for species interactions when explaining species occurrences, while we emphasised the caveats related to the interpretation of residual associations as interactions. In Section 7.7, we discussed under which circumstances the estimated species associations are likely (or not) to help in predicting species occurrences. As a practical tool, we introduced predictions that are conditional on the known occurrences or abundances of the other species, as well as conditional cross-validation that can be used to specifically ask how much the predictive power increases when species associations are accounted for.

Strength 6. HMSC allows one to partition observed variation in species occurrences into components related to environmental variation in measured vs. random processes (or unmeasured variation) at different spatial scales – both at the species and community levels. We introduced such a variance partitioning in the single-species context in Section 5.5, and have applied it at the community level in many of the examples of this book. In HMSC, this is done by calculating the average proportions over the species that is attributed to (groups of) the covariates included as fixed effects, and to each of the random effects included in the model.

Strength 7. HMSC tackles the fourth-corner problem (the influence of species traits on their occurrences) in a way that accounts for the phylogenetic signal in the data. This is illustrated in Chapter 6, where the full approach (Section 6.5) simultaneously models both the effects of traits and phylogeny in species niches. Thus, when assessing how traits influence species' occurrences through their effect on how species respond to the environmental covariates, the species are not considered as independent data points, as their phylogenetic dependencies are accounted for. In Chapter 11, we illustrated the relationships between HMSC analyses and the fourth-corner method by applying both of them to the same data.

Strength 8. HMSC can be applied to many kinds of study designs (including hierarchical, temporal or spatial) and many types of data (such as presence–absence, counts and continuous measurements). Concerning the study designs, we introduced in the single-species context in Chapter 5 how HMSC can be applied to hierarchical, spatial and temporal study designs. In the chapters thereafter, we have shown how HMSC is applied to community data collected with all of these study designs. Concerning the data types, we introduced in Section 5.3 the normal model for continuously valued data, the probit model for presence–absence data, and the Poisson and lognormal Poisson models for count data. Trivially, these also allow the use of the lognormal model, which is obtained by log-transforming the response data and then applying the normal model. The variety of link functions implemented in the current Hmsc also allow for the use of hurdle models, which are fitted by first modelling the presence–absences with the probit model, and then the abundances conditional on presences using e.g. the Poisson model (see Section 5.3.3). As HMSC is built on the generalised linear modelling framework, it is technically possible to implement many other types of link functions, yet doing so provides technical challenges that we leave for future work (see Section 12.2.1).

Strength 9. HMSC can generate predictions at the species, community or trait levels, while propagating uncertainty in the predicted parameter values. In many parts of this book, we have performed two kinds of predictions. The first type of predictions has been illustrated as gradient plots, which show the predictions of different community features over gradients of the environmental variables that are included as predictors. Such community features include individual species, species richness and community-weighted mean trait values. In all of these cases, we have illustrated not only the mean prediction, but also its credible interval, which shows the range of values within which the true value belongs, with e.g. 95 per cent probability. The second type of predictions are those that are made for new sampling units, as for example when we derived species distribution maps for the community of Finnish birds in Section 11.1.5, including predictions for individual species, species richness and community-weighted mean trait values. In Section 11.2 we discussed the important (but often neglected) issue of uncertainty in predicted species distribution maps, and showed how such uncertainty can be quantified by the posterior predictive distribution.

Strength 10. HMSC is computationally efficient, being able to analyse small datasets (with a few hundred sampling units and tens of species) in seconds, and large datasets (with some tens of thousands of sampling units and a few thousand species) in several days. The computational efficiency of HMSC stems from efficient MCMC sampling algorithms, enabled by the prior structure that allows direct sampling from the full conditional distributions. However, the original assessment of Ovaskainen et al. (2017b) quoted above was perhaps somewhat optimistic, as further and more extensive testing has revealed that obtaining satisfactory mixing can often require running very long MCMC chains. We will return to this point in the next section when we discuss the future development needs. Importantly, tools implemented in Hmsc enable a proper evaluation of MCMC convergence through the R-package coda, as illustrated throughout this book and discussed in more detail in Section 8.6.

12.2 Future Development Needs

Even if the HMSC framework presented in this book merges a large number of methodological developments (Abrego et al. 2017a; Ovaskainen et al. 2010; Ovaskainen & Soininen 2011; Ovaskainen et al. 2016a; Ovaskainen et al. 2016b; Ovaskainen et al. 2017a; Ovaskainen et al. 2017b; Sebastián-González et al. 2010; Tikhonov et al. 2017; Tikhonov et al. 2020b), there is still room for many future developments. In this section, we discuss the methodological challenges related to various types of data models, computational efficiency, as well as research questions in community ecology that cannot yet be addressed with HMSC.

12.2.1 Data Models

As discussed above, HMSC currently implements the normal, probit, Poisson and lognormal Poisson models. While this selection of data models already allows the analysis of many kinds of community data, it is still very restricted compared to what is available, for example for univariate modelling in the generalised linear modelling context. In theory, implementation of different data models in the context of not only univariate but also multivariate generalised linear models is straightforward (Clark et al. 2017). However, doing so in a computationally efficient way becomes technically challenging when extended to JSDMs with hierarchical structures and spatially structured latent variables. Below, we list four specific development needs related to data models.

The first future development need related to data models concerns point pattern data. Such data are very common in ecology, as exemplified by the GBIF (www.gbif.org/) base, which encompasses > 1,000 million records of presence-only data. Yet, thus far we have lacked rigorous methods for utilising such data to resolve which processes govern community assembly. Recent work has shown that the frequently used MAXENT (Phillips et al. 2006) approach to point pattern data (e.g. in the context of presence-only data) is equivalent to inhomogeneous Poisson point process models (Renner & Warton 2013; Renner et al. 2015), which are a standard approach in statistical literature. Newly emerging methods (Adams et al. 2009; Gonçalves & Gamerman 2018) for Gaussian Cox models (Møller et al. 1998) allow implementing point process models with increased computational efficiency. Implementing such models to HMSC would allow model-based analyses of multispecies point process data, and in particular bringing trait-based and phylogenetic perspectives into analyses.

The second future development need related to data models concerns proportional data, such as sequence data (e.g. Section 7.9) or pin-point plant cover data (e.g. Damgaard 2008). In Section 7.9, where we analysed sequence data using both a hurdle modelling and lognormal Poisson modelling approach, we already mentioned that the most natural model for such data would be Dirichlet-Multinomial. While straightforward implementations exist for such models (Damgaard 2015; Holmes et al. 2012), it remains a challenge to implement them into the multivariate context of HMSC in a computationally efficient manner.

The third future development need related to data models is the ability to account for uncertainty in species identification. This would be critically important for making robust inferences from modern biomonitoring data, which are often based on high-throughput sequencing of DNA data (Bush et al. 2017), the reliable identification of which (for example, to the species level) is a challenge. While methods of probabilistic taxonomic placement offer quantification of species identification uncertainty (e.g. Somervuo et al. 2017), such uncertainty is still seldom accounted for in downstream analyses. In the Bayesian framework of HMSC, a probability distribution of community data matrices could be considered as a discrete prior for the latent (true but unknown) community matrix, and thus propagation of species identification uncertainty could be implemented using standard MCMC sampling schemes.

The fourth future development need related to data models concerns the recent calls for unifying two active areas of statistical ecology, namely joint species distribution models and occupancy models (Beissinger et al. 2016; Guillera-Arroita 2017), which developments were initiated by

Tobler et al. (2019). Occupancy models explicitly separate the true species occurrence or abundance from the observation process, and thus may predict for example that a species is present with high probability even if it was not observed. The application of occupancy models generally requires repeated visits to the same sampling units, as these are needed to estimate detection probability. The current implementation of HMSC allows the observation process to be accounted for in two ways. First, covariates related to the observation process can be incorporated in the fixed effects, as we illustrated in Section 7.9 by including the log-transformed sequencing depth as one of the predictors. Second, in the case of repeated visits to the same locations, the study design of HMSC can be set so that the sampling unit (i.e. each row of the **X** matrix) corresponds to a particular visit to a particular location, and the dependency structure in the data is accounted for by setting the location as a random effect. This approach allows one to quantify the variation among the repeated visits. Thus, if making the usual assumption of occupancy models that the true occupancy does not change between the repeated visits, it provides an estimate of the importance of detection uncertainty. However, this does not allow the probabilities of species presence to be explicitly separated from the probabilities of detection, and doing so would require a more mechanistic implementation of occupancy models in HMSC.

12.2.2 Computational Efficiency

One important technical limitation of the current HMSC is that applying it to big data can be computationally very intensive. This is because both the time required for a single MCMC step and the number of MCMC steps needed to achieve a satisfactory MCMC convergence increase as a function of the size of the data, as well as the complexity of the model (Section 8.8).

Decreasing the time required for performing a single MCMC step can be trivially achieved by utilising hardware with greater computational power. However, this does not fully solve the problem, as some parts of the current Hmsc implementation scale very badly with data size, and thus even a 1,000-fold increase of computational power would not necessarily allow the model to be fitted to some big dataset. Thus, further progress in devising more computationally efficient algorithms is needed. One example of such recent developments is the work by Tikhonov et al. (2020), which resolved the adverse scaling of the spatial latent factor approach as a function of spatial locations (Section 8.8) by approximating the full Gaussian process prior either by the Gaussian predictive process or by the nearest neighbour Gaussian process. Similar developments with respect to other computationally intensive parts of the HMSC model might be possible as well.

Concerning MCMC convergence, and thus the number of MCMC iterations needed, a particular problem is that satisfactory MCMC convergence is more difficult to achieve with the probit and the Poisson models than with the normal models. This is because the data augmentation step used to implement the probit model, for example, can cause problems with MCMC convergence. As such, solving this problem is currently an active area in statistical research (e.g. Duan et al. 2018). Resolving problems with MCMC convergence can be achieved by 'brute force', that is, by running more MCMC iterations, or by devising MCMC sampling algorithms that lead to faster MCMC convergence. Given the ongoing methodological revolution in Bayesian analyses of big data – on which the current implementation of HMSC is partially built – we are optimistic that in the near future it will be possible to broaden the applicability of HMSC to even bigger data than is currently possible.

12.2.3 Model Structures Related to Ecological and Evolutionary Processes

Throughout this book, we have described how HMSC promotes a synthesis between statistical analysis and ecological theory by incorporating statistical structures that relate, either directly or indirectly, to theoretical concepts from community ecology. As with the data models discussed above, there is much room for future development in this area. Below, we list four specific development needs related to devising model structures that would broaden the set of ecological and evolutionary questions that can be addressed by HMSC.

First, current HMSC assumes that species are homogenous units, in the sense that both the environmental responses and the traits are assumed to be species-specific, thus neglecting any intra-specific variation. It has been widely recognised that intra-specific variation can have a profound influence on community ecology, for example due to local adaptation within species and coadaptation among species (e.g. Jung et al. 2010; Laughlin et al. 2012). Bringing an individual-level perspective to joint species distribution modelling would enable addressing community-level questions related to local adaptation, such as the link between niche breadth and genetic/phenotypic variability. It might also increase predictive power through accounting for intra-specific variation in how species respond to the abiotic environment and to each other. One way to utilise within-species variation in measured traits in the current HMSC model is to include both species occurrences and population-level mean traits simultaneously in the response variables. This would link trait variation not only to environmental variation, but also to variation in species abundance. However, bringing an individual-level perspective to HMSC more mechanistically would require major developments, such as a further hierarchical layer that would model population- or individual-level responses as deviations of species-level responses.

Second, while the current HMSC models the link from species traits and phylogenies to their environmental niches, it does not model the link from species traits and phylogenies to biotic interactions, nor does it allow incorporation of *a priori* information about interaction networks. The reason for this asymmetry in how the fixed and random effects are modelled is not ecological, as clearly both the abiotic and biotic responses of the species are equally influenced by their traits, and for both cases the phylogenetic relationships can work equally well as a proxy for missing traits. The reason why the association matrix $\mathbf{\Omega}$ is not yet modelled as a function of traits and phylogenies is that establishing a good model structure is not straightforward. One option might be to apply covariance regression developed for factor models (Fox & Dunson 2015) to model the species loadings (the λ parameters) as a function of species traits and phylogenies in a similar way as the species niches (the β parameters) are currently modelled. However, this would lead to the situation where species with similar traits would necessarily be predicted to have a positive association. While it is natural to assume that species with similar traits have similar abiotic responses, this is not necessarily the case with biotic interactions, as species with similar traits can be assumed to compete for the same resources and thus have negative competitive interactions. Thus, a complementary approach would be needed to capture how negative associations depend on species traits. Another asymmetry in how species traits influence their abiotic and biotic niches is that with abiotic niches, one may consider one species at a time, however with species interactions the combination of the traits of the two interacting species is important. In cases where there is direct information about the species interaction network (e.g. food-web structure), including such information as an additional data type to HMSC could allow finer dissection of species associations in order to assign them more specifically to those caused by species interactions and to those caused by something else, such as missing covariates or dispersal limitation. In summary, developing ways of modelling species associations as a function of traits, phylogenies and network structures poses an important challenge for future developments.

Third, the current implementation of HMSC allows the modelling of species associations as a function of environmental covariates (Section 7.5), which is important because the direction and strength of interspecific interactions in ecological communities are known to covary with environmental conditions (Pellissier et al. 2018). However, modelling the species associations as a function of environmental covariates requires the relevant environmental covariates to be known and measured. Often this is not the case, and thus one might wish to ask the more general question of how species-to-species association matrices vary over space or time with any *a priori* assumptions about the underlying drivers. Model structures allowing such extensions might be possible to implement, for example with the help of Bayesian covariance regression combined with a Gaussian process approach (Durante et al. 2014; Fox & Dunson 2015).

Fourth, while the current HMSC implements the additive effects of spatial and temporal latent variables, joint species distribution models with spatio-temporal correlation structures have thus far been implemented only for small species communities (Thorson et al. 2016). Such models would be needed for spatio-temporal data – these kinds of data are becoming increasingly available with the fast developments of modern biomonitoring methods (Bush et al. 2017). Thus, there is a great demand for building computationally efficient spatio-temporal joint species distribution models. Here the challenge is not so much in devising an appropriate model structure, but in implementing computationally efficient MCMC sampling algorithms that would allow parameterising such models with large data.

In addition to these technical development needs, it would be at least equally important to better understand how robustly the statistical results can be related to the underlying ecological and evolutionary processes, and what kind of confounding factors may hinder the interpretation of the statistical results. One way of addressing this question is to conduct virtual ecologist studies, as we have done in Chapter 10 and in Ovaskainen et al. (2019). Another way of addressing this question is to combine observational studies with experimental approaches.

Epilogue

In this book, we have described the field of joint species distribution modelling, particularly from the point of view of Hierarchical Modelling of Species Communities (HMSC). While much of our focus has been on statistical aspects and software implementation of HMSC, we started the book with a brief review of community ecological theory. Being able to relate the statistical results to ecological theory is critical, as it enables one to place the results of a specific case study in a broader context. Thus, let us finish the book by returning to the links between joint species distribution modelling and ecological theory.

We recall that single species distribution models are also called niche models, indicating that they are heavily linked to the concept of a species niche and the related Niche Theory. This is because the parameters of a single-species model can be viewed to describe the niche of the focal species, i.e. how environmental conditions, such as resource availability or climatic suitability, translate to predictions of species occurrence or abundance. However, much of the community ecology theory becomes meaningful only in the context of multiple species, and thus joint species distribution models have many more links to community ecology theory than single species distribution models. As joint species distribution models include species distribution models for all the species comprising the community, they naturally also relate directly to Niche Theory. But beyond this, they also relate to many other parts of ecological theory, as we will summarise in the following paragraphs.

Environmental filtering derives from niche theory, but focuses on the variation in species niches among multiple species. Variation in species niches further relates to variation in species traits, which is ultimately generated by evolutionary processes. As a result, it can be expected that such variation is phylogenetically structured. As we have discussed in many parts of this book, joint species distribution models are built to explicitly account for these relationships, which, in our opinion, is one of their most powerful feature for linking data to ecological theory. Biotic filtering relates to species interactions, which are modelled in joint species distribution models by residual association structures. The development of latent variable approaches has greatly expanded the applicability of joint species distribution models to estimate residual associations for large species communities, as well as to use these models to make predictions that account for biotic interactions. In spite of these methodological developments, the links from joint species distribution models to ecological theory are, in our opinion, still less developed concerning biotic filtering than concerning environmental filtering. One reason for this is that residual associations typically result not only from biotic interactions but also from other confounding factors, such as the effect of missing covariates. Another reason is that current joint species distribution models do not explicitly model the residual species associations as a function of species traits, phylogenetic relationships or food-web structures.

In addition to environmental filtering and biotic filtering, joint species distribution models also link to neutral processes, such as ecological drift and dispersal limitation. Both of these are captured by random effects, which model variation in community structure that cannot be related to measured variation in environmental conditions. Concerning dispersal limitation, spatial joint species distribution models are built on spatially explicit random effects; the respective spatial scale parameters can be viewed to model the spatial scales of dispersal. However, as biotic interactions are also modelled through random effects, biotic interactions, dispersal and neutral processes are to some extent confounded in current joint species distribution models, even if they are very different processes in terms of ecological theory. Here the limitation is perhaps not primarily in joint species distribution modelling techniques *per se*, but in the nature of non-manipulative observational data. These data do not typically have sufficient resolution to disentangle the different processes that may generate similar patterns of biodiversity variation over environmental gradients, space and time. Thus, one big challenge for the future is to acquire direct data on the assembly processes at the community level and combine these data with observational data using process-based joint species distribution models – to be possibly developed in the future.

We hope that this book has helped the reader to delve into the world of joint species distribution modelling, including what it is really about, how to apply it in practice, what its pros and cons are and in which way parameter estimates may or may not be interpreted. From our side, the process of writing this book has surely clarified all of these aspects in our own minds, as well as identified a myriad of development needs, concerning how to link joint species distribution models to theory, what kind of new model structures would be needed and the limitations of current software implementations. Given that the field of joint species distribution modelling has already been shown to hold much promise – while still clearly in its infancy – we predict that exciting times are ahead, for both those who wish to contribute to the future model development and for those who plan to apply joint species distribution models to their data!

References

Abarenkov, K., Somervuo, P., Nilsson, R. H., Kirk, P. M., Huotari, T., Abrego, N. & Ovaskainen, O. (2018). Protax-fungi: A web-based tool for probabilistic taxonomic placement of fungal internal transcribed spacer sequences. *New Phytol.*, 220, 517–525.

Abrego, N., Christensen, M., Bässler, C., Ainsworth, A. M. & Heilmann-Clausen, J. (2017b). Understanding the distribution of wood-inhabiting fungi in European beech reserves from species-specific habitat models. *Fungal Ecol.*, 27, 168–174.

Abrego, N., Norberg, A. & Ovaskainen, O. (2017a). Measuring and predicting the influence of traits on the assembly processes of wood-inhabiting fungi. *J. Ecol.*, 105, 1070–1081.

Adams, R. P., Murray, I. & MacKay, D. J. C. (2009). Tractable nonparametric Bayesian inference in Poisson processes with Gaussian process intensities. *Proceedings of the 26th International Conference on Machine Learning*, 9–16.

Akaike, H. (1974). A new look at the statistical model identification. *IEEE Trans. Autom. Contr.*, 19, 716–723.

Algar, A. C., Kharouba, H. M., Young, E. R. & Kerr, J. T. (2009). Predicting the future of species diversity: Macroecological theory, climate change, and direct tests of alternative forecasting methods. *Ecography*, 32, 22–33.

Amarasekare, P. & Nisbet, R. (2001). Spatial heterogeneity, source-sink dynamics, and the local coexistence of competing species. *Am. Nat.*, 158, 572–584.

Anderson, M. J. (2001). A new method for non-parametric multivariate analysis of variance. *Austral Ecol.*, 26, 32–46.

Anderson, M. J., Crist, T. O., Chase, J. M., Vellend, M., Inouye, B. D., Freestone, A. L., Sanders, N. J., Cornell, H. V., Comita, L. S., Davies, K. F., Harrison, S. P., Kraft, N. J. B., Stegen, J. C. & Swenson, N. G. (2011). Navigating the multiple meanings of β diversity: A roadmap for the practicing ecologist. *Ecol. Lett.*, 14, 19–28.

Anderson, T. W. (1951). Estimating linear restrictions on regression coefficients for multivariate normal distributions. *Ann. Math. Statist.*, 22, 327–351.

Araújo, M. B. & Guisan, A. (2006). Five (or so) challenges for species distribution modelling. *J. Biogeogr.*, 33, 1677–1688.

Araújo, M. B. & Luoto, M. (2007). The importance of biotic interactions for modelling species distributions under climate change. *Global Ecol. Biogeogr.*, 16, 743–753.

Araújo, M. B. & Peterson, A. T. (2012). Uses and misuses of bioclimatic envelope modeling. *Ecology*, 93, 1527–1539.

Araújo, M. B., Anderson, R. P., Márcia Barbosa, A., Beale, C. M., Dormann, C. F., Early, R., Garcia, R. A., Guisan, A., Maiorano, L., Naimi, B., O'Hara, R. B., Zimmermann, N. E. & Rahbek, C. (2019). Standards for distribution models in biodiversity assessments. *Sci. Adv.*, 5(1), eaat4858.

Arrhenius, O. (1921). Species and area. *J. Ecol.*, 9, 95–99.

Asher, J., Warren, M., Fox, R., Harding, P., Jeffcoate, G. & Jeffcoate, S. (2001). *Millennium Atlas of Butterflies in Britain and Ireland*. Oxford University Press.

Barberán, A., Ladau, J., Leff, J. W., Pollard, K. S., Menninger, H. L., Dunn, R. R. & Fierer, N. (2015). Continental-scale distributions of dust-associated bacteria and fungi. *Proc. Natl. Acad. Sci. USA*, 112, 5756–5761.

Barry, S. C. & Welsh, A. H. (2002). Generalized additive modelling and zero inflated count data. *Ecol. Model.*, 157, 179–188.

Bates, D., Mächler, M., Bolker, B. & Walker, S. (2015). Fitting linear mixed-effects models using lme4. *J. Stat. Soft.*, 67(1), 1–48.

Bauman, D., Drouet, T., Fortin, M. & Dray, S. (2018). Optimizing the choice of a spatial weighting matrix in eigenvector-based methods. *Ecology*, 99, 2159–2166.

Beale, C. M., Lennon, J. J., Elston, D. A., Brewer, M. J. & Yearsley, J. M. (2007). Red herrings remain in geographical ecology: A reply to Hawkins et al. (2007). *Ecography*, 30, 845–847.

Beaulieu, J. M., Jhwueng, D., Boettiger, C. & O'Meara, B. C. (2012). Modeling stabilizing selection: Expanding the Ornstein–Uhlenbeck model of adaptive evolution. *Evolution*, 66, 2369–2383.

Beck, J., Böller, M., Erhardt, A. & Schwanghart, W. (2014). Spatial bias in the GBIF database and its effect on modeling species' geographic distributions. *Ecol. Inform.*, 19, 10–15.

Beissinger, S. R., Iknayan, K. J., Guillera-Arroita, G., Zipkin, E. F., Dorazio, R. M., Royle, J. A. & Kéry, M. (2016). Incorporating imperfect detection into joint models of communities: A response to Warton et al. *Methods Ecol. Evol.*, 31, 736–737.

Bhattacharya, A. & Dunson, D. B. (2011). Sparse Bayesian infinite factor models. *Biometrika*, 98, 291–306.

Blonder, B. (2018). Hypervolume concepts in niche- and trait-based ecology. *Ecography*, 41, 1441–1455.

Boakes, E. H., McGowan, P. J. K., Fuller, R. A., Chang-qing, D., Clark, N. E., O'Connor, K. & Mace, G. M. (2010). Distorted views of biodiversity: Spatial and temporal bias in species occurrence data. *PLoS. Biol.*, 8, 1–11.

Bohmann, K., Evans, A., Gilbert, M. T. P., Carvalho, G. R., Creer, S., Knapp, M., Yu, D. W. & de Bruyn, M. (2014). Environmental DNA for wildlife biology and biodiversity monitoring. *Trends Ecol. Evol.*, 29, 358–367.

Bokma, F., Bokma, J. & Mönkkönen, M. (2001). Random processes and geographic species richness patterns: Why so few species in the North?. *Ecography*, 24, 43–49.

Bolker, B. M. (2008). *Ecological Models and Data in R*. Princeton University Press.

Booth, T. H., Nix, H. A., Busby, J. R. & Hutchinson, M. F. (2014). Bioclim: The first species distribution modelling package, its early applications and relevance to most current MaxEnt studies. *Diversity Distrib.*, 20, 1–9.

Borcard, D. & Legendre, P. (2002). All-scale spatial analysis of ecological data by means of principal coordinates of neighbour matrices. *Ecol. Model.*, 153, 51–68.

Borcard, D., Gillet, F. & Legendre, P. (2011). *Numerical Ecology with R*. Springer.

Boulangeat, I., Gravel, D. & Thuiller, W. (2012). Accounting for dispersal and biotic interactions to disentangle the drivers of species distributions and their abundances. *Ecol. Lett.*, 15, 584–593.

Box, G. E. P. (1976). Science and statistics. *J. Am. Stat. Assoc.*, 71, 791–799.

Box, G. E. P. (1979). Robustness in the strategy of scientific model building. Eds. Robert L. Launer & Graham N. Wilkinson. In: *Robustness in Statistics*, Academic Press, 201–236.

Bray, J. R. & Curtis, J. T. (1957). An ordination of the upland forest communities of southern Wisconsin. *Ecol. Monogr.*, 27, 325–349.

Breiner, F. T., Guisan, A., Bergamini, A. & Nobis, M. P. (2015). Overcoming limitations of modelling rare species by using ensembles of small models. *Methods Ecol. Evol.*, 6, 1210–1218.

Brooker, R. W., Maestre, F. T., Callaway, R. M., Lortie, C. L., Cavieres, L. A., Kunstler, G., Liancourt, P., Tielbörger, K., Travis, J. M. J., Anthelme, F., Armas, C., Coll, L., Corcket, E., Delzon, S., Forey, E., Kikvidze, Z., Olofsson, J., Pugnaire, F., Quiroz, C. L., Saccone, P., Schiffers, K., Seifan, M., Touzard, B. & Michalet, R. (2008). Facilitation in plant communities: The past, the present, and the future. *J. Ecol.*, 96, 18–34.

Brooks, S. P. & Gelman, A. (1998). General methods for monitoring convergence of iterative simulations. *J. Comput. Graph. Stat.*, 7, 434–455.

Brotons, L., Herrando, S. & Pla, M. (2007). Updating bird species distribution at large spatial scales: Applications of habitat modelling to data from long-term monitoring programs. *Divers. Distrib.*, 13, 276–288.

Brown, A. M., Warton, D. I., Andrew, N. R., Binns, M., Cassis, G. & Gibb, H. (2014). The fourth-corner solution: Using predictive models to understand how species traits interact with the environment. *Methods Ecol. Evol.*, 5, 344–352.

Bruno, J. F., Stachowicz, J. J. & Bertness, M. D. (2003). Inclusion of facilitation into ecological theory. *Trends Ecol. Evol.*, 18, 119–125.

Brus, D. J., Hengeveld, G. M., Walvoort, D. J. J., Goedhart, P. W., Heidema, A. H., Nabuurs, G. J. & Gunia, K. (2012). Statistical mapping of tree species over Europe. *Eur. J. Forest. Res.*, 131, 145–157.

Burnham, K. P. & Anderson, D. R. (2002). *Model Selection and Multimodel Inference: A Practical Information-Theoretic Approach*. Springer.

Burns, A. R., Stephens, W. Z., Stagaman, K., Wong, S., Rawls, J. F., Guillemin, K. & Bohannan, B. J. M. (2015). Contribution of neutral processes to the assembly of gut microbial communities in the zebrafish over host development. *ISME J.*, 10, 655.

Bush, A., Sollmann, R., Wilting, A., Bohmann, K., Cole, B., Balzter, H., Martius, C., Zlinszky, A., Calvignac-Spencer, S., Cobbold, C. A., Dawson, T. P., Emerson, B. C., Ferrier, S., Gilbert, M. T., Herold, M., Jones, L., Leendertz, F. H., Matthews, L., Millington, J. D. A., Olson, J. R., Ovaskainen, O., Raffaelli, D., Reeve, R., Rödel, M., Rodgers, T. W., Snape, S., Visseren-Hamakers, I., Vogler, A. P., White, P. C. L., Wooster, M. J. & Yu, D. W.

(2017). Connecting Earth observation to high-throughput biodiversity data. *Nature Ecol. Evol.*, 1, 0176.

Cadotte, M. W., Arnillas, C. A., Livingstone, S. W. & Yasui, S. E. (2015). Predicting communities from functional traits. *Trends Ecol. Evol.*, 30, 510–511.

Calabrese, J. M., Certain, G., Kraan, C. & Dormann, C. F. (2014). Stacking species distribution models and adjusting bias by linking them to macroecological models. *Global Ecol. Biogr.*, 23, 99–112.

Cale, W. G., Henebry, G. M. & Yeakley, J. A. (1989). Inferring process from pattern in natural communities. *Bioscience*, 39, 600–605.

Carpenter, S. R. (1996). Microcosm experiments have limited relevance for community and ecosystem ecology. *Ecology*, 77, 677–680.

Ceballos, G., Ehrlich, P. R., Barnosky, A. D., García, A., Pringle, R. M. & Palmer, T. M. (2015). Accelerated modern human-induced species losses: Entering the sixth mass extinction. *Sci. Adv.* 1(5), e1400253.

Charrad, M., Ghazzali, N., Boiteau, V. & Niknafs, A. (2014). NbClust: An R package for determining the relevant number of clusters in a data set. *J. Stat. Softw.*, 61(6), 1–36.

Chase, J. M. & Leibold, M. A. (2003). *Ecological Niches: Linking Classical and Contemporary Approaches*. University of Chicago Press.

Chase, J. M., McGill, B. J., Thompson, P. L., Antão, L. H., Bates, A. E., Blowes, S. A., Dornelas, M., Gonzalez, A., Magurran, A. E., Supp, S. R., Winter, M., Bjorkman, A. D., Bruelheide, H., Byrnes, J. E. K., Cabral, J. S., Elahi, R., Gomez, C., Guzman, H. M., Isbell, F., Myers-Smith, I., Jones, H. P., Hines, J., Vellend, M., Waldock, C. & O'Connor, M. (2019). Species richness change across spatial scales. *Oikos*, 128, 1079–1091.

Chave, J., Muller-Landau, H. & Levin, S. (2002). Comparing classical community models: Theoretical consequences for patterns of diversity. *Am. Nat.*, 159, 1–23.

Chen, T., He, T., Benesty, M., Khotilovich, V., Tang, Y., Cho, H., Chen, K., Mitchell, R., Cano, I., Zhou, T., Li, M., Xie, J., Lin, M., Geng, Y. & Li, Y. (2018). Xgboost: Extreme gradient boosting. R package version 0.71.2.

Clark, J. S., Gelfand, A. E., Woodall, C. W. & Zhu, K. (2014). More than the sum of the parts: Forest climate response from joint species distribution models. *Ecol. Appl.*, 24, 990–999.

Clark, J. S., Nemergut, D., Seyednasrollah, B., Turner, P. J. & Zhang, S. (2017). Generalized joint attribute modeling for biodiversity analysis: Median-zero, multivariate, multifarious data. *Ecol. Monogr.*, 87, 34–56.

Clements, F. E. (1916). *Plant Succession: Analysis of the Development of Vegetation. Publication no. 242* . Carnegie Institute of Washington.

Clements, F. E. (1936). Nature and structure of the climax. *J. Ecol.*, 24, 252–284.

Cornelissen, J. H. C., Lavorel, S., Garnier, E., Díaz, S., Buchmann, N., Gurvich, D. E., Reich, P. B., Steege, H., Morgan, H. D., Heijden, M. G. A., Pausas, J. G. & Poorter, H. (2003). A handbook of protocols for standardised and easy measurement of plant functional traits worldwide. *Aust. J. Bot.*, 51, 335–380.

Cramp, S., Simmons, K. E. L. & Perrins, C. M. (1977–1994). *Handbook of the Birds of Europe, the Middle East, and North Africa: The Birds of the Western Palearctic*. Oxford University Press.

Crookston, N. & Finley, A. (2008). yaImpute: An R package for kNN imputation. *J. Stat. Softw.*, 23(10), 1–16.

D'Amen, M., Rahbek, C., Zimmermann, N. E. & Guisan, A. (2017). Spatial predictions at the community level: From current approaches to future frameworks. *Biol. Rev.*, 92, 169–187.

Damgaard, C. (2015). Modelling pin-point cover data of complementary vegetation classes. *Ecol. Inform.*, 30, 179–184.

Damgaard, C. (2008). Modelling pin-point plant cover data along an environmental gradient. *Ecol. Model.*, 214, 404–410.

Damschen, E. I., Harrison, S. & Grace, J. B. (2010). Climate change effects on an endemic-rich edaphic flora: Resurveying Robert H. Whittaker's Siskiyou sites (Oregon, USA). *Ecology*, 91, 3609–3619.

Damuth, J. (1981). Population density and body size in mammals. *Nature*, 290, 699–700.

De'ath, G., Therneau, T. M., Atkinson, B., Ripley, B. & Oksanen, J. (2014). Mvpart: Multivariate partitioning. R package version 1.6-2.

Dengler, J. (2009). Which function describes the species–area relationship best? A review and empirical evaluation. *J. Biogeogr.*, 36, 728–744.

Diamond, J. M. (1975). Assembly of Species Communities. Eds. M. L. Cody, & J. M. Diamond. In: *Ecology and Evolution of Communities*. Harvard University Press, 342–444.

Dolédec, S., Chessel, D., ter Braak, C. J. F. & Champely, S. (1996). Matching species traits to environmental variables: A new three-table ordination method. *Environ. Ecol. Stat.*, 3, 143–166.

Dormann, C. F. (2007a). Effects of incorporating spatial autocorrelation into the analysis of species distribution data. *Global Ecol. Biogeogr.*, 16, 129–138.

Dormann, C. F. (2007b). Promising the future? global change projections of species distributions. *Basic Appl. Ecol.*, 8, 387–397.

Dormann, C. F., Bobrowski, M., Dehling, D. M., Harris, D. J., Hartig, F., Lischke, H., Moretti, M. D., Pagel, J., Pinkert, S., Schleuning, M., Schmidt, S. I., Sheppard, C. S., Steinbauer, M. J., Zeuss, D. & Kraan, C. (2018). Biotic interactions in species distribution modelling: 10 questions to guide interpretation and avoid false conclusions. *Global Ecol. Biogeogr.*, 27, 1004–1016.

Dormann, C. F., McPherson, J. M., Araújo, M. B., Bivand, R., Bolliger, J., Carl, G., Davies, R. G., Hirzel, A., Jetz, W., Daniel Kissling, W., Kühn, I., Ohlemüller, R., Peres-Neto, P. R., Reineking, B., Schröder, B., Schurr, F. M. & Wilson, R. (2007). Methods to account for spatial autocorrelation in the analysis of species distributional data: A review. *Ecography*, 30, 609–628.

Dormann, C. F., Schymanski, S. J., Cabral, J., Chuine, I., Graham, C., Hartig, F., Kearney, M., Morin, X., Römermann, C., Schröder, B. & Singer, A. (2012). Correlation and process in species distribution models: Bridging a dichotomy. *J. Biogeogr.*, 39, 2119–2131.

Dray, S., Dufour, A. & Thioulouse, J. (2018). Ade4: Analysis of ecological data: Exploratory and Euclidean methods in environmental sciences.

Dray, S., Bauman, D., Blanchet, G., Borcard, D., Clappe, S., Guenard, G., Jombart, T., Larocque, G., Legendre, P., Madi, N. & Wagner, H. H. (2019). Adespatial: Multivariate multiscale spatial analysis. R package version 0.3-7.

Dray, S. & Legendre, P. (2008). Testing the species traits–environment relationships: The fourth-corner problem revisited. *Ecology*, 89, 3400–3412.

Dray, S., Legendre, P. & Peres-Neto, P. R. (2006). Spatial modelling: A comprehensive framework for Principal Coordinate Analysis of Neighbour Matrices (PCNM). *Ecol. Model.*, 196, 483–493.

Drummond, A. J., Suchard, M. A., Xie, D. & Rambaut, A. (2012). Bayesian phylogenetics with BEAUti and the BEAST 1.7. *Mol. Biol. Evol.*, 29, 1969–1973.

Duan, L. L., Johndrow, J. E. & Dunson, D. B. (2018). Scaling up data augmentation MCMC via calibration. *J. Mach. Learn. Res.*, 19, 2575–2608.

Dunson, D. B. (2018). Statistics in the big data era: Failures of the machine. *Stat. Probabil. Lett.*, 136, 4–9.

Dunstan, P. K., Foster, S. D. & Darnell, R. (2011). Model based grouping of species across environmental gradients. *Ecol. Model.*, 222, 955–963.

Durante, D., Scarpa, B. & Dunson, D. B. (2014). Locally adaptive factor processes for multivariate time series. *J. Mach. Learn. Res.*, 15, 1493–1522.

Elith, J. & Leathwick, J. R. (2009). Species distribution models: Ecological explanation and prediction across space and time. *Annu .Rev. Ecol. Evol. Syst.*, 40, 677–697.

Elton, C. (1966). The Patterns of Animal Communities. 1st edn. John Wiley & Sons, London; Methuen; New York.

Elton, C. (1927). Animal Ecology. University of Chicago Press.

European Environment Agency. (2016). Corine land cover (CLC) 2012, version 18.5.1. available at http://Land.copernicus.eu/pan-european/corine-land-cover/clc-2012/view

Fauth, J. E., Bernardo, J., Camara, M., Resetarits, W. J., Van Buskirk, J. & McCollum, S. A. (1996). Simplifying the jargon of community ecology: A conceptual approach. *Am. Nat.*, 147, 282–286.

Ferrier, S. & Guisan, A. (2006). Spatial modelling of biodiversity at the community level. *J. Appl. Ecol.*, 43, 393–404.

Fick, S. E. & Hijmans, R. J. (2017). WorldClim 2: New 1-km spatial resolution climate surfaces for global land areas. *Int. J. Climatol.*, 37, 4302–4315.

Fisher, R. A., Corbet, A. S. & Williams, C. B. (1943). The relation between the number of species and the number of individuals in a random sample of an animal population. *J. Anim. Ecol.*, 12, 42–58.

Fithian, W., Elith, J., Hastie, T. & Keith, D. A. (2015). Bias correction in species distribution models: Pooling survey and collection data for multiple species. *Methods Ecol. Evol.*, 6, 424–438.

Foster, S. D., Givens, G. H., Dornan, G. J., Dunstan, P. K. & Darnell, R. (2013). Modelling biological regions from multi-species and environmental data. *Environmetrics*, 24, 489–499.

Fox, E. B. & Dunson, D. B. (2015). Bayesian nonparametric covariance regression. *J. Mach. Learn. Res.*, 16, 2501–2542.

Fox, G. A., Negrete-Yankelevich, S. & Sosa, V. J. (2015). *Ecological Statistics: Contemporary Theory and Application*. Oxford University Press.

Franklin, J. (2009). *Mapping Species Distributions, Spatial Inference and Prediction*. Cambridge University Press.

Garbelotto, M. & Gonthier, P. (2013). Biology, epidemiology, and control of *Heterobasidion* species worldwide. *Annu. Rev. Phytopathol.*, 51, 39–59.

Gause, G. F. (1934). *The Struggle for Existence*. Williams and Wilkins.

GBIF (2018) The Global Biodiversity Information Facility. Available at: www.gbif.org/what-is-gbif

Gelfand, A. E., Fuentes, M., Hoeting, J. A. & Smith, R. L. (2019). *Handbook of Environmental and Ecological Statistics*. Chapman & Hall/CRC Press.

Gelman, A., Carlin, J. B., Stern, H. S., Dunson, D. B., Vehtari, A. & Rubin, D. B. (2013). *Bayesian Data Analysis*. 3rd edn. Chapman & Hall.

Gelman, A. & Rubin, D. B. (1992). Inference from iterative simulation using multiple sequences. *Statist. Sci.*, 7, 457–472.

Geweke, J. (1996). Bayesian reduced rank regression in econometrics. *J. Econom.*, 75, 121–146.

Geweke, J. & Zhou, G. (1996). Measuring the price of the arbitrage pricing theory. *Rev. Financ. Stud.*, 9, 557–587.

Gleason, H. A. (1926). The individualistic concept of the plant association. *J. Torrey Bot. Soc.*, 53, 7–26.

Gneiting, T. & Raftery, A. E. (2007). Strictly proper scoring rules, prediction, and estimation. *Am. Stat.*, 102, 359–378.

Goldberg, D. E. (1990). Components of Resource Competition in Plant Communities. Eds. J. B. Grace & D. Tilman). In: *Perspectives on Plant Competition* , Academic Press, 27–49.

Golding, N. & Harris, D. J. (2015). BayesComm: Bayesian community ecology analysis. R package version 0.1-1.

Gonçalves, F. B. & Gamerman, D. (2018). Exact Bayesian inference in spatiotemporal cox processes driven by multivariate Gaussian processes. *J. R. Stat. Soc. B*, 80, 157–175.

Gorelick, N., Hancher, M., Dixon, M., Ilyushchenko, S., Thau, D. & Moore, R. (2017). Google Earth Engine: Planetary-scale geospatial analysis for everyone. *Remote Sens. Environ.*, 202, 18–27.

Gotelli, N., Hart, E. & Ellison, A. (2015). EcoSimR: Null model analysis for ecological data. R package version 0.1.0.

Gotelli, N. J. (2000). Null model analysis of species co-occurrence patterns. *Ecology*, 81, 2606–2621.

Gotelli, N. J., Anderson, M. J., Arita, H. T., Chao, A., Colwell, R. K., Connolly, S. R., Currie, D. J., Dunn, R. R., Graves, G. R., Green, J. L., Grytnes, J., Jiang, Y., Jetz, W., Kathleen Lyons, S., McCain, C. M., Magurran, A. E., Rahbek, C., Rangel, T. F. L. V. B., Soberón, J., Webb, C. O. & Willig, M. R. (2009). Patterns and causes of species richness: A general simulation model for macroecology. *Ecol. Lett.*, 12, 873–886.

Götzenberger, L., de Bello, F., Bråthen, K. A., Davison, J., Dubuis, A., Guisan, A., Lepš, J., Lindborg, R., Moora, M., Pärtel, M., Pellissier, L., Pottier, J., Vittoz, P., Zobel, K. & Zobel, M. (2012). Ecological assembly rules in plant communities-approaches, patterns and prospects. *Biol. Rev.*, 87, 111–127.

Gower, J. C. (1966). Some distance properties of latent root and vector methods used in multivariate analysis. *Biometrika*, 53, 325–338.

Grafen, A. & Hails, R. (2002). *Modern Statistics for the Life Sciences*. Oxford University Press.

Grenouillet, G., Buisson, L., Casajus, N. & Lek, S. (2011). Ensemble modelling of species distribution: The effects of geographical and environmental ranges. *Ecography*, 34, 9–17.

Grime, J. P. (1973). Competitive exclusion in herbaceous vegetation. *Nature*, 242, 344–347.

Grinnell, J. (1917). The niche-relationships of the California thrasher. *The Auk*, 34, 427–433.

Grueber, C. E., Nakagawa, S., Laws, R. J. & Jamieson, I. G. (2011). Multimodel inference in ecology and evolution: Challenges and solutions. *J. Evol. Biol.*, 24, 699–711.

Guillera-Arroita, G. (2017). Modelling of species distributions, range dynamics and communities under imperfect detection: Advances, challenges and opportunities. *Ecography*, 40: 281–295.

Guisan, A., Edwards, T. C. & Hastie, T. (2002). Generalized linear and generalized additive models in studies of species distributions: Setting the scene. *Ecol. Model.*, 157, 89–100.

Guisan, A. & Rahbek, C. (2011). SESAM – A new framework integrating macroecological and species distribution models for predicting spatio-temporal patterns of species assemblages. *J. Biogeogr.*, 38, 1433–1444.

Guisan, A. & Thuiller, W. (2005). Predicting species distribution: Offering more than simple habitat models. *Ecol. Lett.*, 8, 993–1009.

Guisan, A., Thuiller, W. & Zimmermann, N. (2017). *Habitat Suitability and Distribution Models: With Applications in R*. Cambridge University Press.

Guisan, A. & Zimmermann, N. E. (2000). Predictive habitat distribution models in ecology. *Ecol. Model.*, 135, 147–186.

Hanski, I. (2016). *Messages from Islands: A Global Biodiversity Tour*. 1st edn. University of Chicago Press.

Harris, D. J. (2015). Generating realistic assemblages with a joint species distribution model. *Methods Ecol. Evol.*, 6, 465–473.

Harvey, P. H. & Pagel, M. D. (1991). *The Comparative Method in Evolutionary Biology*. 1st edn. Oxford University Press.

Hassell, M. P. (1975). Density-dependence in single-species populations. *J. Anim. Ecol.*, 44, 283–295.

Hawkins, B. A., Field, R., Cornell, H. V., Currie, D. J., Guégan, J., Kaufman, D. M., Kerr, J. T., Mittelbach, G. G., Oberdorff, T., O'Brien, E. M., Porter, E. E. & Turner, J. R. G. (2003). Energy, water, and broad-scale geographic patterns of species richness. *Ecology*, 84, 3105–3117.

Haylock, M. R., Hofstra, N., Klein Tank, A. M. G., Klok, E. J., Jones, P. D. & New, M. (2008). A European daily high-resolution gridded data set of surface temperature and precipitation for 1950–2006. *J. Geophys. Res.*, 113.

He, Q., Bertness, M. D. & Altieri, A. H. (2013). Global shifts towards positive species interactions with increasing environmental stress. *Ecol. Lett.*, 16, 695–706.

Hebert, P. D., Alina, C., Ball, S. L. & deWaard, J. R. (2003). Biological identifications through DNA barcodes. *Proc. R. Soc. Lond. Ser. B Biol. Sci.*, 270, 313–321.

Hijmans, R. J., Cameron, S. E., Parra, J. L., Jones, P. G. & Jarvis, A. (2005). Very high resolution interpolated climate surfaces for global land areas. *Int. J. Climatol.*, 25, 1965–1978.

Hijmans, R. J. & Graham, C. H. (2006). The ability of climate envelope models to predict the effect of climate change on species distributions. *Global Change Biol.*, 12, 2272–2281.

Hijmans, R. J., Phillips, S. J., Leathwick, J. R. & Elith, J. (2017). Dismo: Species distribution modeling. R package version 1.

Hill, N. A., Foster, S. D., Duhamel, G., Welsford, D., Koubbi, P. & Johnson, C. R. (2017). Model-based mapping of assemblages for ecology and conservation management: A case study of demersal fish on the Kerguelen plateau. *Divers. Distrib.*, 23, 1216–1230.

Holmes, I., Harris, K. & Quince, C. (2012). Dirichlet multinomial mixtures: Generative models for microbial metagenomics. *PLoS ONE*, 7, 1–15.

Holyoak, M., Leibold, M. A. & Holt, R. D. (2005). *Metacommunities: Spatial Dynamics and Ecological Communities.* 1st edn. The University of Chicago Press.

Hubbell, S. P. (2001). *The Unified Neutral Theory of Biodiversity and Biogeography.* Princeton Univ. Press.

Hui, F. K. C. (2017). Boral: Bayesian ordination and regression AnaLysis. R package version 1.4.

Hui, F. K. C., Taskinen, S., Pledger, S., Foster, S. D. & Warton, D. I. (2015). Model-based approaches to unconstrained ordination. *Methods Ecol. Evol.*, 6, 399–411.

Hui, F. K. C., Warton, D. I., Foster, S. D. & Dunstan, P. K. (2013). To mix or not to mix: Comparing the predictive performance of mixture models vs. separate species distribution models. *Ecology*, 94, 1913–1919.

Hutchinson, G. E. (1959). Homage to Santa Rosalia or why are there so many kinds of animals. *Am. Nat.*, 93, 145–159.

iNaturalist. (2019) iNaturalist research-grade observations [WWW document]. www.inaturalist.org.

Ives, A. R. & Helmus, M. R. (2011). Generalized linear mixed models for phylogenetic analyses of community structure. *Ecol. Monogr.*, 81, 511–525.

Izenman, A. J. (1975). Reduced-rank regression for the multivariate linear model. *J. Multivariate Anal.*, 5, 248–264.

Jarzyna, M. A. & Jetz, W. (2018). Taxonomic and functional diversity change is scale dependent. *Nat. Commun.*, 9, 2565.

Jetz, W., Thomas, G. H., Joy, J. B., Hartmann, K. & Mooers, A. O. (2012). The global diversity of birds in space and time. *Nature*, 491, 444.

Johnson, J. B. & Omland, K. S. (2004). Model selection in ecology and evolution. *Trends Ecol. Evol.*, 19, 101–108.

Johnston, A., Ausden, M., Dodd, A. M., Bradbury, R. B., Chamberlain, D. E., Jiguet, F., Thomas, C. D., Cook, A. S. C. P., Newson, S. E., Ockendon, N., Rehfisch, M. M., Roos, S., Thaxter, C. B., Brown, A., Crick, H. Q. P., Douse, A., McCall, R. A., Pontier, H., Stroud, D. A., Cadiou, B., Crowe, O., Deceuninck, B., Hornman, M. & Pearce-Higgins, J. (2013). Observed and predicted effects of climate change on species abundance in protected areas. *Nat. Clim. Change*, 3, 1055.

Jordano, P. (1987). Patterns of mutualistic interactions in pollination and seed dispersal: Connectance, dependence asymmetries, and coevolution. *Am. Nat.*, 129, 657–677.

Jung, V., Violle, C., Mondy, C., Hoffmann, L. & Muller, S. (2010). Intraspecific variability and trait-based community assembly. *J. Ecol.*, 98, 1134–1140.

Kassambara, A. & Mundt, F. (2019). Factoextra: Extract and visualize the results of multivariate data analyses. R package version 1.0.6.

Kattge, J., Díaz, S., Lavorel, S., Prentice, I. C., Leadley, P., Bönisch, G., Garnier, E., Westoby, M., Reich, P. B., Wright, I. J., Cornelissen, J. H. C., Violle, C., Harrison, S. P., van Bodegom, P. M., Reichstein, M., Enquist, B. J., Soudzilovskaia, N. A., Ackerly, D. D., Anand, M., Atkin, O., Bahn, M., Baker, T. R., Baldocchi, D., Bekker, R., Blanco, C. C., Blonder, B., Bond, W. J., Bradstock, R., Bunker, D. E., Casanoves, F., Cavender-Bares, J., Chambers, J. Q., Chapin Iii,F.S., Chave, J., Coomes, D., Cornwell, W. K., Craine, J. M., Dobrin, B. H., Duarte, L., Durka, W., Elser, J., Esser, G., Estiarte, M., Fagan, W. F., Fang, J., Fernández-Méndez, F., Fidelis, A., Finegan, B., Flores, O., Ford, H., Frank, D., Freschet, G. T., Fyllas, N. M., Gallagher, R. V., Green, W. A., Gutierrez, A. G., Hickler, T., Higgins, S. I., Hodgson, J. G., Jalili, A., Jansen, S., Joly, C. A., Kerkhoff, A. J., Kirkup, D., Kitajima, K., Kleyer, M., Klotz, S., Knops, J. M. H., Kramer, K., Kühn, I., Kurokawa, H., Laughlin, D., Lee, T. D., Leishman, M., Lens, F., Lenz, T., Lewis, S. L., Lloyd, J., Llusià, J., Louault, F., Ma, S., Mahecha, M. D., Manning, P., Massad, T., Medlyn, B. E., Messier, J., Moles, A. T., Müller, S. C., Nadrowski, K., Naeem, S., Niinemets, Ü, Nöllert, S., Nüske, A., Ogaya, R., Oleksyn, J., Onipchenko, V. G., Onoda, Y., Ordoñez, J., Overbeck, G., Ozinga, W. A., Patiño, S., Paula, S., Pausas, J. G., Peñuelas, J., Phillips, O. L., Pillar, V., Poorter, H., Poorter, L., Poschlod, P., Prinzing, A., Proulx, R., Rammig, A., Reinsch, S., Reu, B., Sack, L., Salgado-Negret, B., Sardans, J., Shiodera, S., Shipley, B., Siefert, A., Sosinski, E., Soussana, J. -., Swaine, E., Swenson, N., Thompson, K., Thornton, P., Waldram, M., Weiher, E., White, M., White, S., Wright, S. J., Yguel, B., Zaehle, S., Zanne, A. E. & Wirth, C. (2011). TRY – a global database of plant traits. *Global Change Biol.*, 17, 2905–2935.

Kearney, M. (2006). Habitat, environment and niche: What are we modelling?. *Oikos*, 115, 186–191.

Keddy, P. A. (1992). Assembly and response rules: Two goals for predictive community ecology. *J. Veg. Sci.*, 3, 157–164.

Kessler, M. (2009). The impact of population processes on patterns of species richness: Lessons from elevational gradients. *Basic Appl. Ecol.*, 10, 295–299.

Kingsland, S. E. (1986). Modeling nature: Episodes in the history of population ecology. *J. Hist. Biol.*, 19(2), 313–314.

Kissling, W. D., Dormann, C. F., Groeneveld, J., Hickler, T., Kühn, I., McInerny, G. J., Montoya, J. M., Römermann, C., Schiffers, K., Schurr, F. M., Singer, A., Svenning, J., Zimmermann, N. E. & O'Hara, R. B. (2012). Towards novel approaches to modelling biotic interactions in multispecies assemblages at large spatial extents. *J. Biogeogr.*, 39, 2163–2178.

Köppen, W. (1884). Die wärmezonen der erde, nach der dauer der heissen, gemässigten und kalten zeit und nach der wirkung der wärme auf die organische welt betrachtet [the thermal zones of the Earth according to the duration of hot, moderate and cold periods and to the impact of heat on the organic world]. *Meteorologische Zeitschrift (published 2011)*, 20(3), 351–360.

Krebs, C. J. (1972). *Ecology, the Experimental Analysis of Distribution and Abundance*. Harper & Row.

Kukkala, A. S. & Moilanen, A. (2013). Core concepts of spatial prioritisation in systematic conservation planning. *Biol. Rev.*, 88, 443–464.

Ladle, R. J. & Whittaker, R. J. (2011). *Conservation Biogeography*. Wiley-Blackwell.

Laughlin, D. C., Joshi, C., van Bodegom, P. M., Bastow, Z. A. & Fulé, P. Z. (2012). A predictive model of community assembly that incorporates intraspecific trait variation. *Ecol. Lett.*, 15, 1291–1299.

Lavorel, S. & Garnier, E. (2002). Predicting changes in community composition and ecosystem functioning from plant traits: Revisiting the holy grail. *Funct. Ecol.*, 16, 545–556.

Lawton, J. H. (1999). Are there general laws in ecology?. *Oikos*, 84, 177–192.

le Roux, P. C., Pellissier, L., Wisz, M. S. & Luoto, M. (2014). Incorporating dominant species as proxies for biotic interactions strengthens plant community models. *J. Ecol.*, 102, 767–775.

Legendre, P. (1993). Spatial autocorrelation: Trouble or new paradigm? *Ecology*, 74, 1659–1673.

Legendre, P. & Anderson, M. J. (1999). Distance-based redundancy analysis: Testing multispecies responses in multifactorial ecological experiments. *Ecol. Monogr.*, 69, 1–24.

Legendre, P., Borcard, D. & Peres-Neto, P. (2005). Analyzing beta diversity: Partitioning the spatial variation of community composition data. *Ecol. Monogr.*, 75, 435–450.

Legendre, P., Dale, M. R. T., Fortin, M., Gurevitch, J., Hohn, M. & Myers, D. (2002). The consequences of spatial structure for the design and analysis of ecological field surveys. *Ecography*, 25, 601–615.

Legendre, P., Galzin, R. & Harmelin-Vivien, M. (1997). Relating behavior to habitat: Solutions to the fourth-corner problem. *Ecology*, 78, 547–562.

Legendre, P. & Legendre, L. (2012). *Numerical Ecology*. 3rd edn. Elsevier.

Lehtomäki, J. & Moilanen, A. (2013). Methods and workflow for spatial conservation prioritization using Zonation. *Environ. Modell. Softw.*, 47, 128–37.

Leibold, M. A. (1995). The niche concept revisited: Mechanistic models and community context. *Ecology*, 76, 1371–1382.

Leibold, M. A. & Chase, J. M. (2018). *Metacommunity Ecology*. 1st edn. Princeton University Press.

Leibold, M. A., Holyoak, M., Mouquet, N., Amarasekare, P., Chase, J. M., Hoopes, M. F., Holt, R. D., Shurin, J. B., Law, R., Tilman, D., Loreau, M. & Gonzalez, A. (2004). The metacommunity concept: A framework for multi-scale community ecology. *Ecol. Lett.*, 7, 601–613.

Levine, S. H. (1976). Competitive interactions in ecosystems. *Am. Nat.*, 110, 903–910.

Liaw, A. & Wiener, M. (2002). Classification and regression by randomForest. *R News*, 2, 18–22.

Lindström, Å, Green, M., Husby, M., Kålås, J. A. & Lehikoinen, A. (2015). Large-scale monitoring of waders on their boreal and arctic breeding grounds in northern Europe. *Ardea*, 103(1), 3–15.

Lomolino, M. V., Riddle, B. R., Whittaker, R. J. & Brown, J. H. (2010). *Biogeography*. 4th edn. Sinauer Associates.

Lou, M. & Golding, G. B. (2012). The effect of sampling from subdivided populations on species identification with DNA barcodes using a Bayesian statistical approach. *Mol. Phylogenet. Evol.*, 65(2), 765–773.

MacArthur, R. H. (1972). *Geographical Ecology: Patterns in the Distribution of Species*. Princeton University Press.

MacArthur, R. H. & Levins, R. (1967). The limiting similarity, convergence, and divergence of coexisting species. *Am. Nat.*, 101, 377–385.

MacArthur, R. H. & Wilson, E. O. (1967). *The Theory of Island Biogeography*. Princeton University Press.

Madon, B., Warton, D. I. & Araújo, M. B. (2013). Community-level vs species-specific approaches to model selection. *Ecography*, 36, 1291–1298.

Magurran, A. E. (2004). *Measuring Biological Diversity*. 1st edn. Wiley-Blackwell.

Magurran, A. E. & Henderson, P. A. (2003). Explaining the excess of rare species in natural species abundance distributions. *Nature*, 422, 714.

Margules, C. & Sarkar, S. (2007). *Systematic Conservation Planning*. Cambridge University Press.

Margules, C. R. & Pressey, R. L. (2000). Systematic conservation planning. *Nature*, 405, 243–253.

Marincowitz, S., Crous, P. W., Groenewald, J. Z. & Wingfield, M. J. (2008). Microfungi occurring on proteaceae in the fynbos. CBS Fungal Biodiversity Centre.

May, R. M. (1971). Stability in multispecies community models. *Math. Biosci.*, 12, 59–79.

McArdle, B. H. & Anderson, M. J. (2001). Fitting multivariate models to community data: A comment on distance-based redundancy analysis. *Ecology*, 82, 290–297.

McCune, J. L. (2016). Species distribution models predict rare species occurrences despite significant effects of landscape context. *J. Appl. Ecol.*, 53, 1871–1879.

McGill, B. J. (2003). A test of the unified neutral theory of biodiversity. *Nature*, 422, 881.

McGill, B. J. (2010). Towards a unification of unified theories of biodiversity. *Ecol. Lett.*, 13, 627–642.

McGill, B. J., Enquist, B. J., Weiher, E. & Westoby, M. (2006b). Rebuilding community ecology from functional traits. *Trends Ecol. Evol.*, 21, 178–185.

McGill, B. J., Etienne, R. S., Gray, J. S., Alonso, D., Anderson, M. J., Benecha, H. K., Dornelas, M., Enquist, B. J., Green, J. L., He, F., Hurlbert, A. H., Magurran, A. E., Marquet, P. A., Maurer, B. A., Ostling, A., Soykan, C. U., Ugland, K. I. & White, E. P. (2007). Species abundance distributions: Moving beyond single prediction theories to integration within an ecological framework. *Ecol. Lett.*, 10, 995–1015.

McGill, B. J., Maurer, B. A. & Weiser, M. D. (2006a). Empirical evaluation of neutral theory. *Ecology*, 87, 1411–1423.

Meier, E. S., Edwards Jr, T. C., Kienast, F., Dobbertin, M. & Zimmermann, N. E. (2011). Co-occurrence patterns of trees along macro-climatic gradients and their potential influence on the present and future distribution of *Fagus sylvatica* L. *J. Biogeogr.*, 38, 371–382.

Merow, C., Smith, M. J., Edwards Jr, T. C., Guisan, A., McMahon, S. M., Normand, S., Thuiller, W., Wüest, R. O., Zimmermann, N. E. & Elith, J. (2014). What do we gain from simplicity versus complexity in species distribution models?. *Ecography*, 37, 1267–1281.

Meyer, D., Dimitriadou, E., Hornik, K., Weingessel, A. & Leisch, F. (2017). e1071: Misc functions of the department of statistics, probability theory group (formerly: E1071). R package version 1.6-8.

Midgley, G. F., Hannah, L., Millar, D., Rutherford, M. C. & Powrie, L. W. (2002). Assessing the vulnerability of species richness to anthropogenic climate change in a biodiversity hotspot. *Global Ecol. Biogeogr.*, 11, 445–451.

Milborrow, S. (2017). Earth: Multivariate adaptive regression splines. R package version 4.5.1.

Miller, J., Damschen, E. & Ives, A. (2018). Data from: Functional traits and community composition: A comparison among community-weighted means, weighted correlations, and multilevel models. *Dryad Digital Repository*, doi:10.5061/dryad.7gj0s3b.

Miller, J. E. D., Damschen, E. I. & Ives, A. R. (2019). Functional traits and community composition: A comparison among community-weighted means, weighted correlations, and multilevel models. *Methods Ecol. Evol.*, 10, 415–425.

Mitchell, T. J. & Beauchamp, J. J. (1988). Bayesian variable selection in linear regression. *J. Am. Stat. Assoc.*, 83, 1023–1032.

Mod, H. K., le Roux, P. C. & Luoto, M. (2014). Outcomes of biotic interactions are dependent on multiple environmental variables. *J. Veg. Sci.*, 25, 1024–1032.

Mod, H. K., le Roux, P. C., Guisan, A. & Luoto, M. (2015). Biotic interactions boost spatial models of species richness. *Ecography*, 38, 913–921.

Moilanen, A. (2007). Landscape Zonation, benefit functions and target-based planning: Unifying reserve selection strategies. *Biol. Conserv.*, 134, 571–579.

Møller, J., Syversveen, A. R. & Waagepetersen, R. P. (1998). Log Gaussian Cox processes. *Scand. J. Stat.*, 25, 451–482.

Mönkkönen, M. & Forsman, J. T. (2002). Heterospecific attraction among forest birds: A review. *Ornithol. Sci.*, 1, 41–51.

Morin, P. J. (2011). *Community Ecology*, 2nd Edition. John Wiley & Sons,Ltd.

Morin, X., Viner, D. & Chuine, I. (2008). Tree species range shifts at a continental scale: New predictive insights from a process-based model. *J. Ecol.*, 96, 784–794.

Nascimento, F. F., Reis, M. D. & Yang, Z. (2017). A biologist's guide to Bayesian phylogenetic analysis. *Nat. Ecol. Evol.*, 1, 1446–1454.

Nix, H. A. (1986). A Biogeographic Analysis of Australian Elapid Snakes. Ed. Longmore, R. In: *Australian Flora and Fauna Series* 7 . Bureau of Flora and Fauna, 4–15.

Norberg, A., Abrego, N., Blanchet, F. G., Adler, F. R., Anderson, B. J., Anttila, J., Araújo, M. B., Dallas, T., Dunson, D., Elith, J., Foster, S. D., Fox, R., Franklin, J., Godsoe, W., Guisan, A., O'Hara, B., Hill, N. A., Holt, R. D., Hui, F. K. C., Husby, M., Kålås, J. A., Lehikoinen, A., Luoto, M., Mod, H. K.,

Newell, G., Renner, I., Roslin, T., Soininen, J., Thuiller, W., Vanhatalo, J., Warton, D., White, M., Zimmermann, N. E., Gravel, D. & Ovaskainen, O. (2019). A comprehensive evaluation of predictive performance of 33 species distribution models at species and community levels. *Ecol. Monogr.*, 89(3): e01370.

O'Hara, R. B. & Sillanpää, M. J. (2009). A review of Bayesian variable selection methods: What, how and which. *Bayesian Anal.*, 4, 85–117.

Oksanen, J., Blanchet, F. G., Kindt, R., Legendre, P., Minchin, P. R., O'Hara, B., Simpson, G. L., Solymos, P., Henry, M., Stevens, H. & Wagner, H. (2019). Vegan: Community ecology package. R package version 2.5-5.

Ovaskainen, O., Abrego, N., Halme, P. & Dunson, D. (2016a). Using latent variable models to identify large networks of species-to-species associations at different spatial scales. *Methods Ecol. Evol.*, 7, 549–555.

Ovaskainen, O., Finkelshtein, D., Kutoviy, O., Cornell, S., Bolker, B. & Kondratiev, Y. (2014). A general mathematical framework for the analysis of spatiotemporal point processes. *Theor. Ecol.*, 7, 101–113.

Ovaskainen, O., Hottola, J. & Siitonen, J. (2010). Modeling species co-occurrence by multivariate logistic regression generates new hypotheses on fungal interactions. *Ecology*, 9, 2514–2521.

Ovaskainen, O., Roy, D. B., Fox, R. & Anderson, B. J. (2016b). Uncovering hidden spatial structure in species communities with spatially explicit joint species distribution models. *Methods Ecol. Evol.*, 7, 428–436.

Ovaskainen, O., Rybicki, J. & Abrego, N. (2019). What can observational data reveal about metacommunity processes? *Ecography*, 42, 1877–1886.

Ovaskainen, O., Schigel, D., Ali-Kovero, H., Auvinen, P., Paulin, L., Norden, B. & Norden, J. (2013). Combining high-throughput sequencing with fruit body surveys reveals contrasting life-history strategies in fungi. *ISME J.*, 7, 1696–1709.

Ovaskainen, O. & Soininen, J. (2011). Making more out of sparse data: Hierarchical modeling of species communities. *Ecology*, 92, 289–295.

Ovaskainen, O., Tikhonov, G., Dunson, D., Grøtan, V., Engen, S., Sæther, B. & Abrego, N. (2017a). How are species interactions structured in species-rich communities? A new method for analysing time-series data. *Proc. R. Soc. Lond. Ser. B Biol. Sci.*, 284.

Ovaskainen, O., Tikhonov, G., Norberg, A., Blanchet, F. G., Duan, L., Dunson, D., Roslin, T. & Abrego, N. (2017b). How to make more out of community data? A conceptual framework and its implementation as models and software. *Ecol. Lett.*, 20, 561–576.

Pacifici, K., Reich, B. J., Miller, D. A. W., Gardner, B., Stauffer, G., Singh, S., McKerrow, A. & Collazo, J. A. (2017). Integrating multiple data sources in species distribution modeling: A framework for data fusion*. *Ecology*, 98, 840–850.

Pearce, J. & Ferrier, S. (2000). Evaluating the predictive performance of habitat models developed using logistic regression. *Ecol. Model.*, 133, 225–245.

Pearman, P. B., Randin, C. F., Broennimann, O., Vittoz, P., Knaap, W. O. v. d., Engler, R., Lay, G. L., Zimmermann, N. E. & Guisan, A. (2008). Prediction of plant species distributions across six millennia. *Ecol. Lett.*, 11, 357–369.

Pearson, R. G. & Dawson, T. P. (2003). Predicting the impacts of climate change on the distribution of species: Are bioclimate envelope models useful? *Global Ecol. Biogeogr.*, 12, 361–371.

Pearson, R. G., Thuiller, W., Araújo, M. B., Martinez-Meyer, E., Brotons, L., McClean, C., Miles, L., Segurado, P., Dawson, T. P. & Lees, D. C. (2006). Model-based uncertainty in species range prediction. *J. Biogeogr.*, 33, 1704–1711.

Pellissier, L., Albouy, C., Bascompte, J., Farwig, N., Graham, C., Loreau, M., Maglianesi, M. A., Melián, C. J., Pitteloud, C., Roslin, T., Rohr, R., Saavedra, S., Thuiller, W., Woodward, G., Zimmermann, N. E. & Gravel, D. (2018). Comparing species interaction networks along environmental gradients. *Biol. Rev.*, 93, 785–800.

Pellissier, L., Anne Bråthen, K., Pottier, J., Randin, C. F., Vittoz, P., Dubuis, A., Yoccoz, N. G., Alm, T., Zimmermann, N. E. & Guisan, A. (2010). Species distribution models reveal apparent competitive and facilitative effects of a dominant species on the distribution of tundra plants. *Ecography*, 33, 1004–1014.

Peres-Neto, P. R., Olden, J. D. & Jackson, D. A. (2001). Environmentally constrained null models: Site suitability as occupancy criterion. *Oikos*, 93, 110–120.

Peterson, A. T., Soberón, J., Pearson, R. G., Anderson, R. P., Martínez-Meyer, E., Nakamura, M. & Araújo, M. B. (2011). *Ecological Niches and Geographic Distributions.* 1st edn. Princeton University Press.

Phillips, S. J., Anderson, R. P. & Schapire, R. E. (2006). Maximum entropy modeling of species geographic distributions. *Ecol. Model.*, 190, 231–259.

Plummer, M., Best, N., Cowles, K., Vines, K., Sarkar, D., Bates, D., Almond, R. & Magnusson, A. (2018). Coda: Output Analysis and Diagnostics for MCMC, R Package Version 0.19-2.

Pocheville, A. (2015). Chapter 26. The Ecological Niche: History and Recent Controversies. Eds. Heams, T., Huneman, P. Lecointre, G. & Silberstein, M. In: *Handbook of Evolutionary Thinking in the Sciences.* Springer, 547–586.

Pollock, L. J., Morris, W. K. & Vesk, P. A. (2012). The role of functional traits in species distributions revealed through a hierarchical model. *Ecography*, 35, 716–725.

Pollock, L. J., Tingley, R., Morris, W. K., Golding, N., O'Hara, R. B., Parris, K. M., Vesk, P. A. & McCarthy, M. A. (2014). Understanding co-occurrence by modelling species simultaneously with a joint species distribution model (JSDM). *Methods Ecol. Evol.*, 5, 397–406.

Poorter, L. & Bongers, F. (2006). Leaf traits are good predictors of plant performance across 53 rain forest species. *Ecology*, 87, 1733–1743.

Preston, F. W. (1948). The commonness, and rarity, of species. *Ecology*, 29, 254–283.

Preston, F. W. (1960). Time and space and the variation of species. *Ecology*, 41, 612–627.

R Development Core Team. (2019). *R: A Language and Environment for Statistical Computing.* R Foundation for Statistical Computing.

Randin, C. F., Dirnböck, T., Dullinger, S., Zimmermann, N. E., Zappa, M. & Guisan, A. (2006). Are niche-based species distribution models transferable in space?. *J. Biogeogr.*, 33, 1689–1703.

Rangel, T.F., Diniz-Filho, J. & Colwell, R. (2007). Species richness and evolutionary niche dynamics: A spatial pattern oriented simulation experiment. *Am. Nat.*, 170, 602–616.

Renner, I. W., Elith, J., Baddeley, A., Fithian, W., Hastie, T., Phillips, S. J., Popovic, G. & Warton, D. I. (2015). Point process models for presence-only analysis. *Methods Ecol. Evol.*, 6, 366–379.

Renner, I. W. & Warton, D. I. (2013). Equivalence of MAXENT and Poisson point process models for species distribution modeling in ecology. *Biometrics*, 69, 274–281.

Ricklefs, R. E. (1987). Community diversity: Relative roles of local and regional processes. *Science*, 235, 167–171.

Ricklefs, R. E. (1990). *Ecology*. 3rd edn. W. H. Freeman and Co.

Ricklefs, R. E. (2008). Disintegration of the ecological community. *Am. Nat.*, 172, 741–750.

Ridgeway, G. (2017). Gbm: Generalized boosted regression models. R package version 2.1.3.

Root, R. B. (1967). The niche exploitation pattern of the blue-gray gnatcatcher. *Ecol. Monogr.*, 37, 317–350.

Saint-Germain, M., Buddle, C. M., Larrivée, M., Mercado, A., Motchula, T., Reichert, E., Sackett, T. E., Sylvain, Z. & Webb, A. (2007). Should biomass be considered more frequently as a currency in terrestrial arthropod community analyses? *J. Appl. Ecol.*, 44, 330–339.

Saurola, P., Valkama, J. & Velmala, J. (2013). The Finnish bird ringing atlas. Vol 1.

Schindler, D. W. (1998). Whole-ecosystem experiments: Replication versus realism: The need for ecosystem-scale experiments. *Ecosystems*, 1, 323–334.

Schoch, C. L., Seifert, K. A., Huhndorf, S., Robert, V., Spouge, J. L., Levesque, C. A., Chen, W. & Fungal Barcoding Consortium. (2012). Nuclear ribosomal internal transcribed spacer (ITS) region as a universal DNA barcode marker for fungi. *Proc. Natl. Acad. Sci. USA*, 109, 6241–6246.

Schwarz, G. (1978). Estimating the dimension of a model. *Ann. Stat*, 6, 461–464.

Sebastián-González, E., Sánchez-Zapata, J. A., Botella, F. & Ovaskainen, O. (2010). Testing the heterospecific attraction hypothesis with time-series data on species co-occurrence. *Proc. R. Soc. Lond. B Biol. Sci.*, 277, 2983–2990.

Segurado, P., Araújo, M. B. & Kunin, W. E. (2006). Consequences of spatial autocorrelation for niche-based models. *J. Appl. Ecol.*, 43, 433–444.

Simberloff, D. S. & Wilson, E. O. (1969). Experimental zoogeography of islands: The colonization of empty islands. *Ecology*, 50, 278–296.

Smith, T. W. & Lundholm, J. T. (2010). Variation partitioning as a tool to distinguish between niche and neutral processes. *Ecography*, 33, 648–655.

Somervuo, P., Yu, D. W., Xu, C. C. Y., Ji, Y., Hultman, J., Wirta, H. & Ovaskainen, O. (2017). Quantifying uncertainty of taxonomic placement in DNA barcoding and metabarcoding. *Methods Ecol. Evol.*, 8, 398–407.

Spiegelhalter, D. J., Best, N. G., Carlin, B. P. & Van, D. L. (2002). Bayesian measures of model complexity and fit. *J. R. Statist. Soc. B*, 64, 583–639.

Stroud, J. T., Bush, M. R., Ladd, M. C., Nowicki, R. J., Shantz, A. A. & Sweatman, J. (2015). Is a community still a community? Reviewing definitions of key terms in community ecology. *Ecol. Evol.*, 5, 4757–4765.

Talluto, M. V., Boulangeat, I., Ameztegui, A., Aubin, I., Berteaux, D., Butler, A., Doyon, F., Drever, C. R., Fortin, M., Franceschini, T., Liénard, J., McKenney, D., Solarik, K. A., Strigul, N., Thuiller, W. & Gravel, D. (2016). Cross-scale integration of knowledge for predicting species ranges: A metamodeling framework. Glob. *Ecol. Biogeogr.*, 25, 238–249.

ter Braak, C. J. F. (1986). Canonical correspondence analysis: A new eigenvector technique for multivariate direct gradient analysis. *Ecology*, 67, 1167–1179.

Thorson, J. T., Ianelli, J. N., Larsen, E. A., Ries, L., Scheuerell, M. D., Szuwalski, C. & Zipkin, E. F. (2016). Joint dynamic species distribution models: A tool for community ordination and spatio-temporal monitoring. *Global Ecol.Biogeogr.*, 25, 1144–1158.

Thorson, J. T., Scheuerell, M. D., Shelton, A. O., See, K. E., Skaug, H. J. & Kristensen, K. (2015). Spatial factor analysis: A new tool for estimating joint species distributions and correlations in species range. *Methods Ecol. Evol.*, 6, 627–637.

Thuiller, W., Brotons, L., Araújo, M. B. & Lavorel, S. (2004). Effects of restricting environmental range of data to project current and future species distributions. *Ecography*, 27, 165–172.

Thuiller, W., Münkemüller, T., Lavergne, S., Mouillot, D., Mouquet, N., Schiffers, K. & Gravel, D. (2013). A road map for integrating eco-evolutionary processes into biodiversity models. *Ecol. Lett.*, 16, 94–105.

Tikhonov, G., Abrego, N., Dunson, D. & Ovaskainen, O. (2017). Using joint species distribution models for evaluating how species-to-species associations depend on the environmental context. *Methods Ecol. Evol.*, 8, 443–452.

Tikhonov, G., Duan, L., Abrego, N., Newell, G., White, M., Dunson, D. & Ovaskainen, O. (2020a). Computationally efficient joint species distribution modeling of big spatial data. *Ecology*, e02929.

Tikhonov, G., Opedal, Ø, Abrego, N., Lehikoinen, A., de Jonge, M. M., Oksanen, J. & Ovaskainen, O. (2020b). Joint species distribution modelling with the R-package Hmsc. *Methods Ecol. Evol.*, 11, 442–447.

Tilman, D. (1994). Competition and biodiversity in spatially structured habitats. *Ecology*, 75, 2–16.

Tjur, T. (2009). Coefficients of determination in logistic regression models – A new proposal: The coefficient of discrimination. *Am. Stat.*, 63, 366–372.

Tobler, M. W., Kéry, M., Hui, F. K. C., Guillera-Arroita, G., Knaus, P. & Sattler, T. (2019). Joint species distribution models with species correlations and imperfect detection. *Ecology*, 100, e02754.

Troudet, J., Grandcolas, P., Blin, A., Vignes-Lebbe, R. & Legendre, F. (2017). Taxonomic bias in biodiversity data and societal preferences. *Sci. Rep.*, 7, 9132.

Tuomisto, H. (2010). A diversity of beta diversities: Straightening up a concept gone awry. part 1. defining beta diversity as a function of alpha and gamma diversity. *Ecography*, 33, 2–22.

Valkama, J., Saurola, P., Lehikoinen, A., Lehikoinen, E., Piha, M., Sola, P. & Velmala, W. (2014). The Finnish bird ringing atlas. Vol 2.

Vellend, M. (2016). *The Theory of Ecological Communities*. 1st edn. Princeton University Press.

Vellend, M. (2010). Conceptual synthesis in community ecology. *Q. Rev. Biol.*, 85, 183–206.

Wallace, A. R. (1876). *The Geographical Distribution of Animals, with a Study of the Relations of Living and Extinct Faunas as Elucidating the Past Changes of the Earth's Surface*. Macmillan and Co.

Warton, D. I., Blanchet, F. G., O'Hara, R. B., Ovaskainen, O., Taskinen, S., Walker, S. C. & Hui, F. K. C. (2015). So many variables: Joint modeling in community ecology. *Trends Ecol. Evol.*, 30, 766–779.

Watanabe, S. (2013). A widely applicable Bayesian information criterion. *J. Mach. Learn. Res.*, 14, 867–897.

Watanabe, S. (2010). Asymptotic equivalence of Bayes cross validation and widely applicable information criterion in singular learning theory. *J. Mach. Learn. Res.*, 11, 3571–3594.

Wei, T., Simko, V., Levy, L., Xie, Y., Jin, Y. & Zemla, J. (2017). Package 'corrplot': Visualization of a Correlation Matrix. R package version 0.84.

Whitfeld, T. J. S., Kress, W. J., Erickson, D. L. & Weiblen, G. D. (2012). Change in community phylogenetic structure during tropical forest succession: Evidence from New Guinea. *Ecography*, 35, 821–830.

Whittaker, R. H. (1960). Vegetation of the Siskiyou mountains, Oregon and California. *Ecol. Monogr.*, 30, 279–338.

Whittaker, R. H. (1962). Classification of natural communities. *Bot.Rev.*, 28, 1–239.

Whittaker, R. H. (1972). Evolution and measurement of species diversity. *Taxon*, 21, 213–251.

Whittaker, R. H. (1975). *Communities and Ecosystems*. Macmillan, New York.

Wiens, J. J., Ackerly, D. D., Allen, A. P., Anacker, B. L., Buckley, L. B., Cornell, H. V., Damschen, E. I., Jonathan Davies, T., Grytnes, J., Harrison, S. P., Hawkins, B. A., Holt, R. D., McCain, C. M. & Stephens, P. R. (2010). Niche conservatism as an emerging principle in ecology and conservation biology. *Ecol. Lett.*, 13, 1310–1324.

Wilman, H., Belmaker, J., Simpson, J., d. l. Rosa, C., Rivadeneira, M. M. & Jetz, W. (2014). EltonTraits 1.0: Species-level foraging attributes of the world's birds and mammals. *Ecology*, 95, 2027–2027.

Wilson, K. A., Underwood, E. C., Morrison, S. A., Klausmeyer, K. R., Murdoch, W. W., Reyers, B., Wardell-Johnson, G., Marquet, P. A., Rundel, P. W., McBride, M. F., Pressey, R. L., Bode, M., Hoekstra, J. M., Andelman, S., Looker, M., Rondinini, C., Kareiva, P., Shaw, M. R. & Possingham, H. P. (2007). Conserving biodiversity efficiently: What to do, where, and when. *PLoS Biol.*, 5, 1–12.

Wisz, M. S., Pottier, J., Kissling, W. D., Pellissier, L., Lenoir, J., Damgaard, C. F., Dormann, C. F., Forchhammer, M. C., Grytnes, J., Guisan, A., Heikkinen, R. K., Høye, T. T., Kühn, I., Luoto, M., Maiorano, L., Nilsson, M., Normand, S., Öckinger, E., Schmidt, N. M., Termansen, M., Timmermann, A., Wardle, D. A., Aastrup, P. & Svenning, J. (2013). The role of biotic interactions in shaping distributions and realised assemblages of species: Implications for species distribution modelling. *Biol. Rev.*, 88, 15–30.

Wood, S. N. (2011). Fast stable restricted maximum likelihood and marginal likelihood estimation of semiparametric generalized linear models. *J. R. Statist. Soc. B*, 73, 3–36.

Wootton, J. T. (2005). Field parameterization and experimental test of the Neutral Theory of Biodiversity. *Nature*, 433, 309.

Yates, K. L., Bouchet, P. J., Caley, M. J., Mengersen, K., Randin, C. F., Parnell, S., Fielding, A. H., Bamford, A. J., Ban, S., Barbosa, A. M., Dormann, C. F., Elith, J., Embling, C. B., Ervin, G. N., Fisher, R., Gould, S., Graf, R. F., Gregr, E. J., Halpin, P. N., Heikkinen, R. K., Heinänen, S., Jones, A. R., Krishnakumar, P. K., Lauria, V., Lozano-Montes, H., Mannocci, L., Mellin, C., Mesgaran, M. B., Moreno-Amat, E., Mormede, S., Novaczek, E., Oppel, S., Ortuño Crespo, G., Peterson, A. T., Rapacciuolo, G., Roberts, J. J., Ross, R. E., Scales, K. L., Schoeman, D., Snelgrove, P., Sundblad, G., Thuiller, W., Torres, L. G., Verbruggen, H., Wang, L., Wenger, S., Whittingham, M. J., Zharikov, Y., Zurell, D. & Sequeira, A. M. M. (2018). Outstanding challenges in the transferability of ecological models. *Trends Ecol. Evol.*, 33, 790–802.

Zhang, X. & Vincent, A. C. J. (2018). Predicting distributions, habitat preferences and associated conservation implications for a genus of rare fishes, seahorses (Hippocampus spp.). *Divers. Distrib.*, 24, 1005–1017.

Zimmermann, N. E., Edwards Jr, T. C., Graham, C. H., Pearman, P. B. & Svenning, J. (2010). New trends in species distribution modelling. *Ecography*, 33, 985–989.

Zobel, M. (1997). The relative of species pools in determining plant species richness: An alternative explanation of species coexistence?. *Trends Ecol. Evol.*, 12, 266–269.

Zurell, D., Berger, U., Cabral, J. S., Jeltsch, F., Meynard, C. N., Münkemüller, T., Nehrbass, N., Pagel, J., Reineking, B., Schröder, B. & Grimm, V. (2010). The virtual ecologist approach: Simulating data and observers. *Oikos*, 119, 622–635.

Zurell, D., Pollock, L. J. & Thuiller, W. (2018). Do joint species distribution models reliably detect interspecific interactions from co-occurrence data in homogenous environments?. *Ecography*, 41, 1812–1819.

Zurell, D., Thuiller, W., Pagel, J., Cabral, J. S., Münkemüller, T., Gravel, D., Dullinger, S., Normand, S., Schiffers, K. H., Moore, K. A. & Zimmermann, N. E. (2016). Benchmarking novel approaches for modelling species range dynamics. *Glob. Change Biol.*, 22, 2651–2664.

Zuur, A. F., Hilbe, J. M. & Ieno, E. N. (2013). *Beginner's Guide to GLM and GLMM with R: A Frequentist and Bayesian Perspective for Ecologists*. Highland Statistics Ltd.

Zuur, A., Ieno, E. N., Walker, N., Saveliev, A. A. & Smith, G. M. (2009). Mixed Effects Models and Extensions in Ecology with R. Springer.

Zuur, A. F., Saveliev, A. A. & Ieno, E. N. (2012). *Zero Inflated Models and Generalized Linear Mixed Models with R*. Highland Statistics Ltd.

Index

www.ingramcontent.com/pod-product-compliance
Lightning Source LLC
LaVergne TN
LVHW011517200926
848484LV00001B/59
* 9 7 8 1 1 0 8 4 9 2 4 6 1 *